Peptides

Biology and Chemis

Peptides

Biology and Chemistry

Proceedings of the 1994 Chinese Peptide Symposium
June 13–17, 1994, Beijing, China

Edited by

Gui-Shen Lu
Institute of Materia Medica
Chinese Academy of Medical Sciences
Beijing 100050, China

James P. Tam
Department of Microbiology and Immunology
Vanderbilt University
Nashville, TN 37232, U.S.A.

and

Yu-Cang Du
Shanghai Institute of Biochemistry
Chinese Academy of Sciences
Shanghai 200031, China

ESCOM ▪ Leiden ▪ 1995

CIP-data Koninklijke Bibliotheek, Den Haag

Peptides

Peptides: Biology and Chemistry: Proceedings of the 1994 Chinese Peptide Symposium, June 13–17, Beijing, China / ed. by Gui-Shen Lu, James P. Tam and Yu-Cang Du – Leiden: ESCOM – Ill.
With index, ref.
Subject headings: Peptides/Proteins.

ISBN 90-72199-20-0 (hardbound)

Published by:

ESCOM Science Publishers B.V.
P.O. Box 214
2300 AE Leiden
The Netherlands

Printed in The Netherlands

Preface

The third Chinese Peptide Symposium, held in Beijing, China on June 13–17, 1994, attracted 156 delegates representing 11 countries. Nobel Laureate Professor Bruce R. Merrifield was among the 51 international participants, which included many other eminent peptide scientists. Our goal for CPS-94 was to provide a forum for the exchange of knowledge, cooperation and friendship between the international and Chinese scientific communities, and I believe this goal was met.

The program consisted of 10 sessions, with 43 oral and 42 poster presentations. Topics included synthetic methods, molecular diversity and peptide library, design of neuroactive and other active peptides, conformation and protein modeling, peptide immunology and challenging problems in peptides. There were 75 articles selected for publication in these proceedings.

CPS-94 was hosted by the Institute of Materia Medica, Chinese Academy of Medical Sciences and Beijing Medical University. Professor James Tam, Professor Meng-shen Cai and I were honored to serve as Vice-chairmen and Chairperson, respectively, as recommended by the Program Committee of the CPS-92.

The enthusiastic cooperation and excellent contributions were gratifying, and the active response of the invited speakers guaranteed the success of the symposium. The presentations were of excellent caliber and represented the most current and significant aspects of peptide science.

Dr. Su Sun Wang and Dr. Yu-Cang Du were the recipients of 'The Cathay Award', offered for the first time at CPS-94 to recognize outstanding scientists who have made seminal contributions in peptide sciences. Four outstanding young scientists were selected by the Organizing Committee to receive awards sponsored by the H.H. Liu Education Foundation.

Dr. Elizabeth Schram and ESCOM Science Publishers kindly agreed to publish the CPS-94 Proceedings a second time, and I thank them for their contributions and continued support of the symposium.

As Chairperson, I greatly appreciate the generous financial assistance of the sponsors and donors. On behalf of the Organizing Committee, I recognize the administrative assistance of the Institute of Materia Medica, Chinese Academy of Medical Sciences and Beijing Medical University. I especially want to thank Professor James P. Tam and his group for their assistance in organization abroad.

Finally, I am grateful for the technical assistance of my colleagues, especially Dr. Han-Lin Chi and Ms. Ying Sun, who devoted much time and effort to insure the success of CPS-94.

Gui-Shen Lu

Chinese Peptide Symposium – 1994

June 13–17, 1994, Beijing
Institute of Materia Medica, Chinese Academy of Medical Sciences
Beijing Medical University

Chairperson

Gui-Shen Lu, *Institute of Materia Medica, CAMS, Beijing, China*

Vice Chairmen

Meng-Shen Cai and James P. Tam

Organizing Committee

Meng-Shen Cai, *Beijing Medical University, Beijing, China*
Dun-Hua Dai, *Institute of Pharmaceutical Chemistry, Beijing, China*
Richard DiMarchi, *Lilly Pharmaceutical Co., U.S.A.*
Zheng-Kai Ding, *Chinese Academy of Military Medical Science, Beijing, China*
Yu-Cang Du, *Shanghai Institute of Biochemistry, Shanghai, China*
Arthur M. Felix, *Hoffmann-La Roche Inc., U.S.A.*
Xiao-Yu Hu, *Lanzhou University, Lanzhou, China*
Robert Liu, *H.H. Liu Education Foundation*
Gui-Shen Lu, *Institute of Materia Medica, CAMS, Beijing, China*
S. Sakakibara, *Peptide Institute, Osaka, Japan*
James P. Tam, *Vanderbilt University, U.S.A.*
Daniel F. Veber, *SmithKline Beecham Pharmaceuticals, U.S.A.*
De-Xin Wang, *Institute of Materia Medica, CAMS, Beijing, China*
Jie-Cheng Xu, *Shanghai Institute of Organic Chemistry, Shanghai, China*
Xiao-Jie Xu, *Peking University, Beijing, China*
Yun-Hua Ye, *Peking University, Beijing, China*

Awarding Committee of Cathay Award

Bruce Merrifield, *Rockefeller University, U.S.A.*
Gui-Shen Lu, *Institute of Materia Medica, CAMS, Beijing, China*
James P. Tam, *Vanderbilt University, U.S.A.*
Robert C. Liu, *H.H. Liu Education Foundation*

Awarding Committee of H.H. Liu Peptide Awards

Meng-Shen Cai, *Beijing Medical University, Beijing, China*
Dun-Hua Dai, *Institute of Pharmaceutical Chemistry, Beijing, China*
Zheng-Kai Ding, *Chinese Academy of Military Medical Science, Beijing, China*
Yu-Cang Du, *Shanghai Institute of Biochemistry, Shanghai, China*
Arthur M. Felix, *Hoffmann-La Roche Inc., U.S.A.*
Xiao-Yu Hu, *Lanzhou University, Lanzhou, China*
Gui-Shen Lu, *Institute of Materia Medica, CAMS, Beijing, China*
James P. Tam, *Vanderbilt University, U.S.A.*
De-Xin Wang, *Institute of Materia Medica, CAMS, Beijing, China*
Xiao-Jie Xu, *Peking University, Beijing, China*
Yun-Hua Ye, *Peking University, Beijing, China*
Ceng-Hu Zhu, *H.H. Liu Education Foundation*

Sponsors

National Natural Science Foundation of China
H.H. Liu Education Foundation
Lilly Pharmaceutical Co.
Dr. and Mrs. Su Sun Wang
Hoffmann-La Roche, Inc.
Bachem, Inc., USA
Glaxo Holdings, Inc.

Donors

Chiron Mimotopes Pty., Ltd.
Phoenix Laboratory, Inc.
Dr. Ding Cheng
Merck Research Laboratories
Peptide Institute, Inc. (Osaka)

Abbreviations

Abbreviations used in the proceedings volume are defined below:

AAA	amino acid analysis
ABR	anterior byssus retractor
ACE	angiotensin-converting enzyme
ACh	acetylcholine
Acm	acetamidomethyl
ACTH	corticotropin
AIDS	acquired immune deficiency syndrome
AKH	adipokinetic hormone
ALS	amyotrophic lateral sclerosis
ANG	angiotensin
AOA	antiovulatory assay
Aph	aminophenylalanine
Aph(atz)	4-(N-5'-(3'-amino-1H-1',2',4'-triazolyl) phenylalanine
ara-C	cytosine arabinoside
AVP	arginine-8-vasopressin
BBB	blood brain barrier
BBC	benzotriazo(-1-yloxy-bis(pyrrolidino) carbonium hexa-fluorophosphate
BDB	bis-diazotized benzidine
BK	bradykinin
BmK	buthus martensii
Boc	tert-butyloxycarbonyl
BOP	benzotriazolyloxy tris-(dimethylamino) phosphonium hexa-fluorophosphate
T-t-Bos	tri-tert-butoxysilyl ester
BPTI	bovine trypsin inhibitor
BSA	bovine serum albumin
bST	bovine somatotropin
BTG	bovine thyroglobulin
Bu	butyl derivative
Bzl	benzyl
CADD	computer-aided drug design
cAMP	cyclic adenosyl monophosphate
Cbz	carbobenzoxy
CD	circular dichroism
CE,CZE	capillary zone electrophoresis
CHE	cholinesterase
CL	corpus luteum
CNS	central nervous system
CP	carboxypeptidase
CRF	corticotropin releasing factor
CRF-BP	CRF binding protein
CZE	capillary zone electrophoresis
DCC	dicyclohexylcarbodiimide
DCM	dichloromethane
DDSi	dichlorodimethylsilane
DEAE	diethylaminoethanol
DELT	deltorphin
DEPBT	3-diethylphosphoryloxy-1,2,3-benzotriazin-4(3H)-one
DHBT	2,3-dihydro-3-hydroxy-4-oxobenzotriazine
DIPEA	diisopropylethylamine
DMA	dimethylacetamide
DMF	dimethylformamide
DMSO	dimethyl sulfoxide
Dmt	2',6'-dimethyltyrosine
DPDPE	cyclo[DPen^2-DPen^5]-enkephalin
DTT	dithiothreitol
EDT	ethanedithio

EGF	epidermal growth factor
ELISA	enzyme-linked immunosorbent assay
Enk	enkephalin
EPSP	exitatory postsynaptic potential
Et_3N,TEA	triethylamine
FABMS	fast atom bombardment mass spectrometry
Fc	crystallizable Ig fragment
Fg	fibrinogen
Fmoc	fluorenylmethoxycarbonyl
FSH	follicle stimulating hormone
FTIC	fluorescein isothiocyanate
FTIR	Fourier transform infrared
GAP	gonadotropin-releasing hormone associated protein
GC	gas chromatography
GLP	glucagon-like peptide
GnRH	gonadotropin releasing hormone
gP	glycoprotein
GPCR	G-protein coupled receptor
GPI	guinea pig ileum
GRF	growth hormone releasing factor
GRP	gastrin releasing peptide
GSTs	glutathione s-transferases
GTP	guanosine triphosphate
hCG	human chorionic gonadotropin
HCV	hepatitis C virus
Hfe	homophenylalanine
hF-GRP	human follicular gonadotropin peptide
HIV	human immunodeficiency virus
HMBC	heteronuclear multiple-bond correlation
HMQC	heteronuclear multiple-quantum correlation
HOOBt	hydroxyoxodihydrobenzo-triazine
HOSu	*N*-hydroxysuccinimide
h-PCs	homophytochelatins
hPRP	human PACAP-related peptide
HRA	histamine releasing activity
HVTLE	high voltage thin layer electrophonesis
Hyp	hydroxyproline
IC	inhibitory concentration
Ig	immunoglobulin
IL	interleukin
ILys	lysine(N^{ε}-isopropyl)
IPSP	inhibitory postsynaptic potential
IR	infrared
LH	luteinizing hormone
LHRH	luteinizing hormone releasing hormone
Lm-PLL	lactosylated poly L-lysine
LPC	lyso-phosphatidylcholine
LSIMS	liquid secondary ion mass spectrometry
MAb	monoclonal antibody
MALDMS	matrix assisted laser desorption mass spectrometry
MAp	multiple antigen peptide
MBHA	methylbenzhydrylamine
MBS	*m*-maleimidobenzoyl-*N*-hydroxysuccinimide ester
MCPS	multiple constrained peptide synthesis
MCS	*b*-maleimido capric acyl-*N*-hydroxy succinimide ester
MDP	muramyl dipeptide

Meb 4-methylbenzyl
M-Enk methionine enkephalin
MEP memory-enhancing peptide
MHC major histocompatibility complex
MIC minimal inhibitory concentration
Mop morpholinomethyl phenylalanine
MS mass spectrometry; malaria surface antigens
Mtr methoxytrimethylphenyl-sulphonyl
MVD mouse vas deferens
MVP molecular viewing program

D^2Nal D-β-(2-naphthyl)alanine
NFT neurofibrillary tangles
NGF nerve growth factor
NMM *N*-methylmorpholine
NMR nuclear magnetic resonance
NOE nuclear Overhauser effect

OM oxyntomodulin
ONSu *N*-hydroxysuccinimide ester

PACAP pituitary adenylate cyclase activating polypeptide
D^3-Pal D-β-(3-pyridyl)alanine
PBS phosphate buffered saline
PC phytochelatin
PCR polymerase chain reaction
PDB phorbol 12,13-dibutyrate
PEG polyethylene glycol
PKA protein kinase A
PKC protein kinase C
Pmc 2,2,5,7,8-pentamethyl-chroman-6-sulfonyl
PMS premenstrual syndrome
PPP platelet-poor plasma
PRP platelet-rich plasma
PSPP papaver somniferum pollen peptide
PVP polyvinylprolidone

RGD Arg-Gly-Asp fibrinogen binding sequence
RHD rheumatic heart disease
RIA radioimmunoassay
RMSD root mean square deviation
RPHPLC reversed-phase high pressure liquid chrömatography
RT reverse transcriptase

SA systematic analoging
SMPS simultaneous multiple peptide synthesis
SP substance P
SPDP 3-(2-pyridyldithio) propionic acid N-hydroxy-succinimide ester

Tα_1 thymosin α_1
TASP template assembled synthetic proteins
Tβ_4 thymosin β_4
TCS trypsin-catalyzed semi-synthesis
TEA triethylamine
TF5 thymosin fraction 5
TFA trifluoroacetic acid
TFE trifluroethanol
TFMSA trifluoromethanesulfonic acid
TGF transformation growth factor
THF tetrahydrofuran
Tic 1,2,3,4-tetrahydroquino-line-3-carboxylic acid
TM transmembrane
7-TM/GPCR 7-transmembrane G-protein coupled receptor
TMSBr trimethylsilyl bromide
TMT 2',6'-dimethyl-β-methyltyrosines

TNF	tumor necrosis factor
Tos	tosyl
TPK	tyrosine-specific phosphate kinase
TPPI	time-proportional phase incrementation
TRH	thyrotropin releasing hormone
Tris	tris(hydroxymethyl)aminomethane
Trt	trityl
TT	tetanus toxoid
UTI	urotrypsin inhibitor
UV	ultraviolet
VIP	vasoactive intestinal peptide
VP	vasopressin
Z	carbobenzoxy

Contents

Session III: Neuropeptides

Session IV: Endocrinological peptides

Session I
Synthetic methods

Chairs: James P. Tam
Vanderbilt University
Nashville, Tennessee, U.S.A.

Meng-Shen Cai
Beijing Medical University
Beijing, China

Tetsuo Shiba
Peptide Institute
Osaka, Japan

and

Dun-Hua Dai
Beijing Pharmaceutical Chemistry Institute
Beijing, China

Retro and retroenantio analogs of cecropin-melittin hybrids

Bruce Merrifield[a], Padmaja Juvvadi[a], David Andreu[b], Josep Ubach[b] and Hans Boman[c]

[a]*The Rockefeller University, New York, NY 10021, U.S.A.*

[b]*The University of Barcelona, Marti i Franques, 1-11, E-08028, Barcelona, Spain*

[c]*Stockholm University, S-10691 Stockholm, Sweden*

Introduction

Hybrid analogs of cecropin A (C_A) and melittin (Me), which are potent antibacterial peptides, have been synthesized. To understand the structural requirements for this activity we have synthesized the enantio, retro, and retroenantio isomers and the N^{α}-acetylated derivatives of two of the hybrids. Against five of the test bacteria all analogs of C_A(1-13)Me(1-13)-NH_2 were as active as the parent peptide but one bacterial strain was resistant to the retro or retroenantio derivatives. Similarly all analogs of C_A(1-7)Me(2-9)-NH_2 were active against three strains, while two strains were resistant to the retro and retroenantio analogs containing free NH_3^+ end groups, but acetylation restored the activity against one of them. From these data some conclusions can be drawn about which structural features of the peptides are necessary for antibacterial activity.

Results and Discussion

Our laboratories have been studying a group of antibacterial peptides derived from the animal kingdom. They were originally isolated from the silk moth *Hyalophora cecropia* and were named cecropins [1]. Other antibacterial peptides have now been isolated from a wide range of animal sources, for example, melittin from the honey bee [2], defensins from phagocytes [3], magainins from frog skin [4] and tachyplesin from hemocytes of the horseshoe crab [5]. It was shown that the 35- to 37-residue cecropin peptides are derived by enzymatic processing of larger preproproteins [6]. The mature peptides have amphipathic helical structures, with a basic N-terminal region, a central Gly-Pro hinge and a more hydrophobic C-terminal region ending in an amide [7]. Recent efforts have been directed toward an understanding of their mechanisms of action through the chemical synthesis of selected analogs. All peptides were synthesized by standard solid phase methods. We found that certain hybrids of cecropins with other antibacterial peptides such as melittin could lead to even more active peptides (8). All of these peptides formed ion conducting channel structures in artificial planar lipid bilayers, suggesting that this might be related to the mechanism of their action in the bacterial membrane [9].

It was then found that the enantiomers of all of these hybrids, composed of all D amino acid residues, were fully active against all of the test organisms [10]. This result shows that the peptides do not function by chiral interactions with receptors or enzymes or chiral lipids, and supports the idea that in the presence of a hydrophobic environment (lipid bilayer or cell membrane) they self-aggregate to form pores or channels, which allow the passage of ions under a voltage gradient. This would dissipate the proton gradient, stop ATP synthesis and metabolic activity, and result in the death of the cell.

\+ + ++ -+ +- + +
+KWKLFKKIEKVGQ/NIRDGIIKAGPAVAVVGQATQIAK-NH2
1 23 37
Cecropin A

\+ ++++
+GIGAVLKVLTTGL/PALISWIKRKRQQ-NH2
1 13 26
Melittin

\+ + ++ -+ +
+KWKLFKKIEKVGQ/GIGAVLKVLTTGL-NH2
1 13 1 13
CA(1-13)Me(1-13)-NH2

\+ + ++ +
+KWKLFKK/IGAVLKVL-NH2
1 7 2 9
CA(1-7)Me(2-9)-NH2

Fig. 1. Sequences of antibacterial peptides.

To learn more about the effects of various structural features of cecropin-melittin hybrids on their activity, we have synthesized the parent hybrid, and the enantio, retro, and retroenantio isomers of $C_A(1\text{-}13)Me(1\text{-}13)\text{-}NH_2$ and $C_A(1\text{-}7)Me(2\text{-}9)\text{-}NH_2$, and their N^{α}-acetyl derivatives (Fig. 1). This allows us to predict and measure the importance of sequence, direction of peptide bond, N-terminal charge, chirality of the amino acids and handedness of the helix.

We define the structures as follows: The parent peptides contain all L amino acids linked by normal peptide bonds in the sequence 1, 2, 3---n, numbered from the N-terminus (Fig. 2). The enantio peptides contain all D amino acids linked by normal peptide bonds in the sequence 1, 2, 3---n. They are the mirror images of the parent peptides. The corresponding two N-acetyl peptides are also mirror image enantiomers. The retro peptides contain all L amino acids linked by the normal peptide bonds, but in the reverse sequence n---3, 2, 1. The retroenantio peptides (Fig. 2) are mirror images of the corresponding retro peptides. They contain all D amino acids linked by normal peptide bonds, in the direction n---3, 2, 1. Similarly, the acetyl peptides are also mirror images. In both of these cases it seems reasonable to use the term enantio for these linear, all-D, mirror-image isomers.

The retro and retroenantio peptides can be viewed from the opposite end by rotating in the plane 180°, in which case the sequences will read 1,2,3---n, but the amide bonds will be -NH-CO- instead of the normal -CO-NH-. If the peptide is a linear random chain or ß-sheet conformation, the amino acid side chains of a retro peptide, when viewed in this reverse direction, will project to the opposite side and resemble a D amino acid. The retroenantio all-D peptide (Fig. 2) side chains will, however, project to the same side as those of the parent all-L peptide. It will therefore have the same sequence and the same side chain topology as the parent, but the amide bonds will be reversed and the charges on the end groups will be reversed so it will not be a mirror image of the parent peptide. For this reason Goodman and Chorev [11] have preferred to name such an isomer a retro-inveso peptide. Many papers and reviews on retroenantio and retroinverso peptides have appeared (see 11-14).

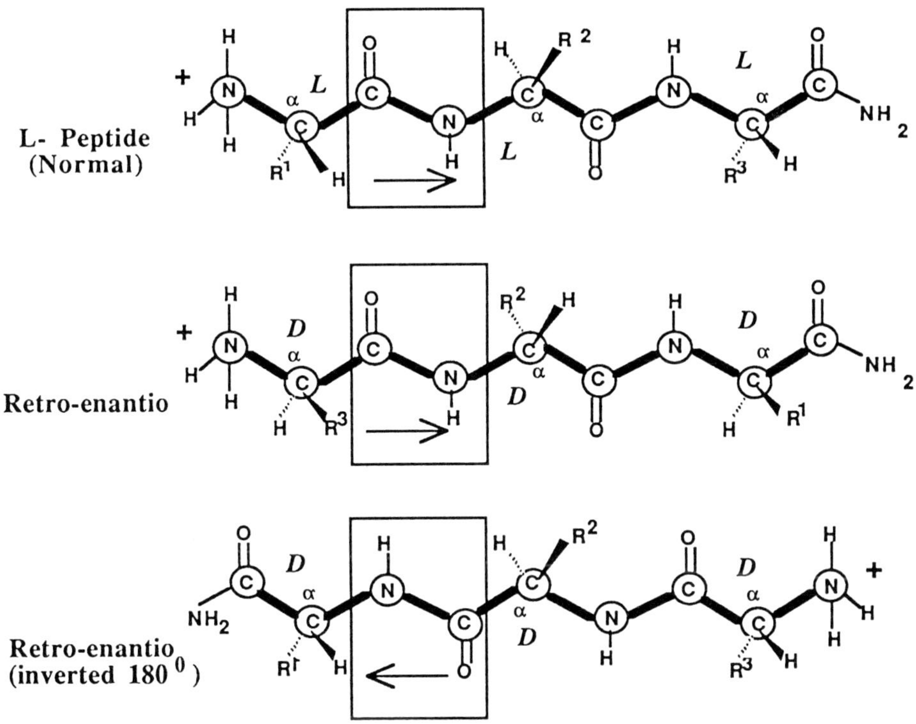

Fig. 2. Linear structures of peptide analogs.

We have recently described some initial experiments on the identification of which structural features of these analogs are critical for the development of antibacterial activity in the hybrid C_A(1-13)Me(1-13)NH_2 [9]. First, assumptions were made as to which features were necessary, i.e. must be the same as in the parent peptide, either singly, in pairs, or in triplets, and then the activity was predicted. This gave a + or -

pattern for the four analogs under study (Table 1). Then the analogs were tested for activity against six representative bacteria (Table 2). This also gave a + or - pattern that could be compared with the predicted patterns. It was found that five of the organisms responded similarly and that they corresponded to the expected activities if, singly, sequence, or amide direction, or chirality was necessary, but not if two or more of these features were required at the same time. Therefore, the essential feature was either sequence, chirality or amide bond direction, but from these limited data we cannot decide which of the three is critical. We already had presented evidence [4] that chirality need not be the same as the all-L parent peptide, because in 10 separate examples, the corresponding all-D peptide was equally as active as the all-L peptide. However, this simple conclusion is open to some doubt based on the above discussion. In addition all of the data from our laboratory and from others indicate that sequence does play an important role and all sequences are not active. We conclude that sequence is the important factor and therefore that chirality and direction of amide bond need not be critical. Thus, for these five organisms, both -NHCO- and -CONH- bonds in all-D and all-L structures lead to active peptides.

Table 1 *Predicted activity of analogs assuming active peptides are random*

Required structure	Normal	Enantio	Retro	Retro-enantio
1. Sequence	+	+	+	+
2. Amide bond direction	+	+	+	+
3. Chirality	+	+	+	+
4. Sequence+amide	+	+	-	-
5. Sequence+ chirality	+	-	-	+
6. Amide + chirality	+	-	+	-
7. Sequence + amide + chirality	+	-	-	-

The sixth organism, *B. subtilis*, was more demanding. Its pattern (Table 2) was ++-- and only fit the predicted pattern for a requirement for both correct sequence and correct amide bond direction. In this case, -NH-CO- bond did not allow formation of active structures. For all of these peptides the acetylated and unacetylated peptides had equivalent activities, indicating that the N amine could be charged or not, with retention of activity.

In contrast to the random configuration, if we consider these antibacterial peptides to be α-helical structures in their active conformations, a different relationship will result (Fig. 3).

Table 2 *Observed activity of $C_A(1\text{-}13)Me(1\text{-}13)NH_2$ and analogs*

Organism	Normal	Enantio	Retro-enantio	Predicted Fit
E. coli	+	+	+	Sequence
P. aeruginosa	+	+	+	or
B. megaterium	+	+	+	amide
Strep. pyogenes	+	+	+	or
Staph. aureus	+	+	+	chirality
B. subtilis	+	+	-	Seq. + amide

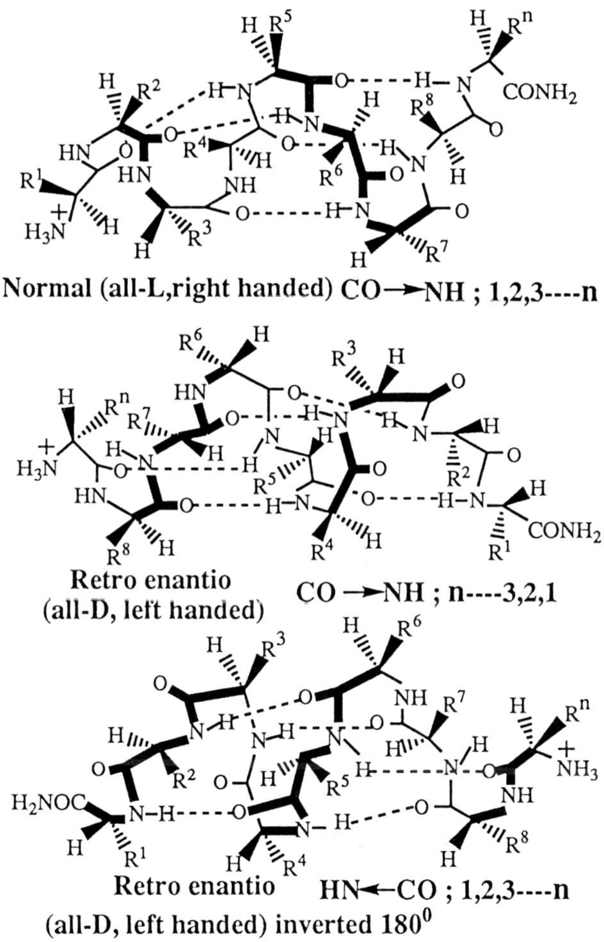

Fig. 3. Helical structures of peptide analogs.

This assumption that the active cecropins and hybrids are amphipathic helices, is well supported by data from several laboratories. The parent peptide will be all-L and a right-handed helix. The enantiomer will be all-D and a left-handed helix. They will be mirror images. This is supported by the CD spectra, which are exact mirror images. The retro peptide will contain all L amino acids in normal peptide bonds and a right-handed helix, but of opposite sequence (n---3,2,1). The retroenantio peptide will be all D amino acids in normal peptide bonds but of opposite sequence (n---3,2,1) and in a left-handed helix. It is a mirror image of the retro peptide, and the N^{α}-acetyl peptides are also mirror images.

Table 3 *Predicted activity of analogs assuming active peptides are helical*

Required Structure	Normal	Enantio	Retro	Retro enantio	Acetyl normal	Acetyl retro	Acetyl retroenantio
1. Sequence	+	+	+	+	+	+	+
2.Amide bond direction	+	+	+	+	+	+	+
3. Chirality/handedness	+	-	-	-	+	+	-
4. Sequence + amide	+	+	-	-	+	-	-
5. Sequence + chirality	+	-	-	-	+	-	-
6. Amide + chirality	+	-	+	-	+	+	-
7.Seq.+amide+chirality	+	-	-	-	+	-	-

However, when the retroenantio peptide is viewed from the opposite end, it will have the same sequence as the parent peptide and the side chains will project from the same side as in the L peptide, but the amide bonds will be reversed, the end group charges will be reversed and in addition the conformation of the helix will remain left-handed in contrast to the right-handed helix of the parent peptide. We therefore make different predictions than when the active structure is considered to be a random linear chain. As shown in Table 3, it is expected that the helical conformation will not allow the chirality or helical handedness of the enantio, retroenantio, or Ac-retroenantio peptides to be the same as the normal parent peptide, even if sequence or amide bonds are inverted. Therefore, we predict that these analogs will be inactive if chirality or handedness is an essential structural feature. Since all of the analogs were fully active towards four of the test organisms, *E. coli, P. aeruginosa, Strep. pyogenes,* and *Staph. aureus*, (Table 4), the fit with prediction is with sequence only or with amide direction only, and chirality is eliminated as an essential feature. With the helical model, *B. subtilis* still fits the prediction that both sequence and amide bond direction must be the same as the parent peptide. For this organism also chirality is eliminated as an essential feature.

The data in Table 5 are the observed activities of C_A(1-7)Me(2-9)NH_2 and analogs. For this shorter, 15-residue, hybrid all isomers were active towards four of the test organisms, *E. coli, B. subtilis, B. megaterium,* and *Staph. aureus.* As with the longer 26-residue peptide, these data match the predictions for either sequence alone or amide direction alone, but not for the two together and not for chirality as a necessary requirement. The data for *P. aeruginosa* agree with the prediction that both sequence and amide direction must be correct.

Table 4 *Observed activity of C_A(1-13)Me(1-13)NH_2 and analogs*

Organism	Normal	Enantio	Retro	Retroenantio	Acetyl normal	Acetyl retro	Acetyl retro enantio	Predicted fit
E. coli *P. aeruginosa* *Strep. pyogenes* *Staph. aureus*	+ + + +	+ + + +	+ + + +	+ + + +	+ + + +	+ + + +	+ + + +	Sequenc e or amide
B. subtilis	+	+	-	-	+	-	-	Seq. + amide

Table 5 *Observed activity of C_A(1-7)Me(2-9)NH_2 and analogs*

Organism	Normal	Enantio	Retro	Retro enantio	Acetyl retro	Acetyl retroenantio	Predicted fit
E. coli *B. subtilis* *B. megaterium* *Strep. pyogenes*	+ + + +	+ + + +	+ + + +	+ + + +	+ + + +	+ + + +	Sequence or amide
Staph. aureus	+	+	-	-	+	+	no NH_3+ on residue n
P. aeruginosa	+	+	-	-	-	-	Seq. +amide

In the case of *Staph. aureus* a novel result was obtained. The retro and retroenantio analogs were both inactive (only 0.5 to 1 % active) but the corresponding acetylated derivatives were active. This indicates that the location of the NH_3^+ charge can have a profound effect. In these two isomers the NH_3^+ is on residue n (Leu) rather than residue 1 (Lys). The latter is in the basic end of the molecule whereas the former is in the more hydrophobic, less charged end. Similar results were not found in the 26-residue series and suggest that the + charge alteration has a much larger effect in the small peptides.

Overall conclusions from the retro and retroenantio analogs:

1. All organisms do not respond alike.
2. All organisms show that chirality need not be the same as in the parent peptide.
3. Most organisms show that either sequence or amide must be the same as in the parent peptide (not both).
4. Some organisms do require both sequence and amide to be the same as in the parent peptide.
5. One strain was resistant if the NH_3^+ is moved from residue 1 (Leu) to residue n (Lys), and activity was recovered by neutralizing with an amide.
6. We know a little more about what makes peptides antibacterial, but there is more to learn.

Overall, the data support the idea that these peptides function by self aggregation into ion conducting pores and not by binding to receptors or other specific chiral structures. It then is easier to understand that peptides with all D amino acids or reversed amide bonds or reversed sequences can aggregate with themselves and form pores that will pass ions and lead to the death of a cell.

In summary, the areas where chemical synthesis has been of value for the study of cecropin and analogs are:

1. Structure proof for sequence and C-terminal amide.
2. Synthesis of more active and broader spectrum analogs.
3. Processing of precursors and activity of intermediates.
4. Evidence for ion channels in membranes.
5. Determination of the effect of chirality on activity and ion channels by synthesis of D enantiomers.
6. Determination of the effect of peptide sequence, amide bond direction and end group charges by synthesis of retro and retroenantio isomers.

Acknowledgements

This work was supported by U.S. Public Health Service Grant DK 01260 (to B.M.), The Swedish Natural Science Council Grant BU 2453 (to H.B.) and CICYT (PB 89-0257), CIRIT, and the U.S Spanish Joint Committee for Cultural and Scientific Cooperation (to D.A.).

References

1. Hultmark, D., Steiner, H., Rasmuson, T. and Boman, H.G., Eur. J. Biochem., 106(1980)7.
2. Habermann, E. and Jentsch, J., Hoppe-Seyler's Z. Physiol. Chem., 348(1967)37.
3. Lehrer, R.I., Lichenstein, A.K. and Ganz, T., Ann. Rev. Immunol. 11(1993)105.
4. Zasloff, M., Proc. Natl. Acad. Sci. USA, 84(1987)5449.
5. Kawano, K., Yonego, T., Miyato, T., Yoshikawa, K., Tokunaga, F., Terada, Y. and Iwanaga, S., J. Biol. Chem. 265(1990)15365.
6. Boman, H.G., Boman, I.A., Andreu, D., Li, Z.-Q., Merrifield, R.B., Schlenstedt, G. and Zimmermann, R., J. Biol. Chem. 264(1989)5852.
7. Merrifield, R.B., Vizioli, L.D. and Boman, H.G., Biochemistry, 21(1982)5020.

8. Boman, H.G., Wade, D., Boman, I.A., Wåhlin, B. and Merrifield, R.B., FEBS Lett., 259(1989)103.
9. Christensen, B., Fink, J., Merrifield, R.B. and Mauzerall, D., Proc. Natl. Acad. Sci. USA, 85(1988)5073.
10. Wade, D., Boman, A., Wahlin, B., Drain, C. M., Andreu, D., Boman, H. G. and Merrifield, R.B., Proc. Natl. Acad Sci. USA, 87(1990)4761.
11. Goodman, M. and Chorev, M., Acc. Chem. Res., 12(1979)1.
12. Prelog, V. and Gerlach, H., Helv. Chim. Acta, 47(1964)2288.
13. Shemyakin, M.M., Ovchinnikov, Y.A. and Ivanov, V.T., Angew. Chem. Int. Ed. Engl., 8(1968)492.
14. Chorev, M. and Goodman, M., Acc. Chem. Res., 26(1993)266.
15. Merrifield, R.B., Merrifield, E.L., Juvvadi, P., Andreu, D. and Boman, H.G., Ciba Foundation Symposium 186 -- Antibacterial Peptides., 1994, In press.

Approach to enzymatic synthesis of lantibiotics

Tetsuo Shiba[a], Kaoru Inami[a], Kiichiro Nakajima[a], Michihiko Kataoka[b] and Sakayu Shimizu[b]

[a]Peptide Institute, Protein Research Foundation, 4-1-2 Ina, Minoh, Osaka 562, Japan

[b]Faculty of Agriculture, Kyoto University, Sakyo-ku, Kyoto 606, Japan

Introduction

Lantibiotic is a name for biologically active peptides containing a lanthionine loop. Molecular basis of the chemistry of lantibiotics was established by our total synthesis of nisin, an antibiotic peptide produced by *Lactococcus lactis*, in 1987 [1-3]. It was the first chemical synthesis of a natural lanthionine peptide. Nisin is now widely used as food preservative.

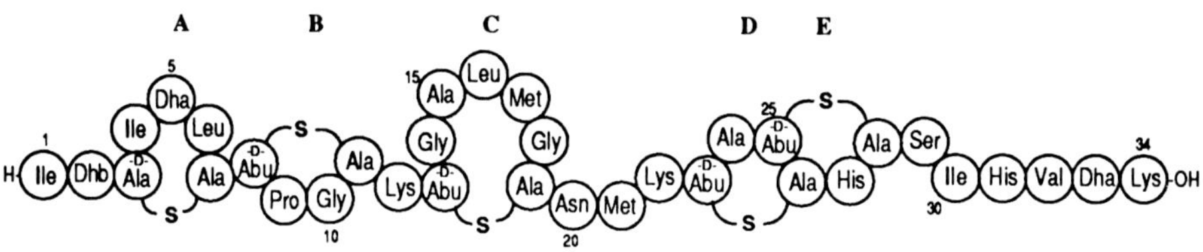

nisin

Application of lanthionine peptides has been recently extended to food preservatives, antibiotics, enzyme inhibitors, and other multiple biologically active substances. Structural determination of a unique triply cyclized lanthionine peptide for ancovenin as an enzyme inhibitor, and a lanthiopeptin (cinnamycin) as an antiviral agent has also been achieved in our group [4,5].

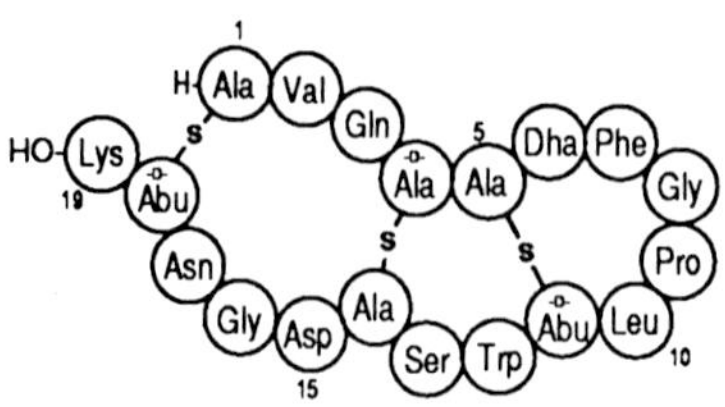

ancovenin

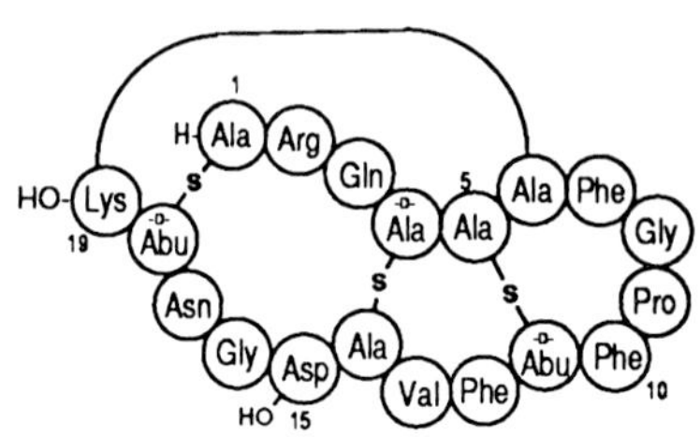

lanthiopeptin (cinnamycin)

Many lantibiotics besides our peptides have been found and studied. Furthermore, the biosynthetic mechanism of lanthionine peptides has been extensively studied [6]. The growing activity in this field has necessitated a supply of authentic samples of various lanthionine peptides which may be better prepared by chemical synthesis. However, the time-consuming and tedious procedures for chemical synthesis of lantibiotics as seen in the case of nisin synthesis may inhibit further attempts at total syntheses of other natural lanthionine peptides. Particularly, the difficulty of preparing synthon amino acids, such as sterically pure methylcysteine, as well as the desulfurization step, and introduction of dehydroamino acid residues, might discourage plans for attempting new syntheses of lantibiotics.

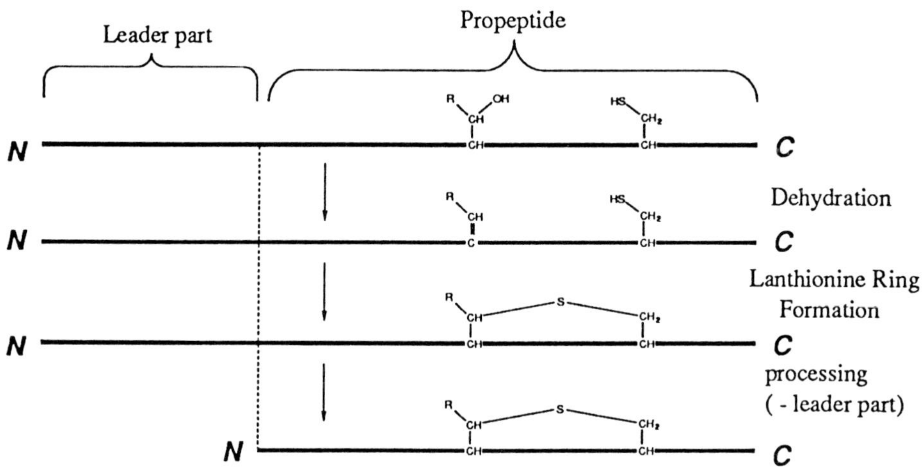

Fig. 1.

We have, therefore, attempted a novel strategy for semisynthesis of lantibiotics through enzymatic formation of a lanthionine ring from open-chain peptide comprising serine or threonine and cysteine. Biosynthetic study of lantibiotics has taught us that in a modified prepeptide connecting with leader peptide at *N*-terminus of propeptide, enzymatic dehydration at the hydroxy amino acid residue followed by lanthionine ring formation between unsaturated amino acid and cysteine residues may proceed, although details of ring formation are not yet completely known. Finally, processing to remove *N*-terminal leader peptide part produces a desired lanthionine peptide. Such a biosynthetic process may be designated schematically as depicted in Fig. 1.

Apart from the biosynthesis mechanism, if one can find a special enzyme which might act as a catalyst to form a lanthionine ring from a shortened peptide corresponding to a propeptide involving hydroxy amino acid and cysteine residues without a leader peptide, it would provide a new strategy for the formation of lanthionine peptide. This would entail a one-step enzymatic reaction from an open-chain peptide prepared easily by conventional chemical synthesis. In other words, an appropriate enzyme may be used as a tool, like a chemical reagent, in this strategy.

Results and Discussion

Substrates for enzymatic reaction were prepared by ABI Peptide Synthesizer 430A on Merrifield resin through Boc strategy followed by purification by reverse-phase HPLC. The following two peptides, corresponding to the propeptides of ancovenin and nisin, were synthesized. Lanthionine and methyllanthionine residues were replaced by L-Ser/ L-Cys and L-Thr / L-Cys residues, respectively.

I **C**VQ**SC***S*FGPL**T**W*S***C**DGN**T**K

II I*T***S**I*S*L**CT**PG**C**K**T**GALMG**C**NMK**T**A**TC**H**C**SIHV*S*K

Enzyme solutions tested for the formation of lanthionine peptides from the above propeptides were investigated and prepared from the culture broth of many microorganisms which were preserved at Faculty of Agriculture, Kyoto University, as shown in Fig. 2 [7].

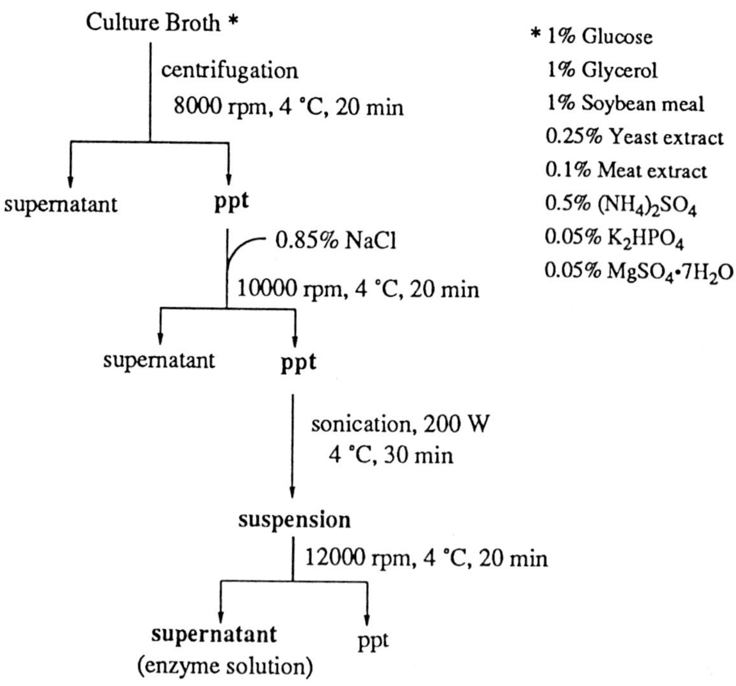

Fig. 2.

An enzymatic reaction was performed in 0.5M potassium phosphate buffer (pH 7.0) solution of the substrate in the presence of pyridoxal phosphate at 37° C for 15 hours. After centrifugation, the supernatant was directly hydrolyzed by 6N HCl and applied to amino acid analyzer in order to check for formation of either lanthionine or methyllanthionine as a preliminary screening test.

Out of about fifty species of microorganisms tested, some of *Rhodococcus* as well as *Streptoverticillium* species clearly showed the formation of lanthionine and methyllanthionine in the product hydrolyzate [8].

The addition of pyridoxal 5'-phosphate known as a coenzyme of cystathionine synthase may prevent formation of unnatural isomers of 3-methyllanthionine and yield predominantly the natural threo form. In one experiment using *Streptoverticillium* for nisin propeptide II, high-performance liquid chromatography after enzymatic reaction clearly showed the formation of nisin molecule, as shown in Fig. 3.

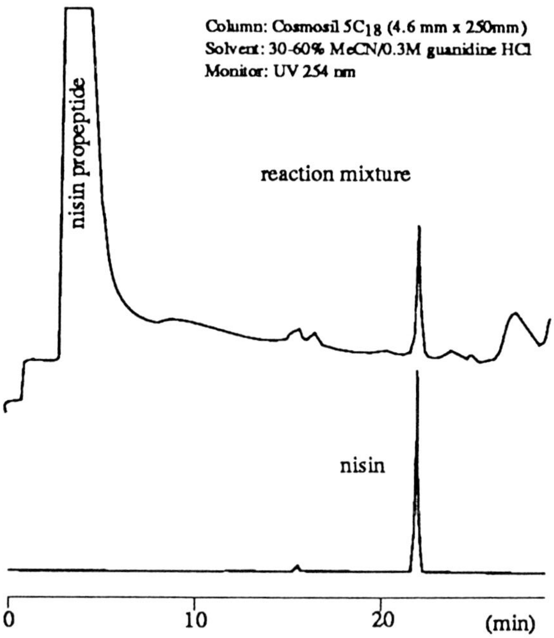

Fig. 3.

For isolation of the enzyme from the cultivation broth of the microorganism, protein obtained through centrifugation followed by precipitation with ammonium sulfate was dialyzed by means of cellulose tubing of Union Carbide Corp. against 20 mM of potassium phosphate buffer and then fractionated by Pharmacia DEAE-Sephacel in a gradient concentration of 0 M to 0.6 M sodium chloride solution. Further purification of the enzyme is now in progress.

References

1. Fukase, K., Kitazawa, M., Sano, A., Shimbo, K., Fujita, H., Horimoto, S., Wakamiya, T., and Shiba, T., Tetrahedron Lett., 29(1988)795.
2. Wakamiya, T., Fukase, K., Sano, A., Shimbo, K., Kitazawa, M., Horimoto, S., Fujita, H., Kubo, A., Maeshiro, Y., and Shiba, T., In Jung, G. and Sahl, H.-G. (Eds.) Nisin and Novel Lantibiotics, ESCOM, Leiden, 1991, pp. 189–203.
3. Fukase, K., Kitazawa, M., Sano, A., Shimbo, K., Horimoto, S., Fujita, H., Kubo, A., Wakamiya, T., and Shiba, T., Bull. Chem. Soc. Jpn., 65(1992)2227.
4. Wakamiya, T., Ueki, Y., Shiba, T., Kido, Y., and Motoki, Y., Bull. Chem. Soc. Jpn., 63(1990)1032.
5. Wakamiya, T., Fukase, K., Naruse, N., Konishi, M., and Shiba, T., Tetrahedron Lett., 29(1988)4771.
6. Jung, G. and Sahl, H.-G. (Eds.) Nisin and Novel Lantibiotics, ESCOM, Leiden, 1991.
7. Kido, Y., Hamakado, T., Yoshida, T., Anno, M., Motoki, Y., Wakamiya, T., and Shiba, T., J. Antibiot., 36(1983)1295.
8. The preliminary result was reported in a review paper: Tetsuo Shiba, in Chemistry of Lantibiotics, New Aspects of Organic Chemistry II, Z. Yoshida and Y. Ohshiro-Eds., Kodansha, Tokyo, 1992, pp. 429.

Self-assembly of N-phosphoamino acids

Yong Ju, Yu-Fen Zhao and Yi Chen
Bioorganic Phosphorus Chemistry Laboratory, Department of Chemistry, Tsinghua University, Beijing 100084, China

Introduction

Phosphorus and amino acids play an important roles in life chemistry. Many reports on peptide formation reactions in the process of chemical evolution have been published [1-4]. Although, the proteins formed on the primitive earth are still under discussion, there is not much investigation of the intrinsic relationship between phosphorus and amino acids. It is not easy to understand the chemistry of these macromolecules, and a systematic investigation of phosphoryl amino acids might help elucidate the above question. Therefore, amino acids were phosphorylated and their chemical properties were investigated [5-8].It seems that the existence of a phosphoryl group leads to many interesting chemical properties. For example, the N-phosphoryl serine tends to self-activation and to result in peptides, esters and phosphoryl esterification, N-O migration of the phosphoryl group, etc, in which the phosphoryl-participating reactions are the basis of bio-catalysis [8–11].

Results and Discussion

Phosphoryl group promoting effect Because the biological reactions usually take place in aqueous media, the properties of N-phosphoryl amino acids were studied in aqueous solution at 38°C. It was found that the N-phosphoryl amino acids can self-catalyze to form homopeptides or phosphorylated peptides with other amino acids and amino acid esters.

$$(i\text{-}PrO)_2P(O)NHCHR^1COOH + H_2NCHR^2COOR^3 \xrightarrow{-H_2O} (i\text{-}PrO)_2P(O)NHCHR^1CONHCHR^2COOR^3$$

In our previous papers [12,13], FAB mass spectrometry was a very useful method for determining the structure of phosphoryl peptides. In this way, possible peptide structure can be established on the basis of the molecular ion or pseudomolecular ion and the corresponding fragment ions in the FAB-MS (such as: M+, M-42, M-42x2 in positive ion FAB-MS and M-1, M-42-1 in negative ion FAB-MS,etc)(Table 1), and the amides bond absorption (about $1650cm^{-1}$) in the IR spectra.

It is also found that self-activation is a unique character of N-phosphoryl amino acids. In the absence of any catalyst or activators, N-phosphoryl amino acids could perform the peptide formation reaction at low temperature. It seems that all these reactions are attributed to the presence of the N-phosphoryl group. Since under similar conditions, no peptide formation reaction occurred for other types of N-protected amino acids, such as Boc-Pro, Z-Gly, (where Boc=$(CH_3)_3OCO$-, Z=Ph-CH_2OCO-) have also been incubated in n-butanol for 15h. at 40-60°C but which no peptides were detected (IR, FAB-MS)[5].

Table 1 *The peptides formation of some N-phosphoryl amino acids with amino acids and amino acid esters in water at 38°C for a week*

Reagent	Possible peptide	M+ and fragment ions in positive FAB-MS (m/z)	M- and fragment ions in negative FAB-MS (m/z)
DIPPHis	DIPPHisHis	457(M+1, 3.5%), 372(M-42X2, 10%)	455(M-1), 413(M-42-1)
	DIPPHis-$(His)_2$	594(M+1, 0.1%), 511(M-42X2+1, 4.5%)	492(M-1), 508(M-42X2-1)
	His-His-His		428(M-1)
	DIPPHis-$(His)_3$		729(M-1), 646(M-42X2-1)
	His-$(His)_3$		586(M), 585(M-1)
DIPPCys	DIPPCysCys	388(M, 1%), 346(M-42, 1%)	388, 346, 304, 342(M-46)
DIPPLys	DIPPLysLys	439(M+1, 0.3%)	438, 396, 354
DIPPAla	DIPPAlaAla	325(M+1, 1%), 282(M-42, 8%)	323, 324, 279, 236
DIPPGlu	DIPPGluGlu		440(M), 398(M-42), 356(M-42X2)
DIPPHis	DIPPHisCys	423(M+1, 1%)	422(M), 380(M-42)
+ Cys		381(M-42+1, 2%)	348(M-42X2)
DIPPHis	DIPPHisCysOBu	478(M, 2%)	
+ CysOBu			
DIPPHis	DIPPHisGlyGly	433(M, 1%), 391(M-42, 2%)	
+ GlyGly	DIPPHisHisGlyGly	571(M, 1%)	432(M-1), 433(M), 390(M-42-1)
DIPPHis	DIPPHisGlyGlyOBu	490(M, 0.5%)	
+ GlyGlyOBu		405(M-42X2, 1%)	489(M), 447(M-42)
DIPPGlu			405
+ GlyGlyOBu	DIPPGluGlyGlyOBu	482(M+1, 0.2%), 440(M-42+1, 1%)	
DIPPAla			
+ GlyGlyOBu	DIPPAlaGlyGlyOBu	424(M, 0.3%)	
DIPPCys			
+ GlyGlyOBu	DIPPCysGlyGlyOBu	456(M+1, 0.2%), 371(M-42X2, 4.8%)	

Phosphoryl group differentiation effect It is known that histidine is present in the active site of many enzymes, such as chymotrypsin, pancreatic ribonuclease, and carboxyl peptidase, etc. and that phosphoryl histidine is present as an intermediate in reactions involving acid phosphatase, phosphofructokinase, and histamine kinase [11]. Thus, it is a very interesting to study the properties of phosphoryl histidine. Studies on phosphoryl transfer reactions of phosphoryl histidine have been reported [7].

Compared to other amino acids, phosphoryl histidine in aqueous solution can form phosphoryldipeptide and tripeptide, other phosphoryl amino acids while yield only dipeptide. The high reactivity of N-phosphorylhistidine could be explained by the intramolecular catalytic effect of the imidazole group. Because of the participation of the imidazole group, phosphorus might be transformed into a hexacoordinate intermediate. The direct P-N bond from the participation of imidazole has previously been proposed [7].

Lipmann F.[14] proposed that the biosynthesis of protein proceeded through a mixed carboxylic-phosphoric anhydride intermediate. Thus, based on these experimental results and some other findings [5,8], it seems that an intramolecular mixed carboxylic phosphoric anhydride intermediate may exist in the reaction process.

Conclusion Our investigation found that the N-phosphoryl amino acids (especially N-phosphoryl- histidine) could not only activate themselves to yield peptides, esters and phosphoryl ester-exchanged products in a non-aqueous environment [5,6], but also could have similar properties in aqueous solution. The biological reaction usually occurs in water solution. Thus, this result might be responsible for the properties of phosphoryl amino acids in living chemistry. The participation of the phosphoryl group may be the clue to the function of the phosphorylated enzyme in many important bioprocesses. It also might be important for prebiotic protein synthesis.

Acknowledgements

The Authors thanks the China Postdoctoral Science Foundation (YJ) and National Natural Science Foundation of China (YFZ) for the financial support.

References

1. Miller, S.L. and Orgel, L.E., The Origins of Life on the Earth, Prentic Hall Inc., Englewood Cliffs, New Jersey, U.S.A., 1974.
2. Fox, S. W. and Dose, K., Molecular Evolution and the Origins of Life, 2nd, Marcel Dekker, New York, 1972.
3. Kovacs, J. and Nagy, H., Nature, 19(1961)531.
4. Munegumi, T., Suzuki, N., Tanikawa, N. and Harada, K., Chem. Lett., (1992)1676.
5. Li, Y.M., Yin, Y.W. and Zhao, Y.F., Int.J.Peptide Protein Res., 39(1992)375.
6. Ma, X.B. and Zhao, Y.F., Synthesis, 8(1992)759.
7. Li, Y.C., Tan, B. and Zhao, Y.F., Heteroatom Chemistry, 4(1992)415.
8. Zhao, Y.F., Li, Y.M., Yin, Y.W. and Li, Y.C., Science in China(Series B), 36(1993)1451.
9. Zaug, A.J. and Cech, T.R., Science, 236(1987)1532.
10. Bass, B.L. and Cech, T.R., Nature, 308(1986)820.
11. Burnel, N. and Natch, M.D., Arch Biochem Biophys, 239(1984)175.
12. Ma, X.B. and Zhao, Y.F., Bio Mass Spectrometry, 20(1991)498.
13. Yin, Y.W., Ma, Y., Zhao, Y.F., Xia, B. and Wang, G.H., Org. Mass Spectrum, 29(1994)201.
14. Lipmann, F., Advances in Enzymology and Related Subject, International Science Publishers, New York, 1941, pp97-152.

Hemolytic and antibacterial activity of synthetic deletion peptides of melittin

Li-Ping Liu, Zhi-Dong Xie, Hu-Sheng Yan, Xiao-Hui Cheng and Bing-Lin He
Institute of Polymer Chemistry, Nankai Uiversity, Tianjin 300071, China

Introduction

Melittin, the principal component of honey bee (*Apis mellifera*) venom, is an amphi-pathic peptide consisting of 26 amino acids: IGAVLKVLTTGLPALISWIKRKR-QQ-NH2. It has many important biological functions, such as hemolytic, antibacterial and anti-inflammatory activity [1]. Melittin can also raise the survival rate of irradiated animals from 0.5% to 50%. At present, melittin is used to treat diseases sunch as rheumatism and psoriasis. Although there is a great difference between the treatment doses and the lethal dose, melittin still has side-effects, including hemolytic anaemia and kidney damage because of its high hemolytic activity. These side-effects limit the use of melittin. If we can determine its active center, we will try to decrease its hemolytic activity without changing its other effects.

Results and Discussion

(1) Synthesis and purification of the peptides: The peptides synthesized in this work were a series of melittin deletion peptides, whose amino acid sequences are listed in Table 1. The glycyl residue at the C-terminus was added to facilitate synthesis. For deprotecting N^{in}-formyl Trp, EDT was added during the final anhydrous liquid HF cleavage treatment (HF:EDT=95:5) [2]. Each purified peptide showed only one peak on the analytical HPLC column. Amino acid analyses of the purified peptides revealed residue ratios identical with those expected.

Table 1 *Amino acid sequences of the synthetic peptides*

Peptide	Sequence
Mel 12	A-L-I-S-W-I-K-R-K-R-Q-G-NH_2
Mel 13	P-A-L-I-S-W-I-K-R-K-R-Q-G-NH_2
Mel 14	L-P-A-L-I-S-W-I-K-R-K-R-Q-G-NH_2
Mel 15	G-L-P-A-L-I-S-W-I-K-R-K-R-Q-G-NH_2
Mel 16	T-G-L-P-A-L-I-S-W-I-K-R-K-R-Q-G-NH_2

(2) Comparison of the hemolytic activity of melittin and the synthetic peptides: Melittin and the five deletion peptides were tested for hemolytic activity using human blood cells. Melittin shows distinct hemolysis at low concentrations. From the hemolysis curve, we can determine that the HD50 of melittin is 0.88μM (Fig. 1). All of the

synthetic deletion peptides show slight hemolysis until their concentrations reach 100μM and the HD_{50} values are above 300μM.

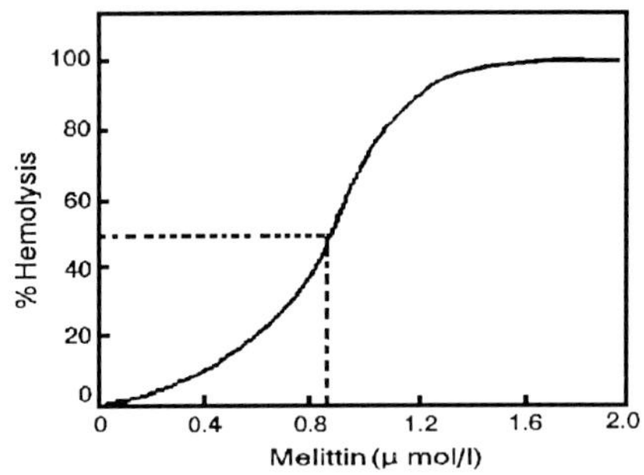

Fig. 1. Hemolysis curve of melittin.

(3) Antibacterial activity: Melittin inhibits the growth of Gram-positive (+) bacteria more effectively than that of Gram-negative(-) bacteria. Therefore we used *S. aureus* and *B. subtilis* as the tested bacteria. The lowest concentrations (MIC) of melittin and the synthetic deletion peptides that can inhibit the growth of bacteria are shown in Table 2.

Table 2 *Minimum inhibitory concentration (MIC in μg/ml) of melittin and synthetic deletion peptides against bacteria*

Peptide	MIC against S. aureus (μg/ml)	MIC against B.subtilis (μg/ml)
Melittin	50	100
Mel 12	100	200
Mel 13	50	50~100
Mel 14	50	100
Mel 15	50	100
Mel 16	50	100

From Table 2, we can see that among the five deletion peptides, only Mel12 has lower antibacterial activity than melittin. The other four peptides have the same activity as melittin. Blondelle *et al* have synthesized twenty-four individual omission analogues of melittin to study the relation between the structure and bioactivity [3]. In these analogues, they found that when hemolytic activity is high, antibacterial activity will also be high. Contrary to their results, the hemolytic activity of the five deletion peptides we synthesized is very low, but antibacterial activity changes little. We can therefore draw the following conclusions: (1) potent antibacterial activity seems a less sequence-dependent than the lysis of red blood cells; (2) there are some differences between hemolytic and antibacterial mechanisms, and the latter may be more complicated. These results lay a foundation for investigating further the other pharmcologic activities of melittin fragments. It is possible to develop highly-effective, low-toxic melittin preparations based on the studies.

Acknowledgement

This work is supported by the Science and Technology Development Foundation of Nankai University.

References

1. Habermann, E. and Jentsch, J. , Hoppe-Seyler's Z. Physiol. Chem., 348(1967)37.
2. Matsueda, G.R., Int. J. Peptide & Protein Res., 20(1982)26.
3. Blondelle, S.E. and Houghten, R.A., Biochemistry, 30(1991)4671.

Synthesis of a fully active HIV-1 protease analogue by a chemical ligation approach

Chuan-Fa Liu, Chang Rao and James P. Tam
Department of Microbiology and Immunology, A-5119 Medical Center North, Vanderbilt University, Nashville, TN 37232-2363, U.S.A.

Introduction

Despite recent advances in recombinant technology, the chemical synthesis of peptides and proteins has distinct advantages and remains an active field in chemistry and biochemistry [1]. The chemical approach can overcome the limitations of the biological protein assembly because of its great generality and flexibility. The ability to alter and create protein functions is greatly enhanced and expanded through the incorporation of non-natural structure units and through the design of unusual architectures [2].

With the perfection of the chemistry for peptide bond formation and for functional group protection and deprotection, small to moderate size peptides ranging from 10 to 50 residues are now routinely prepared by the automated solid phase synthesis method. Strategically, segment condensation is an obviously appealing approach to obtaining more homogeneous products in the synthesis of large peptides, since the intermediate segments in use can be thoroughly purified and characterized. However, the conventional segment condensation approach is generally not applicable to large, protected peptide fragments due to the intrinsic entropic barrier of a bimolecular reaction between large molecules, which is further greatly aggravated by the presence of protecting groups. The development of mild and selective activation methods for the carboxyl components has produced some limited improvements resulting from a decreased need for protecting groups [3, 4]. Accordingly, it is clear that the final solution to this problem must depend on the development of new approaches which can overcome the entropic problem of the bimolecular reaction. For this purpose, we have recently developed a new chemical ligation approach in which the entropy problem of the usual intermolecular peptide-bond-forming process is bypassed through the use of an efficient intramolecular reaction [5, 6]. In our approach, the respective C^{α}-carboxyl and N^{α}-amino groups of two unprotected peptides are first brought together through a highly specific ring-forming reaction betweeen an alkyl aldehyde and a β-functionalized N-terminal amino group. An amide bond is then formed between the two groups through an intramolecular O,N-acyl transfer reaction. The resulting thiazolidine ring product is a proline-like structure, which can then be used as a proline surrogate. The feasibility and potential of this scheme has been verified by studies on small model compounds and by the synthesis of peptides of up to 50 residues. In this report, we describe the synthesis of an analogue of the HIV-1 protease to further demonstrate the usefulness and effectiveness of this strategy. We chose HIV-1 protease as our model synthetic protein because of its critical role in viral replication, its suitable size, and its established three-dimentional structure.

Results and Discussion

As reported previously, the synthetic scheme of our ligation approach consists of three steps: (1) aldehyde initiation, in which a masked glycolaldehyde ester is linked to the carboxyl terminus of an unprotected peptide; (2) ring formation, in which the unmasked aldehyde reacts with the N-terminal cysteine residue of the second peptide to form a thiazolidine ring at acidic pH; (3) O,N-acyl rearrangement, in which O-acyl ester linkage is transferred to N-acyl amide linkage to form a peptide bond with a pseudoproline structure.

The HIV-1 protease, a member of the aspartic protease family, acts as a homodimer in which each of the 99-residue monomers contributes one of the two aspartic acid residues to constitute the active site of the enzyme. The crystal structures of the free enzyme as well as its complexes with specific inhibitors have been solved at the atomic level [7, 8]. There are two proline residues around the middle of its sequence that are suitable ligating junctions: Pro39 situated in the broad loop (sequence 36-42) prior to the flap region and Pro44 located within the first β strand of the flap (Fig. 1). In this study, we chose Pro39 as the junction to accomodate the pseudoproline structure in the resulting product.

H-P^{1} Q I T L W Q R P L V T I R I G G Q L K E A L L *D T G* A D D T V L E E M N L P^{39} G K W <u>K P K M I G G I G G F I K V R Q</u> Y D Q I P V E I C G H K A I G T V L V G P T P V N I I G R N L L T Q I G C T L N F^{99}-OH

Fig. 1. Amino acid sequence of HIV-1 protease. The underlined sequence represents the flap region; the highly conserved active site sequence is in italics.

The overall strategy for the synthesis of this HIV-1 protease analogue is outlined in Fig. 2. As described above, the first step of our ligation scheme involves the introduction of a protected glycolaldehyde functionality onto the C-terminus of the first fragment through an ester linkage. To achieve this, we have devised a new method which allows the chemical introduction of a special functional group onto the C-terminus of an unprotected peptide. We make use of the distinct activation feature of a thioester by a silver ion, which distinguishes the C-terminal C^{α}-thioester from other side-chain carboxylic groups. The activated thioester then couples predominantly to a small amino component carrying the glycolaldehyde ester moiety, of which a large excess is used (usually over 100-fold). The coupling to side chain amines and other nucleophiles can be ignored because of their relatively low molarity compared to the largely present small amino component. The reaction was performed in DMSO in the presence of HOSu, which served as additive, and completed within 1 hr with a > 90% yield. Compared to the enzymatic method used in our previous studies, this chemical method overcomes the limitations on the C-terminal sequences that are imposed by the substrate specificity of a given enzyme, and is thus a universal method of modifying peptide C-termini. The thioester fragment, HIV-1 PR(1-37)-CO-S(CH2)2CO-Gly-OH, 2a, was synthesized by SPPS as reported by Aimoto et al. [4]. The N-component, [Cys^{39}, Abu^{67,95}]-HIV-1PR(39-99), 2b, was also prepared in satisfactory quality. Both peptides were obtained using the conventional Boc-Benzyl type protection strategy and the BOP coupling protocol. The coupling between the thioester fragment and the small H-Leu-$OCH_2CH(OCH_3)_2$ component gave rise to HIV-1 PR(1-38)-CO-$OCH_2CH(OCH_3)_2$, 2c. Deprotection of the acetal was easily performed using 95% TFA in H_2O at 0°C for 5 min. After HPLC purification, the free peptide aldehyde (2d), was immediately reacted with 2b, which was used in an approximate 1.5 to 2-fold excess to form the thiazolidine product (2e). This reaction was carried out in acetate buffer at pH 5-6, and was 80% completed within 5 hr at very low molar concentrations of both components (0.1 - 0.2 mM). The high reactivity of an aldehyde with β-aminothiol contributes to the efficiency of this reaction. Another reason is that there probably is a complementarity between the two components, given that the reaction medium is very close to the natural solution in which the enzyme functions. This complementarity then holds the two reacting ends into close proximity, producing a high effective local concentration.

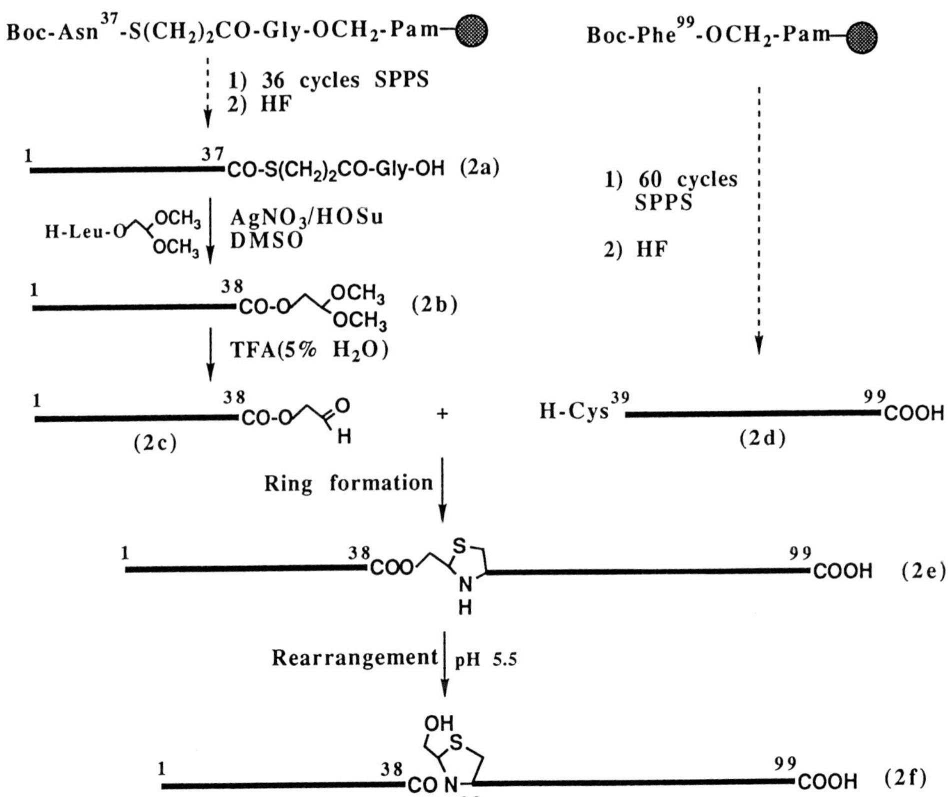

Fig. 2. Overall strategy for the synthesis of a HIV-1 protease analogue.

The subsequent O,N-acyl transfer reaction also occurs right after ring product had been formed to give the pseudoproline structure (2f). Because the HIV-1 protease is autoproteolytic, a general aspartic protease inhibitor, acetyl-pepstatin, which has been shown to inhibit the HIV-1 protease at a K_i = 20 nM, must be added to the ligation solution. The ligation product was separated after the first ring formation was complete and after freeze drying, the material was then dissolved in 6M guanidine·HCl (pH 5.5) / 40% glycerol solution. This solution serves as a storage buffer for the protease. It preserves enzymatic activity and also is a suitable solvent for O,N-acyl rearrangement. The protease attained its maximal enzymatic activity after 3-4 days incubation during which > 90% N-acyl product (2f) has been formed. The enzyme sample was acitve immediately when diluted into the assay buffer. A qualitative assay showed that the resulting HIV-1 protease analogue efficiently catalyze the hydrolysis of the fluorogenic substrate derived from the native Gag(p24/p17) sequence (Fig. 3). Further comparative kinetic studies confirmed that this analogue had the identical kinetic characteristics as the native protease (unpublished data). These results indicated that the protease analogue with a pseudoproline structure in its sequence was able to fold, dimerize correctly into its active conformation, and fully assume catalytic activity. In addition, the obtained protease was analyzed by Matrix-Assisted Laser Desorption Mass Spectrometry (MALDMS), which gave the expected molecular weight (10805 found; 10804 calcd mean mass). All intermediate fragments were also fully characterized by MALDMS and amino acid analysis.

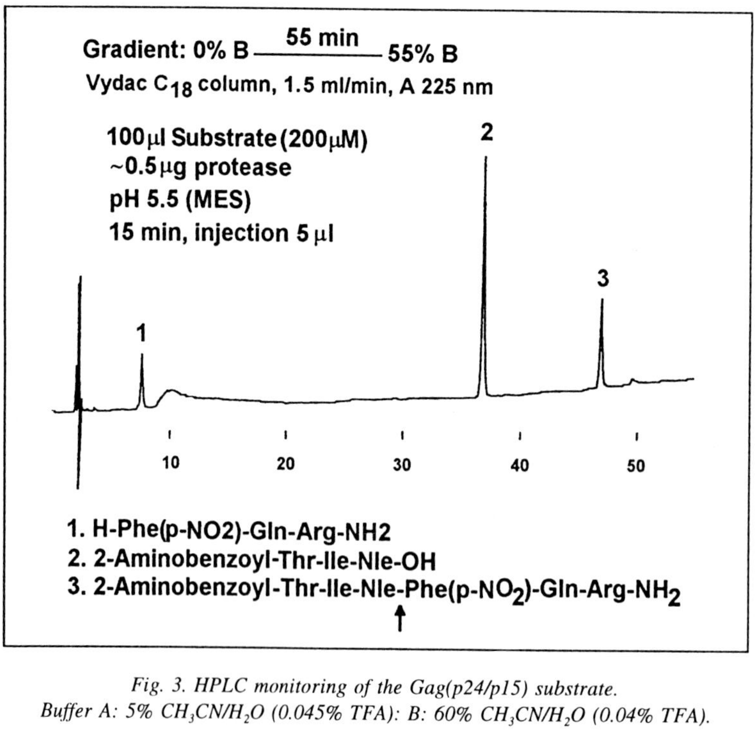

Fig. 3. HPLC monitoring of the Gag(p24/p15) substrate.
Buffer A: 5% CH_3CN/H_2O (0.045% TFA): B: 60% CH_3CN/H_2O (0.04% TFA).

The above example of synthesizing the HIV-1 protease analogue clearly demonstrated that our ligation approach is a useful method for the construction of large peptides and proteins using unprotected segments as building blocks. In the case of HIV-1 protease, it also demonstrated that the thiazolidine pseudoproline structure can be used to replace a proline residue in the sequence of a protein without altering its biological activity. Our strategy combines the advantages of stepwise solid phase peptide synthesis and segment coupling in solution. Because the intramolecular acyl transfer reaction overcomes the problems of the conventional segment condensation method, it is therefore the first choice for future developments in forming peptide bonds between very large fragments.

Acknowledgements

This work was in part supported by grants from USPHS AI 28701 and CA 36544.

References

1. Merrifield, R.B., In Hodges, R.S. and Smith, J.A. (Eds.) Peptides: Chemistry, Structure and Biology (Proceedings of the 13th American Peptide Symposium), ESCOM, Leiden, 1994, pp. 21–30.
2. Tam, J.P., Proc. Natl. Acad. Sci. USA, 85(1988)5409.
3. Blake, J. and Li, C.H., Proc. Natl. Acad. Sci. USA, 78(1981)4055.
4. Hojo, H. and Aimoto, S., Bull. Chem. Soc. Jpn., 64(1991)111.
5. Liu, C.F. and Tam, J.P., J. Am. Chem. Soc., 116(1994)4149.
6. Liu, C.F. and Tam, J.P., Proc. Natl. Acad. Sci. USA, 91(1994)6584.
7. Navia, M.A., Fitzgerald, P.M.D., McKeever, B.M., Leu, C.-T., Heimbach, J.C., Herber, W.K., Sigal, I.S., Darke, P.L. and Springer, J.P., Nature, 337(1989)615.
8. Miller, M., Schneider, J., Sathyanaryana, B.K., Toth, M.V., Marshall, G.R., Clawson, L., Selk, L., Kent, S.B.H. and Wlodawer, A., Science, 246(1989)1149.

Studies on the synthesis and DNA-binding activity of zinc finger motif

Xiao-Yu Hu, Xiang-Qun Li and Rui Wang
Department of Biology, State Key Laboratory of Applied Organic Chemistry, Lanzhou University, Gansu, Lanzhou 730000, China

Introduction

Transcription factors are a type of protein that bind to DNA at some specific base pair box promoter element and selectively activate mRNA reproduction by the gene. A major focus of molecular biology is the identification of mechanisms by which eukaryotic gene transcription is regulated in a precise, temporal and spatial fashion. The identification of transcriptional factors with ability to recognize and bind to specific DNA sequences have provided important clues for elucidating these mechanisms.

Protein recognition of specific DNA binding sites is critical for the biological function of all cells. Transcription, replication and restriction are but a few of the cellular activities that are under the control of binding proteins.

Over the past several years substantial progress has been made in elucidating the structures and molecular mechanisms by which proteins recognize and bind specific DNA sequences.

Eukaryotic genes are regulated differentially in response to a complex set of environmental and development cues. In terms of transcriptional regulation, eukaryotic promoters are large, complex arrangements of short DNA sequences that are recognized by a wide variety of specific DNA-binding proteins that activate or repress transcription. The distinct transcriptional regulatory patterns of individual genes or sets of genes are determined primarily by the specific protein-DNA interactions that occur at the promoter [1].

Structure sites and function analyses of eukaryotic DNA-binding proteins indicate that domains containing less than 100 amino acid residues are sufficient for specific DNA-binding activity. Although detailed structural information is not available, the primary sequences and biochemical properties of various DNA- binding domains are suggestive of at least three basic structural motifs. To date, we know the three basic structural motifs are helix-turn-helix, leucine zipper and zinc finger [2].

Transcription factor SP1 is a protein present in mammalian cells that binds to GC box promoter elements and selectively activates mRNA synthesis from genes that contains functional recognition sites. It contains three continuous zinc finger motifs, which are believed to be metalloprotein structures that interact with DNA [3]. We synthesized the second zinc finger of SP1 (SP1-ZF2) and its mutant (SP1-ZF2/HT. $E^{20}\rightarrow H$. $R^{23}\rightarrow T$). We also synthesized the Cys-Cys loop (ZF6) and the His-His loop (ZF5) of SP1, and linked the two loops together using a β-turn structure to obtain a finger mimic analogue (ZF-15) by stepwise solid-phase technique.

We attempted to coordinate Zn^{++} to the three peptides we made and to test the ability to combine with DNA a fragment.

Results and Discussion

The AA sequences of the three peptides we prepared are:

SP1-ZF2

NH_2-Glu-Arg-Pro-Phe-Met-Cys-Thr-Trp-Ser-Tyr-Cys-
Gly-Lys-Arg-Phe-Thr-Arg-Ser-Asp-Glu-Leu-Gln-Arg-
His-Lys-Arg-Thr-His-Thr-Gly-CO-NH_2

SP1-ZF2/HT (E^{20}→H, R^{23}→T)

NH_2-Glu-Arg-Phe-Met-Cys-Thr-Trp-Ser-Tyr-Cys-Gly-
Lys-Asn-Phe-Thr-Arg-Ser-Asp-His-Leu-Gln-Thr-His-
Lys-Arg-Thr-His-Thr-Gly-CO-NH_2

ZF-15

NH_2-Cys-Thr-Trp-Ser-Tyr-Cys-Gln-Val-Val-Gly-His-
Lys-Arg-Thr-His-CO-NH_2

The amino acid composition analysis is in agreement with the theoretical value. The purity is 95% and detected by HPLC.

We also made these two pieces of DNA, whose sequences are:

1.	5'	GCGCGCGCGCG	3'	G=guanosine
	3'	CGCGCGCGCGC	5'	C=cytidine
2.	5'	GGCCGGCCG	3'	
	3'	GCCGGCCGG	5'	

We obtained the circular dichroism (CD) spectra of the three peptides SP1-ZF2,SP1-ZF2/HT and ZF-15 with or without $ZnCl_2$ solution. The spectrum of ZF-15 is almost the same before and after treatment with $ZnCl_2$ solution, which means that ZF-15 did not form a α-helix structure. SP1-ZF2 and SP1-ZF2/HT had significant changes in CD spectra when treated with or without $ZnCl_2$ solution. The very strong absorption at 208 nm that was present in the CD spectra of SP1-ZF2 and SP1-ZF2/HT with $ZnCl_2$ solution demonstrated the formation of zinc finger structures.

We treated SP1-ZF2, SP1-ZF2/HT and ZF15 with Cobalt ion ($CoCl_2$) solution instead of Zn ion. The tetrahydral coordination of Co^{++} instead of Zn^{++} had no change in this condition, can be identified by resolving the NMR spectrum. We found the specific absorption at around 635 nm for UV-Visible spectra of SP1- ZF2 and SP1-ZF2/HT when treated with $CoCl_2$ solution, which shows cobalt ion coordinates with these two peptides and forms tetrahydral structures. This is an indirect evidence of the formation of zinc finger structures. From the gel (polyacrylamide) retarding electrophoresis experiment (buffer solution: 89 mM Tris-89 mM boric acid, pH=8.0. Voltage: 8V/cm), we found the peptide SP1-ZF2 when treated with $ZnCl_2$ solution bound DNA fragments 5' GCG 3' strongly while SP1-ZF2/HT did not. There was no other combination.

Conclusions

1. We prepared two peptides, SP1-ZF2 and its mutant SP1-ZF2/HT by solid phase method and found these two peptides form zinc finger structures when treated with $ZnCl_2$ solution. They were identified by their CD spectra and UV-Visible spectra with $CoCl_2$ solution.
2. From the gel retarding electrophoresis experiment, we found that SP1-ZF2 bound DNA fragment 5' GCG 3' strongly and SP1-ZF2/HT did not, which means the Glu and Arg residues are very important for recognition. There was no other combination.
3. We found even a single Zn finger domain possesses the ability to bind DNA fragment. There are some arguments in the literature about the combination between single Zinc finger peptide and DNA sequence [4, 5].
4. ZF15, the peptide we designed and made can't form zinc finger structure and can't bind DNA fragment for Protein-DNA interaction, not only the correct spatial structure is necessary, but also the hydrophobic center and recognition code (distinctive amino acid residues such as Arg and Glu) are required.

Acknowledgements

We thank National Natural Science Foundation of China and the State Education Commission of China for financial support and Dr. Zhang Ruo-heng, Dept. of Chemistry, Peking University, for his help on the preparation of peptides.

References

1. Affolter, M., Smith, A.P., Moller, M., Leupin, W. and Gehring, W.J., Proc. Natl. Acad. Sci. U.S.A., 87(1990)4093.
2. Struhl, K., TIBS, 14(1989)137.
3. Nardelli, J., Gibson, T.J., Vesque, C., and Charnay, P., Nature, 349(1991)175.
4. Parraga, G., Science, 241(1988)1489.
5. Archer, T.K., Proc. Natl. Acad. Sci. U.S.A., 87(1990)7560.

New methods in solid phase peptide synthesis

R.P. Sharma, R.J. Broadbridge, D.A. Jones, D.L. Corina and M. Akhtar
Department of Biochemistry, School of Biological Sciences,
University of Southampton, Bassett Crescent East,
Southampton SO16 7PX, U.K.

Introduction

Solid phase peptide synthesis was first introduced by Merrifield in 1963 [1]. The method is highly versatile and can be accommodated to a variety of chemistries. It has been widely utilized on a routine basis for the synthesis of small to medium size peptides. More recently, the development of Fmoc strategy by Sheppard [2], advances in new coupling reagents, resins, linkers and purification procedures have stimulated renewed interest into the synthesis of large peptides (polypeptides). We recently reported the application of tri-*tert*-butoxysilyl esters of amino acids in peptide synthesis on the solid phase from N→C direction [3-4]. Such an approach was designed to permit the use of mild deprotection conditions thus allowing a wider choice of side chain protecting groups. The important advantage of this methodology is that it allows the *in situ* modification of carboxyl function which may have significant application in drug design. The aims of our research in this area (N→C direction), therefore, have been the following: (1) Peptide Synthesis; (2) modification of amino acids and peptides on the solid phase; (3) incorporation of modified amino acids into the peptide backbone; and (4) to develop methods for fragment condensation reactions on the solid phase.

Results and Discussion

Amino acid tri-*tert*-butoxysilyl ester (T-t-Bos) can be conveniently prepared in a single step [5]. The amino acid is allowed to react with silicon tetrachloride and *tert*-butyl alcohol in the presence of pyridine and the reaction is carried out at room temperature for 2-4 hours. The reaction is rapid, the yields are usually high and minimum purification is required. The amino acid esters are stable between P^H4 and P^H8. These properties make this group very attractive for the purpose of peptide synthesis since the protection is in no danger of hydrolyzing under the conditions of peptide bond formation but could be rapidly and selectively removed when desired. The fact that these esters can be hydrolyzed under both basic and acidic conditions widens the range of peptidyl-resin linkages and side chain protections that can be employed in solid phase peptide synthesis using amino acid tri-*tert*-butoxysilyl esters. All amino acid tri-*tert*-butoxysilyl esters have been prepared and fully characterized by their ^{1}H nuclear magnetic resonance (NMR) spectra, infra-red spectra and by accurate mass measurements.

Attachment of the first amino acid via its amino function has previously been achieved through a benzyloxycarbonyl linkage to a commonly available polystyrene resin (Merrifield resin) [6-7]. This has been our usual approach in obtaining a peptide resin linkage that is stable enough to all reagents during the synthesis but that is easily cleaved at the end of the synthesis by treatment with a strong acid such as hydrogen fluoride (HF) etc. We have prepared a number of tri-*tert*-butoxysilyl amino acid ester resins in this manner. A satisfactory substitution level of these amino acid esters to polystyrene resin was obtained.

A number of peptides and peptide analogues have been synthesized manually in a stepwise fashion on the solid phase from N→C direction (Table 1).

Table 1 *Structure of peptides synthesized by N→C method*

1. Leu-Ala-Gly-Val
2. Tyr-Gly-Gly-Phe-Leu
3. Ala-Gly-Ala-Gly
4. Ala-Pro-Ala-Val-Gln
5. Ala-Pro-Ala-Val-Gln-Ser-Thr-Glu-Thr-Lys
6. Ala-Gly-Ala-Gly-Gly-Gly-Phe-Leu-OMe (from Resin-Ala-Gly-Ala-Gly-OH + Gly-Gly-Phe-Leu-OMe)
7. Tyr-Gly-Gly-Phe-Leu CH_2OH
8. Tyr-Gly-Gly-Phe-Leu-$COCH_2Cl$
9. Ala-Gly-Ala-Gly-Phe[OH]-CH_2OH
10. Tyr-Gly-Gly-Phe-Leu [OH]-CH_2OH
11. Arg-Pro-Pro-Gly-Phe-Ser-Pro-Phe-Arg
12. Lys-Thr-Glu-Thr-Ser-Gln-Val-Ala-Pro-Ala
13. Arg-Pro-Pro-Gly-Phe-Ser-Pro-Phe-Arg-OMe (from Resin-Arg-Pro-Pro-Gly-Phe-OH + Ser-Pro-Phe-Arg-OMe)

A reaction protocol for a peptide assembly on a resin is illustrated in Table 2.

In most cases a single coupling cycle of 1 hour duration was required. The hydroxyl functions of the tyrosine, threonine and serine were not protected during the synthesis. The quality of the product was comparable to the product obtained when the peptides were synthesized in the conventional C→N direction using Fmoc chemistry. All synthetic peptides described in Table 1 were analysed by RP-HPLC, FIB mass spectrometry and high voltage thin layer electrophoresis (HVTLE). The peptide analogues were based on leucine-enkephalin. The leucine-enkephalin diol (entry 10) was synthesized by coupling Leu-CHOH-CH_2OH [8] and Resin-Tyr-Gly-Gly-Phe-OH and gave the desired product in excellent yield and purity. The other analogues were prepared in a similar manner. We have synthesized a number of peptides in which modification of amino acid was carried out on the solid phase (entry 9). We have also demonstrated that N→C methodology is an attractive approach for fragment condensation reactions on the solid phase. Peptides 6 and 13 (Table 1) were prepared by fragment couplings on the solid phase.

The possibility of racemisation during the synthesis was always considered and the chiral purity of the leucine-enkephalin synthesized was extensively investigated by RP-HPLC, enzymatic digestion with carboxypeptidase A and specific optical rotation. No racemisation was detected.

Table 2 *Reaction protocol for N→C procedure. The volumes given are for 0.5 g of resin*

		Volume (ml)	Time (min)	Repeat
Prewash	1. Wash DCM	6	1	2
Deprotect	2. 25% TFA in DCM (v/v)	6	5	
	3. Drain and repeat	6	15	
Wash	4. DCM	6	1	3
	5. DMF	6	1	1
	6. Remove resin sample for assay			
Coupling	7. T-t-Bos amino acid; 3 Eq BOP 3 Eq HOBT 3 Eq Diisopropylethylamine 6 Eq in DMF	6	60	
Wash	8. DMF	6	1	2
Monitoring	9. Remove resin sample for assay			

We have developed a simple method to monitor the coupling efficiency during N→C synthesis. The procedure is based on an acid base titration. The amount of free CO_2H is quantified by the back titration of base with acid. A knowledge of the initial [base] and the [acid] needed for neutralisation allows the coupling efficiency to be calculated. The method is promising and has given excellent results. Further work in this area is continuing.

Acknowledgements

Financial support from the Wellcome Trust is gratefully acknowledged.

References

1. Merrifield, R.B., J. Amer. Chem. Soc., 85(1963)2149.
2. Atherton, E. and Sheppard, R.C., Solid Phase Peptide Synthesis: A Practical Approach, Oxford University Press, Oxford 1987.

3. Sharma, R.P., Jones, D.A., Corina, D.L. and Akhtar, M., In Hodges, R.S. and Smith, J.A. (Eds.) Peptides: Chemistry, Structure and Biology (Proceedings of the 13th American Peptide Symposium), ESCOM, Leiden, 1994, pp. 127–129.
4. Sharma, R.P., Jones, D.A., Broadbridge, R.J., Corina, D.L. and Akhtar, M., In: Epton, R. (Ed.), Innovation and Perspectives in Solid Phase Synthesis, 1994: Biological and Biomedical Applications, Mayflower Worldwide, 1994, Ch 52, pp. 353–356.
5. Gruszecki, W., Gruszecka, M. and Bradaczek, H., In Giralt, E. and Andreu, D. (Eds.) Peptides 1990 (Proceedings of the 21st European Peptide Symposium), ESCOM, Leiden, 1991, pp. 27–28.
6. Letsinger, R.L. and Kornet, M.J., J. Amer. Chem. Soc., 85(1963)3045.
7. Felix, A.M. and Merrifield, R.B., J. Amer. Chem. Soc., 92(1970)1385.
8. Sharma, R.P., Gore, M.G. and Akhtar, M., J. Chem. Soc., Chem. Comm. (1979)875.

Characterization of a synthetic antagonist and agonists of memory-enhancing peptide

Yu-Cang Du, An-Wu Zhou, Ke-Ying Ma, Xiao-Ming Jiang and Peng-Liang Wang
Shanghai Institute of Biochemistry, Chinese Academy of Sciences, Shanghai 200031, China

Introduction

ZNC(C)PR, a memory-enhancing peptide (MEP), has been found in animal brain to be an enzymatic product (AVP_{4-8}) of arginine-vasopressin (AVP). We currently report that ^{35}S-labelled ZNC(C)PR showed specific and tight binding to the limbic system in rat brain [1]. After ligands have bound to the membrane, a series of intracellular events occurs including regulation of gene expression through a branching signalling pathway: R→IP_3-(→GAP_{43}→LTP)→Fos→NGF [2]. Based on the results generated from NMR and molecular dynamics studies [3], five peptides were synthesized and compared for their behavioral activities, binding capabilities and potencies in evoking endogenous NGF expression in rat brain. So far two of these synthetic analogs tetrapeptide, Asn-Leu-Pro-Arg and a pentapeptide, PGA-Asp-Cyt-Pro-Arg have been found to be potent agonist and antagonist of MEP respectively. Both showed tight binding on the hippocampal synatosomal membrane. However they showed very different bioactivities on behavioral response and on NGF gene expression: the former is active but the latter is inactive and somewhat suppressive.

Results and Discussion

The structure and behavioral activity of five synthetic peptides, ZNC(C)PR (I), Ac-YQNC(C)PR-NH_2 (II), ZDC(C)PR (III), NLPR (IV) and $(ZNCPR)_2$ (V) are listed in Table 1. The results generated from passive avoidance behavior tests in rats indicated that peptides I, II and IV (doses > 0.1 μg/rat) had positive effects on facilitating memory, while peptide V (dose > 10 μg/rat) had no effect. However, peptide III was not only void of memory-enhancing effect but also showed significant inhibition on maintening behavioral responses when a high dose (10 ng/rat) was used. As synthetic [^{35}S]-ZNC(C)PR has specific and tight binding to the synaptosomal membrane preparation of rat anterior cortex with K_D 3.12 nM and B_{max} 30.8 fmol/mg protein [1], the displacement of specifically bound [^{35}S]-ZNC(C)PR by synthetic peptides from anterior cortical synaptosomal membrane was performed. The data in Table 2 shows the results of competition assays with [^{35}S]-ZNC(C)PR were markedly different: ZDC(C)PR, Ac-YQNC(C)PR-NH_2, and ZNCPR dimer had 87.1%, 23.2%, and 6.5% relative binding potencies respectively, while AVP in a dose range of 10 ng to 5 μg had less than 0.2% of the binding potency of ZNC(C)PR. It is note worthy that the relative binding potencies of these analogs related to their agonist and antagonist activities. It is clear that ZDC(C)PR (III) is a potent antagonist and Ac-YQNC(C)PR-NH_2 (II) is an agonist to the ZNC(C)PR receptor.

Table 1 *Effect of ZNC(C)PR analogs on the retention of passive avoidance behavior in rats*

Compound & structure	Energy (Kcal/mol)	Mean latency (sec) Dose (μg)	Means	± SE	N
(I) ZNC(C)PR	−104.3	saline	51.8	9.6	44
		0.01	97.3	32.6	11
PGA-Asn-Cyt-Pro-Arg		0.1	119.2[b]	21.3	17
		1.0	180.8[b]	26.8	18
(II) Ac-YQNC(C)PR-NH_2		saline	59.9	10.4	20
		0.01	65.1	10.1	20
Acetyl-Tyr-Gln-Asn-Cyt-Pro-Arg-NH_2		0.03	110.0[a]	17.8	20
		0.1	130.6[b]	18.6	20
(III) ZDC(C)PR	−140.2	saline	176.7	87.3	16
		0.1	187.2	94.5	16
PGA-Asp-Cyt-Pro-Arg		1.0	169.9	77.0	15
		10.0	116.6[a]	59.9	18
(IV) NLPR	−52.6	saline	41.1	4.1	24
		0.1	82.9[a]	24.1	21
Asn-Leu-Pro-Arg		1.0	97.3[b]	12.0	21
		3.0	96.9[b]	31.4	6
(V) $(ZNCPR)_2$		saline	23.5	7.6	11
PGA-Asn-Cys-Pro-Arg		1.0	16.0	6.7	4
PGA-Asn-Cys-Pro-Arg		10.0	78.2	39.2	5

Peptides were subcutaneously injected immediately after the learning trial and retention was tested at 24 h. N: Number of rats. Values are given as mean ± SE. [a] $p < 0.05$; [b] $p < 0.01$ vs. saline control (Kruskal-Wallis test and subsequently Mann-Whitney U-test).

Lacking a pyroglutamyl residue at the N-terminal, tetrapeptide NLPR (IV) did not compete with [^{35}S]-ZNC(C)PR i.e., its competitive potency was less than 1%. Interestingly it still remained rather high behavioral potency (see Table 1). To clarify these anomalous findings, two kinds of experiments have been done: 1, Specific bindings of synthetic [^{3}H]-labeled NLPR to the synaptosomal membrane of rat hippocampus and amygdala have been found. Scatchard analysis of specific binding of [^{3}H]-NLPR indicated that a dual binding site existed in hippocampal membrane with K_{D1} 0.2 nM and K_{D2} 2.6 nM, in contrast to ZNC(C)PR which shows only one binding site with K_D 3nM; 2, By using antisense RNA probe from βNGF cDNA in Northern blot analysis with the rat groups pretreated with agonist, antagonist and/or ZNC(C)PR, enhancement of ZNC(C)PR, Ac-YQNC(C)PR-NH_2 and NLPR on *in vivo* NGF mRNA transcription in rat hippocampus were further identified (Fig. 1A). ZDC(C)PR markedly depressed NGF expression evoked by administration with exogenous ZNC(C)PR approximately 50% but did not significantly suppress the basal level of NGF expression in the tissues (Fig. 1B). It seems that peptide IV is a new type of agonist, as it may bind to an unknown subtype of receptor, of which the selectivity must differ from that of the putative ZNC(C)PR receptor.

Table 2 *Relative receptor binding potencies of analogs to ^{35}S-ZNC(C)PR binding site on the synaptosomal membrane from rat cerebral cortex*

Compound	-log EC_{50}	K_i[a] 10^{-8} M	Relative Binding Potency %	Behavioral Potency (+)[b] ED_{50} (μg/rat)	(–)
ZNC(C)PR	7.00	8.69	100.0	0.10	
ZDC(C)PR	6.94	9.98	87.1	>10	10
Ac-YQNC(C)PR-NH_2	6.37	37.4	23.2	0.08	10
$(ZNCPR)_2$	5.87	134.9	6.5	>10	10
AVP	>4.13	>6520	<0.2	0.46	10

[a] Ki was calculated according to the equation: $K_i = [EC_{50}/(1 + F/K_d)]$, where F is the concentration of free ZNC(C)PR in the absence of competing ligand.
[b] (+) or (–) represents facilitation or attenuation of passive avoidance behavior at 24-hr test.

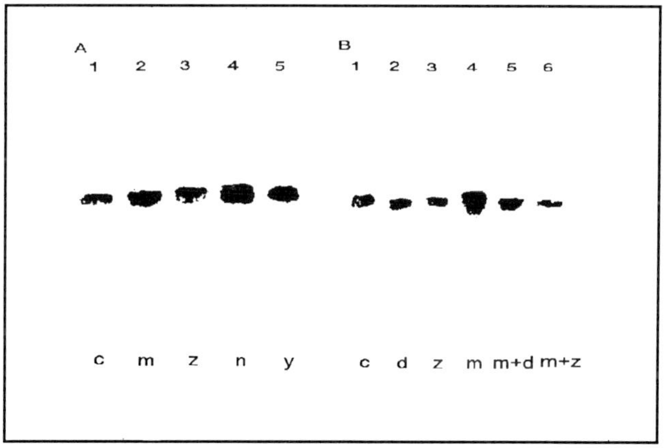

Fig. 1. Effects of agonists and antagonists of ZN(C)CPR on NGF mRNA level in rat brain detected by using the NGF antisense cRNA as probe on Northern blot. (A) The agonistic effects of ZNC(C)PR and its analogs in rat hippocampus (A): column 1, control (c); 2, ZNC(C)PR (m); 3, ZDC(C)PR (z); 4, NLPR (n); 5, Ac-YQNC(C)PR-NH_2 (y). (B) The antagonistic effect of ZDC(C)PR: columns 1, control (c); 2, $(ZNCPR)_2$ (d); 3, ZDC(C)PR (z); 4, ZNC(C)PR(m); 5, ZNC(C)PR and $(ZNCPR)_2$ (m+d); 6, ZNC(C)PR and ZDC(C)PR (m+z).

Acknowledgements

The study was supported by grants from the National Natural Science Foundation and Chinese Academy of Sciences.

References

1. Du, Y.C., Wu, J.H., Jiang, X.M. and Gu, Y.J., Peptides, 15(1994), in press.
2. Du, Y.C., Zhou, A.W., Guo, J., Wang, H.Y., Gu, B.X., Rong, X.W. and Chen, X.F. in Peptide Chemistry 1993, The Proceedings of 31st Japanese Peptide Symposium, Akashi 1993, Ed. by Y. Okada. Protein Research Foundation, Osaka, 1994, pp 373.
3. Wang, P.L., Du, Y.C. Lai, L.H. and Xu, X.J., Acta Pharmacol. Sin., 15(1994)311.

Thymosin α_1, thymosin β_4 and analogs

S.S. Wang, B.S.H. Wang and A.L. Goldstein
Alpha 1 Biomedicals, Inc. Foster City, CA 94404, U.S.A. and
Department of Biochemistry and Molecular Biology, The George Washington University Medical Center, Washington, DC 20037, U.S.A.

Introduction

Thymosin α_1 (Tα_1) was the first biologically active peptide to be isolated from thymosin fraction 5 (TF5), an extract which was prepared from calf thymus glands. Tα1 has been one of the most thoroughly studied immunoregulating hormones. Its

Table 1 *Properties of well-characterized thymosin preparations and purified peptides*

Thymic Preparation	Chemical Properties	Biological Functions
Thymosin Fraction 5	Family of heat-stable acid polypeptides; mol. wt. 1,000-5,000	Induces T-cell differentiation and enhances immunologic function in animal models and in humans; increases ACTH, β-endorphin, and glucocorticoid release; stimulates production of MIF, IL-2, IFN, CSF, and other cytokines
Thymosin α_1(Tα_1)	Peptide of 28 residues, mol. wt. 3,108 pI 4.2; sequence determined	Induces enhancement of cytokines; modulates TdT a ctivity; increases viral, fungal and tumor immunity; amplifies T-cell immunity in humans, and synergizes with IFN and IL-2.
Prothymosin α_1	113 amino acids, Tα_1 at N-terminal, amino acid sequence determined, pI 3.55, mol. wt. 13,500	Biological activity similar to Tα_1 in protecting mice against opportunistic infections with *Candida albicans.*
Thymosin α_7	Peptide of mol. wt. 2,000; pI 3.5	*In vitro* enhancement of suppressor T-cells; expression of Lyt-1,2,3-positive cells.
Thymosin α_{11}	Polypeptide of 35 residues, N-terminal 28 residues identical to Tα_1	Biological activity similar to Tα_1 in protecting mice against infection with *Candida albicans*
Thymosin β_4	Peptide of 43 residues, mol. wt. 4,963, pI 5.1, sequence determined	Actin-sequestering activity induces TdT *in vivo* and *in vitro* in normal and athymic mice; *in vivo* induction of TdT-positive thymocytes of immunosuppressed m ice; stimulates releast of LH-RH (LRF); reduces toxicity of chemotherapy and sepsis.
Thymosin MB-35	Peptide of 35 residues	Stimulates release of growth hormone and prolactin

amino acid sequence was determined [2], and the chemical structure confirmed by total synthesis [3]. The peptide was found to exhibit several interesting biological activities and thus has been studied in several clinical trials for the treatment of lung cancer, primary immunodeficiency syndrome, AIDS, chronic hepatitis B, hepatitis C, and cancer [1, 4, 5]. It has also been found to decrease growth of human lung cancer cells in vitro [6]. Thymosin β_4 (Tβ_4), also isolated from TF5, has been found to be a potent regulator of actin polymerization [8], and can protect mice from the toxicity of cytosine arabinoside (ara-C) [9] by preventing the cycling of hematopoietic stem cells. A summary of some of the chemical and biological properties of the thymosin preparations is given in Table 1.

The amino acid sequences of the synthetic thymosin α_1, thymosin β_4 and their analogs reported here that are currently under study are listed below:

Thymosin α_1
Ac-Ser-Asp-Ala-Ala-Val-Asp-Thr-Ser-Ser-Glu-Ile-Thr-Thr-Lys-Asp-Leu-Lys-Glu-Lys-Lys-Glu-Val-Val-Glu-Glu-Ala-Glu-Asn

Thymosin β_4
Ac-Ser-Asp-Lys-Pro-Asp-Met-Ala-Glu-Ile-Glu-Lys-Phe-Asp-Lys-Ser-Lys-Leu-Lys-Lys-Thr-Glu-Thr-Gln-Glu-Lys-Asn-Pro-Leu-Pro-Ser-Lys-Glu-Thr-Ile-Glu-Gln-Glu-Lys-Gln-Ala-Gly-Glu-Ser

[N_4]-Thymosin$_4$
Ac-Ser-Asp-Lys-Pro

Thymosin β_4-Ala
Ac-**Ala**-Asp-Lys-Pro-Asp-Met-Ala-Glu-Ile-Glu-Lys-Phe-Asp-Lys-Ser-Lys-Leu-Lys-Lys-Thr-Glu-Thr-Gln-Glu-Lys-Asn-Pro-Leu-Pro-Ser-Lys-Glu-Thr-Ile-Glu-Gln-Glu-Lys-Gln-Ala-Gly-Glu-Ser

Thymosin β_4-Xen
Ac-Ser-Asp-Lys-Pro-Asp-Met-Ala-Glu-Ile-Glu-Lys-Phe-Asp-Lys-**Ala**-Lys-Leu-Lys-Lys-Thr-Glu-Thr-Gln-Glu-Lys-Asn-Pro-Leu-Pro-Ser-Lys-Glu-Thr-Ile-Glu-Gln-Glu-Lys-Gln-**Thr-Ser**-Glu-Ser

Thymosin β_9
Ac-**Ala**-Asp-Lys-Pro-Asp-**Leu-Gly**-Glu-Ile-**Asn-Ser**-Phe-Asp-Lys-**Ala**-Lys-Leu-Lys-Lys-Thr-Glu-Thr-Gln-Glu-Lys-Asn-**Thr**-Leu-Pro-**Thr**-Lys-Glu-Thr-Ile-Glu-Gln-Glu-Lys-Gln-Ala-**Lys**

Thymosin β_9-Met
Ac-**Ala**-Asp-Lys-Pro-Asp-Met-**Gly**-Glu-Ile-**Asn-Ser**-Phe-Asp-Lys-**Ala**-Lys-Leu-Lys-Lys-Thr-Glu-Thr-Gln-Glu-Lys-Asn-**Thr**-Leu-Pro-**Thr**-Lys-Glu-Thr-Ile-Glu-Gln-Glu-Lys-Gln-Ala-**Lys**

Thymosin β_{10}
Ac-**Ala**-Asp-Lys-Pro-Asp-Met-Gly-Glu-Ile-**Ala-Ser**-Phe-Asp-Lys-**Ala**-Lys-Leu-Lys-Lys-Thr-Glu-Thr-Gln-Glu-Lys-Asn-**Thr**-Leu-Pro-**Thr**-Lys-Glu-Thr-Ile-Glu-Gln-Glu-Lys-**Arg-Ser-Glu-Ile**-Ser

Thymosin β_{11}

Ac-Ser-Asp-Lys-Pro-**Asn-Leu-Glu**-Glu-**Val-Ala-Ser**-Phe-Asp-Lys-**Thr**-Lys-Leu-Lys-Lys-Thr-Glu-Thr-Gln-Glu-Lys-Asn-Pro-Leu-Pro-**Thr**-Lys-Glu-Thr-Ile-Glu-Gln-Glu-Lys-Gln-Ala-**Ser**

Thymosin β_{12}

Ac-Ser-Asp-Lys-Pro-Asp-**Leu**-Ala-Glu-**Val-Ser-Asn**-Phe-Asp-Lys-**Thr**-Lys-Leu-Lys-Lys-Thr-Glu-Thr-Gln-Glu-Lys-Asn-Pro-Leu-Pro-**Thr**-Lys-Glu-Thr-Ile-Glu-Gln-Glu-Lys-Gln-Ala-**Thr-Ala**

Thymosin β_{13}

Ac-**Ala**-Asp-Lys-Pro-Asp-Met-**Gly**-Glu-Ile-**Ala-Ser**-Phe-Asp-Lys-**Ala**-Lys-Leu-Lys-Lys-Thr-Glu-Thr-Gln-Glu-Lys-Asn-**Thr**-Leu-Pro-**Thr**-Lys-Glu-Thr-Ile-Glu-Gln-Glu-Lys-Gln-Ala-**Lys**

Thymosin β_{14}

Ac-Ser-Asp-Lys-Pro-Asp-**Ile-Ser**-Glu-**Val-Ser-Ser**-Phe-Asp-Lys-**Thr**-Lys-Leu-Lys-Lys-Thr-Glu-Thr-**Ala**-Glu-Lys-Asn-**Thr**-Leu-Pro-**Thr**-Lys-Glu-Thr-Ile-Glu-Gln-Glu-Lys-**Thr**-Ala

Substitutions in amino acids from Tβ_4 primary structure are shown in bold type. Deviations from Tβ_4 range from 2% to 29%. It should be noted that there are many amino acid overlaps between Tα_1 and Tβ_4, especially in the N-terminal regions of the two molecules.

Results and Discussion

Scaled up synthesis of thymosin α_1

Solid phase synthesis of Tα_1 has been described before [10–12]. The process was scaled up to produce thymosin α_1-resin in the batch sizes of up to 1 kg and thymosin α_1 (>99% purity by HPLC) in batch sizes of up to 0.1 kg, under GMP (good manufacturing practice). Identity and purity of the product were established by amino acid analyses, HPLC, capillary electrophoreses, tryptic mapping, molecular weight determination by mass spectrometry, TFA treatment to remove N-terminal acetyl group followed by gas phase Edman sequence analyses [13], C-terminal sequence analysis with carboxypeptidase Y, moisture content determination by Karl Fischer titration, pyrogen tests, GC analysis of residual organic solvent and determination of the contents of acetate ion, phosphate ions, ammonium ion, potassium ion, etc. A typical lot of synthetic thymosin a1 showed 99.8% purity on analytical HPLC, 99.3% on capillary electrophoresis, with amino acid composition of Asp, 4.00; Thr, 3.07; Ser, 2.72; Glu, 6.46; Ala, 3.00; Val, 2.81; Ile, 0.99; Leu, 1.04; and Lys, 4.04. The moisture content is 5.92%; ammonium ion, 2.02%; acetate ion, 2.26%; phosphate ion, 0.42% and acetonitrile, 0.0016%. The over-all yields of thymosin α_1 lots ranged from 25 to 30%, calculated from the first amino acid (Boc-Asp-OBzl) coupled to MBHA-resin [14]. The compound was utilized as lyophilized powder in human clinical trials as well as in animal studies.

Synthesis of thymosin β_4 and analogs

Thymosin β$_4$ and analogs were synthesized on Merrifield-resin by methods analogous to that originally presented previously [15–17].

Effect of thymosin β_4 and N_4-$T\beta_4$ on reduction in cytosine arabinoside-induced toxicity

A total of 160 mice were studied in duplicate experiments. Mice were divided into three groups. Group 1 received normal saline and ara-C (29 mice), and served as control. Group 2 received ara-C and whole synthetic Tβ$_4$ (58 mice). Group 3 received ara-C and N-terminal tetrapeptide, N$_4$-Tβ$_4$ (58 mice). Mice were injected intraperitoneally (i.p.) with a volume of 0.1 mL containing 5 mg ara-C in sterile Dulbecco's phosphate buffered saline (D-PBS) at 0, 7 and 36 hr for a total of 15 mg. They were also injected i.p. at 5 and 34 hr with 0.1 mL of either Tβ$_4$ or N$_4$-Tβ$_4$ in D-PBS at concentrations of 0.1 μg and 1.0 μg. Survival was recorded daily at 10:00 hr. Calculation of z values using a standard confidence test was used to detect differences in survival between test and control groups.

Thymosin β$_4$ and N$_4$-Tβ$_4$ were found to significantly reduce the toxicity of ara-C. In this study, we examined the ability of Tβ$_4$ to abrogate the short-term lethal toxicity of ara-C. In order to compare the abilities of Tβ$_4$ and the N-terminal tetrapeptide, N$_4$-Tβ$_4$ to protect mice from LD50 doses of ara-C, 10~11-week old male C3H/HEN mice were divided into three groups. All mice except a control group of 15 mice (data not shown) were given a standard dose of ara-C and either 0.1 μg or 1.0 μg of the appropriate peptide. This experiment was performed in duplicate with two equal sets of mice on different days to exclude other variables. Data from both sets were added to form one larger data set. As the endpoint of the assay was survival and evaluation of short-term toxicity, mice were followed for 2 weeks (16 days), after which the survivors were sacrificed per institutional protocol. Cumulative survival is shown in Table 2. No change in the number of surviving mice was noted between 8 and 16 days, and these data are shown only through day 8. Statistically significant improvement in survival was noted in groups receiving ara-C plus 0.1 μg Tβ$_4$ or 0.1 μg N$_4$-Tβ$_4$ when compared to controls

Table 2 *Thymosin β_4 and the N_4-$T\beta_4$ analog decrease the toxicity of ara-C in C3H/HEN mice*

Exp. No.	Total dosage of compounds			Number of mice alive							
	Ara-C	Tb4	N4-Tb4	Day 1	Day 2	Day 3	Day 4	Day 5	Day 6	Day 7	Day 8
1	15 mg	0	0	29	29	29	19	16	14	14	14
2	15 mg	2 μg	0	29	29	29	28	25	21	19	19
	15 mg	0.2 μg	0	29	29	29	29	29	28	28	28
3	15 mg	0	2 μg	29	29	28	25	19	18	18	17
	15 mg	0	0.2 μg	29	29	29	27	26	23	23	23

receiving ara-C alone ($p<0.01$). This was not seen in groups receiving ara-C plus higher doses of $T\beta_4$ or N_4-$T\beta_4$. These data suggest a dose- dependent and bifunctional effect of the administered peptides on abrogation of short-term ara-C toxicity. Appreciable side effects were not seen following administration of either peptide, and no deaths were observed in controls receiving doses of 0.1 µg $T\beta_4$ alone.

In conclusion, we have found that $T\beta_4$ protects mice from LD50 doses of ara-C [8]. We have also verified that although the N_4-$T\beta_4$ is active, the parent compound is more active, and is responsible for most of the activities first ascribed by Lenfant et al to the N-terminal tetrapeptide [9].

References

1. Goldstein, A.L., in Combination Therapies 2 (E. Garaci and A.L. Goldstein, eds.) Plenum Press, NY (1993)39.
2. Goldstein, A.L., Low, T.L.K., McAdoo, M., McClure, J., Thurman, G.B., Rossio, J., Lai, C.Y., Chang, D., Wang, S.S., Harvey, C., Ramel, A.H. and Meienhofer, J. Proc. Natl. Acad. Sci. USA, 74(1977)725.
3. Wang, S.S., Kulesha, I.D. and Winter, D.P., J. Amer. Chem. Soc., 101(1979)253.
4. Mutchnick, M.G., Appleman, H.D., Chung, H.T., Aragona, E., Gupta, T.P., Cummings, G.D., Waggoner, J.G., Hoofnagle, J.H. and Shafritz, D.A., Hepatology, 14(1991)409.
5. Schulof, R.S., J. Biolog. Resp. Mod., 3(1986)429.
6. Moody, T.W., Fagarasan, M., Zia, F., Cesnjaj, M. and Goldstein, A.L., Cancer Research, 53(1993) 5214.
7. Sanders, M.C., Goldstein, A.L. and Wang, Y.L., Proc. Natl. Acad. Sci. USA, 89(1992)4678.
8. Moscinski, L.C., Naylor, P.H., Oliver, J. and Goldstein, A.L., Immunopharmacology, 26(1993)83.
9. Lenfant, M., Wdzieczak-Bakala, J., Guittet, E., Prome, J.C., Sotty, D. and Findel, E., Proc. Natl. Acad. Sci. USA, 86(1989)779.
10. Wang, S.S., Makofske, R., Bach, A. and Merrifield, R.B., Int. J. Peptide Protein Res., 15(1980)1.
11. Wang, S.S., US Patent 4,855,407; Aug. 8, 1989.
12. Wang, S.S., Wang, B.S.H., Hughes, J.L., Leopold, E.J., Wu, C.R. and Tam, J.P., Int. J. Peptide Protein Res., 40(1992)344.
13. Wellner, D., Panneerselvam, C. and Horecker, B.L., Proc. Natl. Acad. Sci. USA, 87(1990)1947.
14. Matsueda, G.R. and Stewart, J.M., Peptides, 2(1981)45.
15. Wang, S.S., Wang, B.S.H., Chang, J.K., Low, T.L.K. and Goldstein, A.L., Int. J. Peptide Protein Res., 18(1981)413.
16. Wang, S.S., Wang, B.S.H., Chang, J.K., Low, T.L.K. and Goldstein, A.L., in Peptides: Synthesis, Structure, Function (D.H. Rich and E. Gross, eds.), Pierce Chemical Co., Rockford, IL(1981)189.
17. Low, T.L.K., Wang, S.S. and Goldstein, A.L., Biochemistry, 22(1983)733.

Selective formation of three disulfide bridges in EGF-like peptides

Yan Yang[a], William V. Sweeney[a], Klaus Schneider[b], Brain T. Chait[b] and James P. Tam[c]

[a]*Department of Chemistry, Hunter College of CUNY, 695 Park Ave., New York, NY 10021, U.S.A.*

[b]*Laboratory for Mass Spectrometry and Gaseous Ion Chemistry, The Rockefeller University, 1230 York Ave., New York, NY 10021, U.S.A.*

[c]*Department of Microbiology and Immunology, Vanderbilt University, A5119 M.C.N., Nashville, TN 37232, U.S.A.*

Introduction

One of the challenging problems in peptide synthesis is the formation of correctly paired multiple disulfide bridges. In general there are two strategies: the single-step approach [1, 2] and the sequential approach [3, 4]. In the formation of disulfide bridges in different synthetic epidermal growth factor (EGF)-like peptides by the single-step approach, we have found that several exhibit a significant multiple conformation problem that results in producing a mixture of disulfide isomers. An example of this is the C-terminal EGF-like domain in human blood coagulation factor IX (fIX_{EGF-C}), which did not yield a peptide with the correct disulfide pairing by the single-step approach. On the other hand, formation of the three disulfide bridges in fIX_{EGF-C} by the sequential approach is a difficult undertaking because of the need for three orthogonal sets of thiol protecting groups. We have therefore sought to develop a hybrid two-step approach to resolve this problem [5].

The fundamental idea of this two-step approach is to limit the number of possibilities for disulfide bond formation in the different oxidation stages. There are fifteen possible ways to form three disulfide bonds, but only three ways to form two disulfide bonds. Our approach takes advantage of this, allowing only four cysteines to form sulfur-sulfur pairs in the first step, and the remaining disulfide bridge is formed in the second step. The number of possible disulfide isomers thereby is reduced from fifteen to three.

Results and Discussion

Two different protecting groups for cysteines are required. One is the conventional protecting group, 4-methylbenzyl (Meb) in Boc chemistry or trityl (Trt) in Fmoc chemistry, which will be deblocked from the cysteines concomitant with the cleavage of the peptide from the resin. The other is acetamidomethyl (Acm), which remains on the cysteines after acidic cleavage.

Since fIX_{EGF-C} has three disulfide bridges, the Acm protecting groups could logically be placed on any of the three pairs of cysteines. To determine which pair would be optimal, three peptides of fIX_{EGF-C} sequence were synthesized, each with a different pair of cysteines protected by Acm. In peptide A, Acm was used to block cysteines 5 and 6, in peptide B cysteines 2 and 4, and in peptide C cysteines 1 and 3. Ideally, all three peptides should yield the same final product with an EGF-like disulfide pairing of Cys 1-3, Cys 2-4 and Cys 5-6.

The sequence of fIX_{EGF-C} is as follows, with the human blood coagulation factor IX sequence position [6] shown in parentheses:

H-LDVT C NIKNGR C EQF C KNSADNKVV C S C TEGYRLAENQKS C EPAV-OH
1(84)
45(128)

Air oxidation was used for the formation of two disulfide bridges in the first step. Peptide solution, at pH about 7.5, was slowly stirred at room temperature about 24-48 hr. To remove the Acm and begin the second step oxidation, 5 mM iodine in methanol was added dropwise, under nitrogen and in the dark, to the peptide solution containing 10% acetic acid until a brown color remained. After the solution was cooled in an ice-bath, 10 mM sodium thiosulfate was added to reduce excess iodine.

No disulfide scrambling was expected in the second step, since it was conducted under acidic condition (pH 2). The oxidation process was monitored with HPLC, and special attention was given to identification of the disulfide bridge pairings after the first step folding. Proteolytic digestion in conjunction with mass spectrometric peptide mapping were used for this purpose. Fig. 1 shows an example of how the location of the disulfide bridges was determined for peptide A. After it was cleaved by protease V8, the digestion mixture was subjected to matrix-assisted laser desorption mass spectrometric analysis. If peptide A had adopted an EGF-like pattern (Cys 1-3 and 2-4, while Cys 5 and 6 blocked with Acm), two peptide fragments should have been found, [1-13]+[14-30] and [31-45], but the experimental data did not match this pattern. Instead, three peptide fragments were observed, [1-13], [14-30] and [31-45], corresponding to an undesired isomer having cysteine pairings 1-2 and 3-4 (Fig. 1). In peptide B (data not shown), with Cys 2 and 4 blocked with Acm, the expected folding fragments [1-8]+[9-17] and [24-39]+[40-45] were detected, corresponding to formation of Cys 1-3 and 5-6 bonds. In peptide C, where Cys 1 and 3 were blocked with Acm, a mixture of disulfide isomers was detected (data not shown). The probable reason for this disulfide scrambling is that there is only one residue between Cys 4 and 5, so there is equal likelihood of forming an undesired 2-5 disulfide or the desired 2-4 disulfide linkage. Thus, only peptide B having the Cys 2-4 pair blocked with Acm formed the desired disulfide isomer in high yield.

We have applied these results to the synthesis of a number of other peptides. An example is the N-terminal EGF-like domain in factor X in which the cysteines 2 and 4 are blocked with Acm. The sequence of fX_{EGF-N}, with the corresponding human blood coagulation factor X sequence positions [7] shown in parentheses, is as follows. The synthetic peptide contains aspartic acid in place of the naturally occurring β-hydroxy-aspartic acid in position 20.

H-YKDGDQ C ETSP C QNQGK C KDGLGEYT C T C LEGFEGKN C
ELFTR-OH
1(44)
43(86)

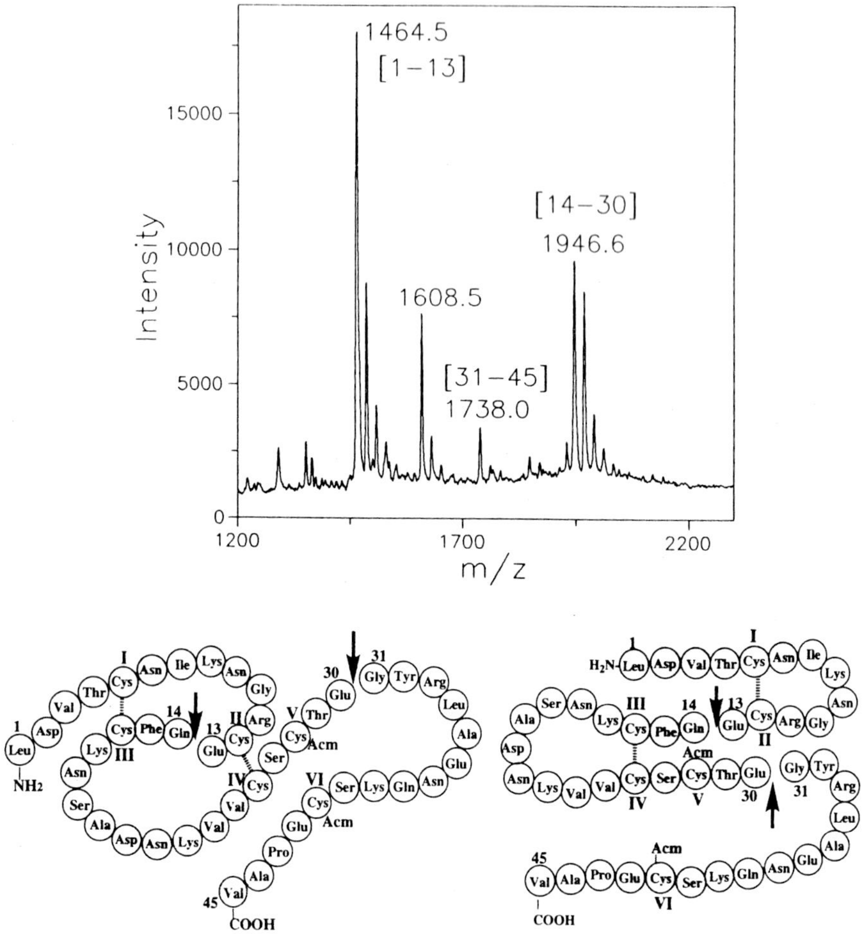

Fig. 1. Assignment of disulfide bridges located in peptide A after the first step oxidation. Arrows indicate the position of Staphylococcus aureus protease V8 cleavage. Top: Matrix-assisted laser desorption mass spectrum; Bottom left: Expected peptide with EGF pairing; Bottom right: Experimentally found disulfide pattern in peptide A.

The peptide mapping results after the first step oxidation of $fX_{EGF\text{-}N}$ are shown in Table 1. It is clear that the N-terminal EGF-like domain in factor X were synthesized with correctly paired disulfide bridges by the two-step approach.

Table 1 *Determination of disulfide pairing after first-step oxidation of* $fX_{EGF\text{-}N}$ *by endoproteinease Lys-C digestion, which cleaves after Lys, and mass spectrometric peptide mapping*

Determined Mass	Calculated Mass	Fragments	Disulfide Pairing
2202.2	2201.3	[1-19]	Cys 1-3
2755.5	2754.9	[20-43]	Cys 5-6
2773.2	2772.9	[20-36]+[37-43]	Cys 5-6

Acknowledgements

The authors would like to thank Dr. R.B. Merrifield, The Rockefeller University, for his generous help. This work was supported in part by US PHS grants HL41935 (J.P.T. and W.V.S.), CA 36544 (J.P.T.) and RR00862 (B.T.C.), and by US PHS grant RR03037 to the Hunter College Synthesis and Sequence Facility.

References

1. Tam, J.P., Wu, C.R., Liu, W. and Zhang, J.W., J. Am. Chem. Soc., 113(1991)6657.
2. Barany, G. and Merrifield, R.B., In Gross, E. and Meienhofer, J. (Eds.), The Peptides: Analysis, Synthesis, Biology. Volume 2. Part A. Academic Press, New York, 1979, p.233.
3. Van Rietschoten, J., Pedroso Muller, E. and Granier, C. In Goodman, M. and Meienhofer, J. (Eds.), Peptides. Proceedings of the Fifth American Peptide Symposium, John Wiley and Sons, New York, 1977, p.522.
4. Akaji, K., Fujino, K., Tatsumi, T. and Kiso, Y., Tetrahedron Lett., 33(1992)1073.
5. Yang, Y., Sweeney, W.V., Schneider, K., Chait, B.T. and Tam, J.P., Protein Sci., 3 (1994), In Press.
6. Yoshitake, S., Schach, B.G., Foster, D.C., Davie, E.W. and Kurachi, K., Biochem., 24(1985)3736.
7. Leytus, S.P., Foster, D.C., Kurachi, K., and Davie, E.W., Biochem., 25(1986)5098.

Study on racemization in thioester condensation by using RP-HPLC

Hong Wang, Hong Ji, Ruo-Heng Zhang and Xiao-Jie Xu
Institute of Physical Chemistry, Peking University, Beijing 100871, China

Introduction

In order to synthesize a highly pure polypeptide, a procedure in which partially protected peptide thioesters are used as building blocks was developed (Fig. 1) [1,2].

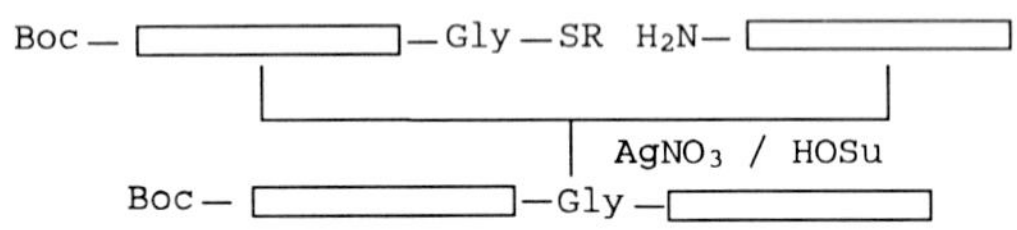

Fig. 1. Reaction route of thioester condensation.

Usually, glycine is chosen as the C-terminal amino acid of segment peptide I is choosed instead of other amino acids to avoid racemization. However, it becomes difficult to apply this strategy in synthesizing a peptide in which there is no glycine in which is near the C- or N-terminal. Hence, we hope to expand possibilities for the C-terminal from glycine to other amino acids so that we can apply our procedure more conveniently. Here, we present a study of the racemization in alanine-thioester condensation examined by RP-HPLC .

Results and Discussion

1. Synthesis of L,D peptide diastereisomers

The pure Boc-(L,D)Phe-GlyOBzl, Boc-Ala-(L,D)-Ala-GlyOBzl, Boc-Ala-(L,D)-Phe-GlyOBzl, Boc-Ala-(L,D)-Leu-GlyOBzl diastereisomers were synthesized by conventional solution method of DCC/HOBt [3]. All were confirmed by FAB-MS, element analyzing, m.p. detecting and RP-HPLC monitoring (Table 1).

2. RP-HPLC separation of L,D peptide diastereisomers

Each pair of the above L,D peptide diastereisomers mixtures was analysed by RP-HPLC (Waters 510, Column: Synchropak RP-P(4.6x250mm), 5μ Bondpak phenyl (3.9x150mm), Cosmosil C18 (10x250mm), Mobile phase: A: 0.1%TFA/H2O, B: 0.1%TFA/MeCN. The results are shown in Table 2.

We observed that only a diastereisometic mixture of Boc-Ala-(L,D)-Ala-GlyOBzl was well separated under the given conditions: Column: Synchropak RP-P(4.6x250mm), gradent: 20~60%B (30min).

Table 1 *Analysis data of L,D peptide diastereisomers*

Peptide diastereisomers	m.p.(°C)	M.W.	$[M+1]^+$	$[\alpha]_{30}$	Calcd %			Found 5		
					C	H	N	C	H	N
Boc-Phe-GlyOBzl	131.6~132.8	412	413	-8.8	61.44	7.85	9.35	61.59	7.90	9.43
Boc-D-Phe-GlyOBzl	131.5~132.0	412	413	+8.8	61.44	7.85	9.35	61.44	7.88	9.37
Mwc-Ala-Ala-GlyOBzl	146.0~147.1	407	408	-17.1	58.93	7.18	10.32	58.96	7.31	10.37
Boc-Ala-D-Ala-GlyOBzl	134.1~135.0	407	408	+5.7	58.83	7.18	10.32	58.67	7.18	10.09
Boc-Ala-Phe-GlyOBzl	103.6~104.8	483	484	-18.9	64.57	6.88	8.69	64.55	6.99	8.69
Boc-Ala-D-Phe-GlyOBzl	137.7~138.2	483	484	+9.4	64.57	6.88	8.69	64.72	6.83	8.74
Boc-Ala-Leu-GlyOBzl	117.6~118.5	449	450	-30.0	66.96	6.84	6.79	66.92	6.95	6.77
Boc-Ala-D-Leu-GlyOBzl	109.0~110.0	449	450	+14.7	66.96	6.84	6.79	67.25	6.98	6.82

Table 2 *RP-HPLC analysis of L,D peptide diastereisomers*

Peptide diastereisomers	Synchropak RP-P (4.6×250 mm)	Resolution μBondpak phenyl (3.9×150 mm)	Cosmosil C18 (10×250 mm)
Boc-(L,D)-Phe-GlyOBzl		0	
Boc-Ala-(L,D)-Ala-GlyOBzl	1	0.7	1.1
Boc-Ala-(L,D)-Phe-GlyOBzl	0.1	0	
Boc-Ala-(L,D)-Leu-GlyOBzl	0		

Table 3 *Racemization during the coupling of Boc-Ala-Ala-SCH_2CH_2CO-GlyOBzl with GlyOBzl TosOH under different conditions*

No.	Solv.	Additive	Base	R%[a]	F% (2hr.)[b]
1	EtOAc	HOSu (15)	NMM (12)	<2%	40%
2	THF	HOSu (15)	NMM (12)	<2%	100%
3	DMF	HOSu (15)	NMM (12)	25%	96%
4	DMSO	HOSu (15)	NMM (12)	22%	100%
5	THF	none	NMM (12)	42%	100%
6	THF	HOSu (1)	NMM (12)	38%	100%
7	THF	HOSu (5)	NMM (12)	26%	100%
8	THF	HOSu (10)	NMM (12)	5%	100%
9	THF	HOSu (20)	NMM (12)	<2%	100%
10	THF	HOSu (25)	NMM (12)	<2%	100%
11	THF	HOSu (30)	NMM (12)	<2%	100%
12	THF	HONp (15)	NMM (12)	46%	30%
13	THF	HOSu (15)	DIEPA (12)	<2%	100%
14	THF	HOSu (15)	TEA (12)	2%	100%
15	THF	HOSu (15)	NMM (6)	4%	95%
16	THF	HOSu (15)	NMM (18)	4%	100%
17	THF	HOSu (15)	NMM (24)	5%	100%

Numbers in parentheses are the ratio of additive (mol)/thioester (mol) or base (mol)/thioester (mol).
[a] R%=D/(L+D); [b] F%=(L+D)/(L+D+S). L,D,S present the peak area of L-peptide, D-peptide and thioester on RP-HPLC chromatograms respectively.

3. Racemization in thioester condensation

A model coupling reaction between Boc-Ala-Ala-SCH2CH2CO-GlyOBzl and GlyOBzl TosOH was monitored by RP-HPLC. To optimize reaction conditions, the effects of solvents, additives and bases were checked. The results were summarized in Table 3.

We found that the addition of HOSu could greatly reduce racemization, but HONp could not. When more than 15 equiv. of HOSu was added, its effect was no longer remarkable. Different bases and different amounts of bases had little influence on racemization. The degree of racemization that occurred for a variety of solvents changed with the following sequence: DMF>DMSO>THF~EtOAc. We conclude that THF should be is recommended as a solvent. When thioester (mol)/NMM (mol)/HOSu (mol) is 1/12/15, there is the least racemization (< 2%) during thioester condensation. Further study is underway.

References

1. Hojo, H. and Aimoto, S.,Bull. Chem. Soc. Jpn., 64(1991)111.
2. Hojo, H. and Aimoto, S.,Bull. Chem. Soc. Jpn., 65(1992)3055.
3. Huang, W.D., Chen, C.Q., Peptides Synthesis,(Ed. Science Publisher), Beijing, 1985.

Synthesis of phosphonotripeptides as ACE inhibitors

He-Ru Chen and Jie-Cheng Xu
Shanghai Institute of Organic Chemistry, Chinese Academy of Sciences, Shanghai 200032, China

Introduction

Since the 1980's orally active angiotensine converting enzyme (ACE) inhibitors have received much attention in the treatment of hypertension [1]. It has been shown that inhibition of this Zn-containing protease by isosteric phosphorus compound inhibitors may be quite promising. Galardy [2] has reported that phospho-L-alanyl-L-proline is a potent ACE inhibitor with a Ki of pH 7.5. We have focused on the design and synthesis of derivatives with the generalstructure CXII (as shown in Scheme 1).

Results and Discussion

N-protected α-aminophosphonic diesters (I) and (II) were synthesized by method A or B. Method A provides a " one pot " procedure with high yield (about 80–95%). The all-proctected title compounds were prepared by condensation of the corresponding n-protected α-aminophosphonic chlorides with H-Ala-Pro-OBzl•HCl with moderate yield (40-70%). We also used a new coupling reagent, benotriazolyloxybis(pyrrolidino) carbonium hexafluoro-phosphate (BBC)[3], but we obtained only bis(pyrrolidino)-Ala-Pro-OBzl carbonium hexafluoro-phosphate. This may provide an indirect evidence for the BBC coupling mechanism.

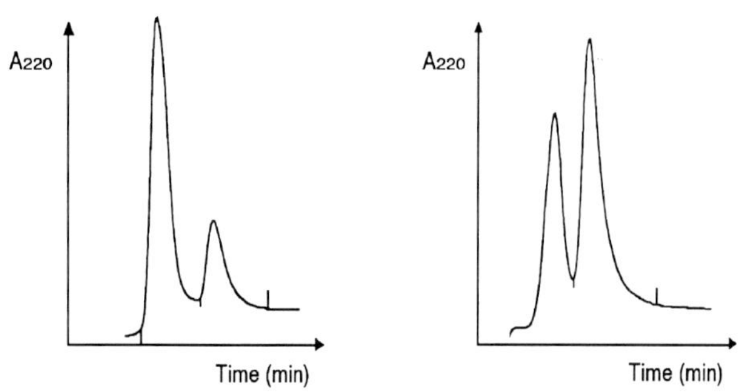

Fig. 1 HPLC profiles of compounds b and c. Mobile phase: 95/5 n-hexane/isopropanol (v/v); Flow rate: 1.3 ml/min; Detection: 220nm.

In order to resolvte the all-protected epimeric tripeptides (CXI), we used a column packed with chiralstationary phase of (L, L)-dipeptide tert-butyl-amide chemically bonded to γ-aminopropylsilanized silica. With chromatography conditions shown in Fig. 1, the value of selectivity (α) lie between 1.22-1.43 (see Table 1). It shows the possibility of separation for these epimeric tripeptides.

Scheme 1 *Synthesis route for the title compounds*

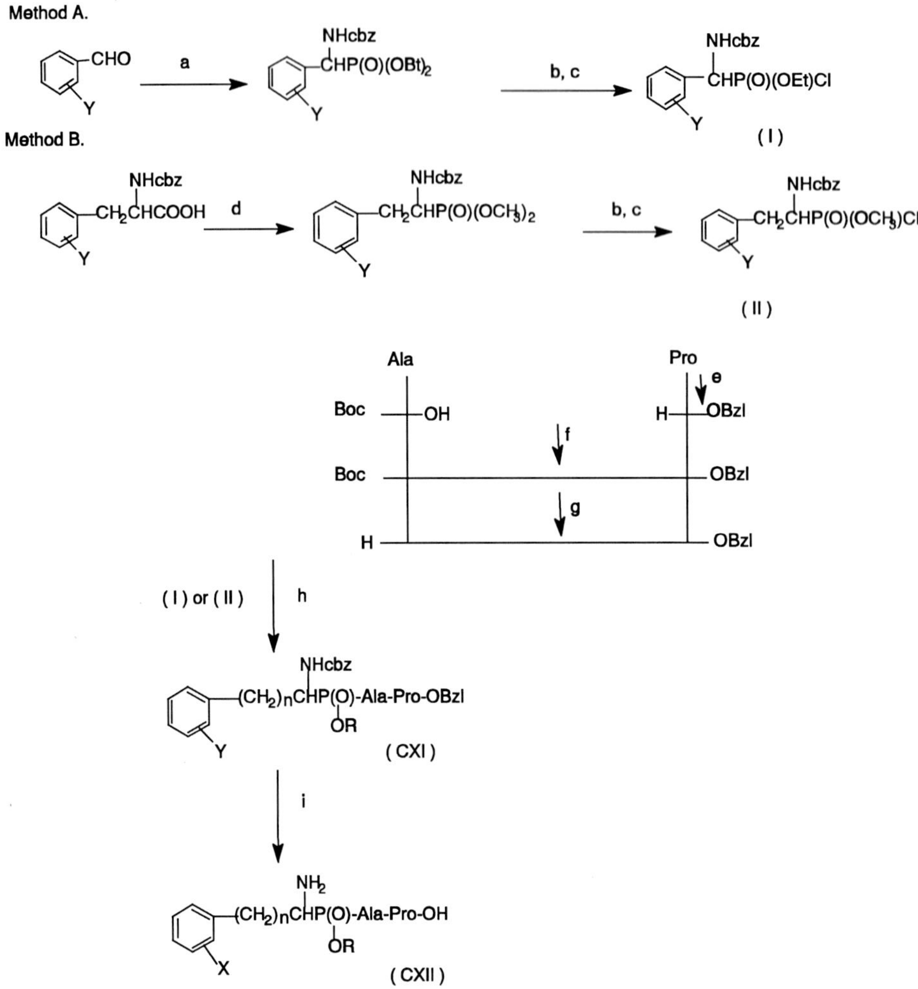

(Y=H, OBzl; X=H, OH; n=0,1)

a. AcCL / cbzNH_2 / HP(O)$(OEt)_2$; b. 1N NaOH / dioxane; c. $SOCl_2$ / $CHCl_3$; d. $Pb(OAc)_4$ / $P(OCH_3)_3$; e. $SOCl_2$ / HOBzl; f. ClCOOBu-i / $CHCl_3$; g. HCl / EtOAc; h. DMAP / $CHCl_3$ / NEt_3; i. H_2 / 10%Pd / C.

Table 1 *Chiral stationary phase resoluting epimeric tripeptides*

No.	compounds			value of parameters		
	Y	n	R	k1	k2	α
a	H	0	Et	5.67	8.13	1.43
b	OBzl(o-)	0	Et	4.73	6.46	1.37
c	OBzl(p-)	0	Et	4.95	6.06	1.22
d	H	1	Me	6.48	8.41	1.30
e	OBzl(p-)	1	Me	5.58	7.07	1.27

Acknowledgements

We thank Prof. Huang, J. X of the Shanghai Institute of Materia, Chinese Academy of Sciences for providing the chiral column.

References

1. Wyvratt, M.J., Patchett, A.A., Medicinal Research Reviews, 5(1985)483.
2. Galardy, R.E., Biochem. Biophys. Commun., 97(1980)94-99
3. Chen, S.Q. and Xu, J.C., Tetrahedron lett., 33(1992)647.

The study on TMSBr/thioanisole/TFA deprotection for the arginine-rich peptides

Xue-Jun Fan, Zheng-Ying Chen and Xiao-Je Dong
Institute of Basic Medical Sciences, AMMS, Beijing 100850, China

Introduction

Hepatitis C is a serious infectious disease, transmitted mainly through blood contamination. Based on computer analysis of the HCV genome and available references [1], we have selectively synthesized two peptides from the structural region of the HCV genome, CP9 (36aa) and CP10 (19aa). Peptide syntheses were accomplished in a Biolynx TM automated peptide synthesizer model 4170 using polyamide as the solid support and pentafluorophenyl active ester of corresponding Fmoc amino acids. For protection, Fmoc group was employed for the α-amino group and t-Butyl group or for other acid-labile groups such as Mtr and Trt side-chain functional groups. The conjugations problem of long peptides was attacked by employing repeated couplings and raising the temperature of the acylating reaction. Peptide was cleaved from resin by the new deprotecting reagent trimethylsilyl bromide (TMSBr) which also removed the sidechain protecting groups of the amino acids including Arg/Mtr groups. Results generated from HPLC and amino acid analysis confirmed that the two desired peptides were obtained. Peptides, CP9 and CP10, showed very good antigenicity in ELISA tests.

Results and Discussion

It is well know that most Fmoc compatible side-chain protecting groups are removed in 95% aqueous TFA at room temperature in 1.5 to 2.5 hours. However, Mtr is a notable exception. Deprotection and cleavage of Arg (Mtr) requires 3 or more hours, which is longer than that required for deprotection of other acid-labile, side-chain protecting groups. CP9 and CP10 bearing twelve and six Arg (Mtr) groups respectively still have attached Mtr groups after cleavage for 24 hours. We therefore introduced a new deprotecting procedure with TMSBr in TFA [2], based on a hard acid principle. This reagent has the ability to cleave readily tert-butyl type protecting groups and Mtr type protecting groups with the aid of a soft nucleophile, such as thioanisole. The arginine-rich 36-residue (CP9) peptidyl-resin and 19-residue (CP10) peptidyl-resin were treated with TMSBr/thioanisole/TFA in the presence of two additional scavengers, m-cresol and EDT, in an icebath for 1.5 h and 1.75 h, respectively, to remove all protecting groups and at the same time cleave the peptides from the resins. Deprotection of these peptides was monitored using RP-HPLC. The identity of the desired peptide was proven by amino acid analysis (shown in Table 1) and RP-HPLC (shown in Figs. 1 and 2).

The above results suggested that TMSBr/thioanisole/TFA as the cleavage mixture can remove Mtr more efficiently and as a deprotecting procedure can play an important role in the synthesis of arginine-rich peptides.

Table 1 *Amino acid analysis of CP9 and CP10*

	CP9	CP10
Asp		3.11(3)
Thr	1.82(2)	1.84(2)
Ser	1.79(2)	
Glu	4.20(4)	2.08(2)
Pro	5.01(5)	2.84(3)
Gly	4.18(4)	
Ala	0.92(1)	
Val	1.93(2)	1.20(1)
Ile	1.13(1)	
Leu	0.94(1)	
Lys	2.27(2)	3.85(4)
Arg	10.5(12)	3.64(4)

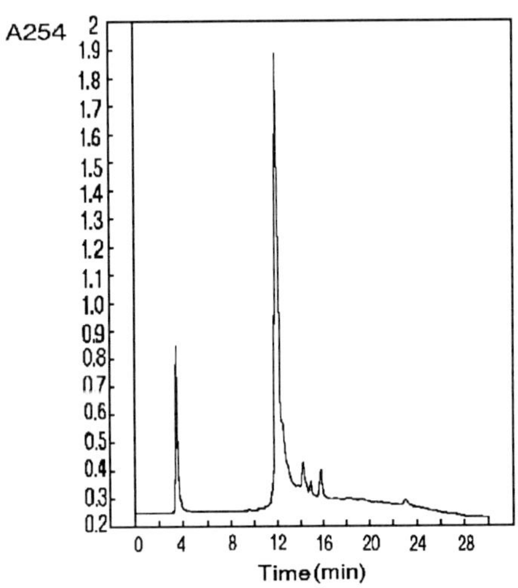

Fig. 1. HPLC Chromatogrm of peptide of CP9 after deprotection and cleavage by TMSBr/thioansole/ TFA in 1.75 hours.

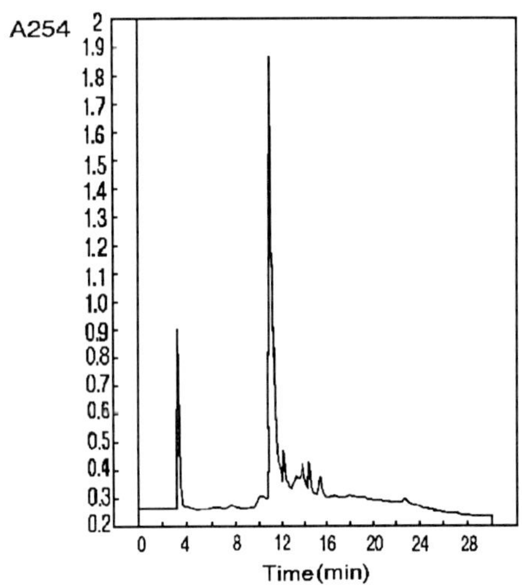

Fig. 2. HPLC Chromatogram of peptide of CP10 after deprotection and cleavage by TMSBr/ thioansole/TFA in 1.5 hours.

References

1. Choo, Q.L., Kuo, G., Weiner A.T., Overby, L.R., Bradley, D.W., Houghton, M., Science, 244(1989)359.
2. Fujii, N., Otaka, A., Sugiyama, N., Harano, M. and Yajima, H., Chem. Pharm. Bull., 35(1987)3880.

Synthesis and ion binding properties of cyclo-[Lys(Z)-Pro-Gly-Lys(Z)-Gly]$_2$ and cyclo-[Phe-Lys(Z)-Pro-Gly-Lys(Z)-Gly]$_2$

Yun Jiang, Zhen-Wei Miao and Xiao-Jie Xu
Department of Chemistry, Peking University, Beijing 100871, China

Introduction

Some cyclopeptides have an interesting feature in that they bind a specific alkali or alkaline earth metal ion. This characteristic can be explained by the contrained structure of the cyclic backbone in the complex state. For example, valinomycin, a macrocyclic dodecadepsipeptide, displays an exceptionally high selectivity for K^+ over Na^+ [1]. Furthermore, cyclopeptide can be used as a supporting template in TASP (template assembled synthetic). In order to build ion channel models with four-helix-bundle structures, 3 templates of cyclic peptides (CP12, CP10 and CP8) have been designed to exhibit ion binding selectivities and conformational constraint. The 3 cyclopeptides are:

```
CP12:  [Phe-Lys(z)-Pro-Gly-Lys(z)-Gly]2
CP10:  [Lys(z)-Pro-Gly-Lys(z)-Gly]2
CP8 :  [Lys(z)-Gly-Lys(z)-Gly]2
```

In CPS-92, we reported the conformation of CP12. In this paper, we will focus on the synthesis of CP12 and CP10, as well as their ion binding properties.

Results and Discussion

Synthesis

The cyclopeptides CP12 and CP10 were synthesized from linear peptides by cyclization or cyclic-dimerization. The linear peptides and their intermediates were synthesized by conventional solution method using DCC/HOBt [2]. The t-butoxy-carbonyl (Boc) and benzyloxy (Z) group were employed for α- and ε-amino protection, respectively, and methyl ester (OMe) was used for carboxyl protection. The Boc group was removed with TFA prior to each coupling, the OMe was removed by NaOH in 1,4-dioxane, and the Z group was not removed. The results of peptide cyclization are shown in Table 1 and Fig. 1.

All of the linear peptides and their intermediates were confirmed by FAB-MS, element analysizing, m.p. detecting and TLC monitoring. From Table 1 we can conclude that the yields of direct cyclization from linear peptides are higher than cyclic-dimerization. This may be interpreted by in cyclic-dimerization, there are always have product from direct cyclization or polymerization.

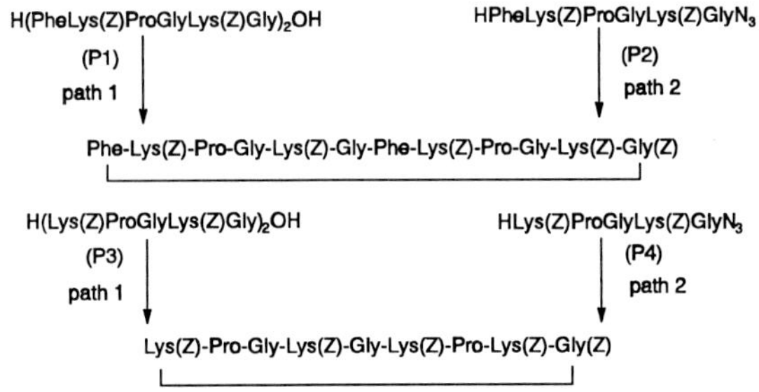

Fig. 1. The procedure of cyclization;
path 1: cyclization of P1 and P3, using DCC/HOBt in THF;
path 2: cyclic-dimerizaton of P2 and P4, using DCC/HOBt in THF.

Table 1 *The yields of cyclization*

peptide	path 1		path 2		$MS^+(+H^+)$
	yield (%)	conc. (mM)	yield (%)	conc. (mM)	
CP12	82.1	1.0	52.0	10	1765
CP10	86.0	1.0	55.0	10	1471

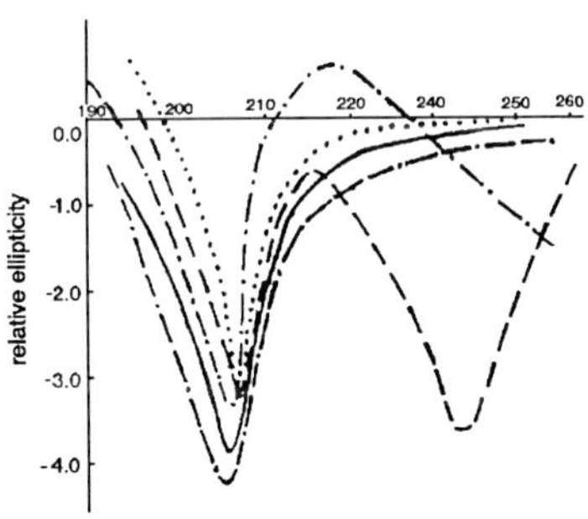

Fig. 2. CD spectra of CP10(-) and its various cation complexes in acetonitnle. The concentration of CP10 was 4.1×10^{-4}. The molar C/P ratio is 1.0: (-----)Mg^{2+}, (——)Ba^{2+},(–•–)K^+, (•••••)Na^+, the cell path is 1mm.

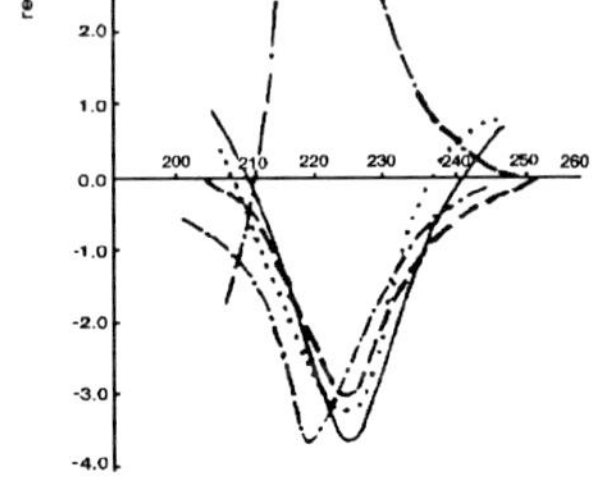

Fig. 3. CD spectra of CP12(-) and its various cation complexes in acetonitnle. The concentration of CP10 was 4.1×10^{-4}. The molar C/P ratio is 1.0: (-----)Mg^{2+}, (——)Ba^{2+},(–•–)K^+, (•••••)Na^+, the cell path is 1mm.

Ion binding properties

The ion binding properties of CP12 and CP10 were studied by CD. The CD spectra were recorded on a Jasco J-500c spectropolarimeter with a DP-500N Data processor at 25°C. The results and conditions were as follows (the cell path is 1mm):

These figures show that CP12 selectively binds for Ba^{2+} and CP10 has high ability to bind Mg^{2+}.

References

1. Shimizu, T., Tarcaka, Y. and Tsuta, K., Bull. Chem. Soc. Jan., 58(1985)3436.
2. Huang, W.D., Chen, C.Q., in Peptides Synthesis, Science Press, 1985, Beijing.

A new approach to chemical synthesis of peptide T analogue

Jin-Huan Shen and Yu-Cang Du
Shanghai Institute of Biochemistry, Chinese Academy of Sciences, Shanghai 200031, China

Introduction

N^{α}-Boc and N^{α}-Fmoc-protecting groups are extensively used in solid-phase peptide synthesis. In Boc-strategy the peptide and all the benzyl-derived side chain protecting groups are detached from the solid support and cleaved by liquid hydrogen fluoride or other strong acid.

The synthetic utility of derivatives of DHBT (2,3-dihydro-3-hydroxy-4-oxobenzotriazine) has been demonstrated in their high yield reactions with hindered α-amino acids [1, 2]. It has been reported that FMOC-amino acid esters of this alcohol are excellent acylating agents.

In an effort to improve and simplify the procedure of solid phase peptide synthesis, we have examined the advantage of DHBT reagent as an active ester in the synthesis of peptide T by Boc-strategy. As an example, octapeptide ^{D}A-S-T-T-T-N-Y-T-(NH_2), a D-Ala[1] substituent of peptide T amide [3], was synthesized on merrifield resin in the presence of Boc-amino acid DHBT esters with free side chains and then the [D-Ala[1]]-peptide T amide was obtained from the resin by NH_3/CH_3OH in high yield.

Results and Discussion

For the synthesis of [D-Ala[1]]-peptide T amide, Boc-threonine was esterified on a 1% cross-linked Divinylbenzen-styrene merrifield resin (0.45 meq/g) and then each coupling cycle was sequentially carried out in DMF with 2.25 equivalent of Boc amino acid DHBT esters, i.e. Boc-tyrosine ester, Boc-asparagine ester, Boc-threonine ester, Boc-serine ester and D-alanine ester etc. In each cycle, the deprotection was performed by using 50% TFA/DCM and the peptide was finally cleaved from the resin with 25% NH_3/CH_3OH. The process of peptide T synthesis is outlined by scheme 1.

In the peptide synthesis process, the coupling efficiency of each amino acid derivative was analyzed step-wise by using salicylaldehyde method. Results in Table 1 show that both the amount of residual free amino group after coupling and the amount of total amino group exposed after deblocking meet the requirements for the next coupling.

After ammonolysis and TFA treatment the crude product was obtained in 89% yield and after purification by HPLC the total yield of the pure [D-Ala[1]]-peptide T amide was 45% based on the starting Boc-threonine resin. In this way the main side-product was found to be [D-Ala[1]]-peptide T without amide group at the C-terminal and with 12.6% yield in pure form. The amino acid composition of this synthesized [D-Ala[1]]-peptide T amide is shown in Table 2.

Scheme 1 *Synthetic procedure of [D-Ala[1]]-peptide T amide*

Boc-Thr-OH + Φ; 50% TFA/CH_2Cl_2; 10%
DIPEA/CH_2Cl_2
Boc-Tyr-DHBT + T-Φ; 50% TFA/CH_2Cl_2; 10%
DIPEA/CH_2Cl_2
Boc-Asn-DHBT + Y-T-Φ; 50% TFA/CH_2Cl_2; 10%
DIPEA/CH_2Cl_2
Boc-Thr-DHBT + N-Y-T-Φ; 50% TFA/CH_2Cl_2; 10%
DIPEA/CH_2Cl_2
Boc-Thr-DHBT + T-N-Y-T-Φ; 50% TFA/CH_2Cl_2; 10%
DIPEA/CH_2Cl_2
Boc-Thr-DHBT + T-T-N-Y-T-Φ; 50% TFA/CH_2Cl_2; 10%
DIPEA/CH_2wl_2
Boc-Ser-DHBT + T-T-T-N-Y-T-Φ; 50% TFA/CH_2Cl_2; 10%
DIPEA/CH_2Cl_2
Boc-[D-Ala]-DHBT + S-T-T-T-N-Y-T-Φ
⇓
Boc-^{D}A-S-T-T-T-N-Y-T-Φ
⇓ 25% NH_3/CH_3OH
⇓ TFA
^{D}A-S-T-T-T-N-Y-T-(NH_2)

Table 1 *Coupling efficiency of amino acid derivatives to the resin*

order of coupling	Thr_8	Tyr_7	Asn_6	Thr_5	Thr_4	Thr_3	Ser_2	$^{D}Ala_1$
amino group liberated(mmol/g)	0.45	0.44	0.38	0.36	0.29	---	0.24	---
amino group calculated(mmol/g)	0.45	0.42	0.38	0.35	0.32	0.30	0.28	0.27
residual amino group(mmol/g)		0.015	0.021	0.018	0.012	0.013	0.011	0.008
coupling yield %		97	95	95	96	96	96	97

The results presented herein show that the analog of peptide T was successfully synthesized in a convenient and economical maner. Although possible interference in the condensation reaction could be made by six side chains including five aliphatic hydroxyl and one phenolic hydroxyl groups, the high yield of synthetic peptide T indicates that the side reactions are avoidable, especially in the mild reaction condition that we used. In our procedure the peptide-resin was subjected to trifluoro-acetic acid treatments and an ammonolysis but not to hydrogen fluoride or other strong acid. In general, we believe that the Boc-amino acid DHBT esters strategy is a useful approach for synthesizing active peptide.

Table 2 *Amino acid compositions of synthetic peptide*

	Asp	Thr	Ser	Ala	Tyr	NH_3
[D-Ala^1]-peptide T amide	1.01	4.36	1.10	0.84	1.30	2.15
[D-Ala^1]-peptide T	1.05	4.40	1.05	0.78	1.17	0.99

References

1. Konig, W. and Giger, R., Chem. Ber., 103(1970)2034.
2. Protein Synthesis Group, Shanghai Institute of Biochemistry, Academia Sinica. Scientia Sinica, 18(1975)745.
3. Pert, C.B., PNAS., 83(1986)9254.

Dichlorodimethylsilane - a useful reagent for the cleavage of peptidyl Wang resin

De-Xin Wang, Ying Sun, Gui-Shen Lu and Qing-Chai Xu
Institute of Materia Medica, Chinese Academy of Medical Sciences, Beijing 100050, China

Introduction

We have previously reported that dichlorodimethylsilane (DDSi) could completely remove, in the presence of phenol, the N^{α}-Boc group [1]. In present study, DDSi was explored as an acidolysis reagent, especially for the cleavage of peptidyl Wang-resin linkage, the usual case of TFA-consuming reaction [2].

The mechanism of this reaction is described by the formula showing the release of acidic phenolic proton due to the formation of a strong Si-O bond.

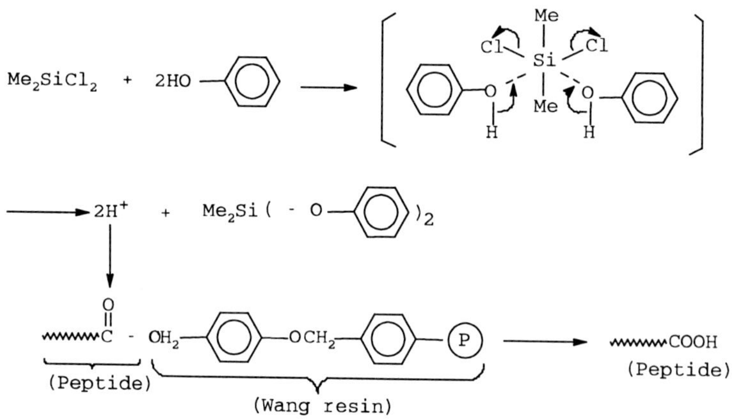

Results and Discussion

Four protected peptidyl Wang-resins were tested as model compounds:

H-Thr(tBu)-Asp(Tbu)-(D)Thr(Tbu)-Ser(Tbu)-Leu-Gln-Asp(Tbu)-His(Trt)-Pro-Thr(Tbu)-Phe-Asn(Trt)-Wang resin .... (1)

H-Thr(Tbu)-Asp(Tbu)-Val-Ser(Tbu)-Leu-Gln-Asp(Tbu)-His(Trt)-Pro-Thr(Tbu)-Phe-Asn(Trt)-Wang resin .... (2)

Ac-Salicylyl-Thr(Tbu)-Thr(Tbu)-Asn(Trt)-Tyr(Tbu)-Thr(Tbu)-Wang resin .... (3)

Ferulyl-Arg(Pmc)-Lys(Boc)-Gly-Asp(Tbu)-Wang resin .... (4)

In order to get convincing results, two parallel acidolyses of compound 4 were carried out with TFA and DDSi. Based on detecting degree of the peptide substitution on the resin before and after cleavage, quite close yields (Table 1) were obtained. Three other peptides also were released by DDSi procedure only with satisfactory yields: 96.5% (1 and 2), 89.2% (3).

Table 1 *Comparison of cleaving yields between TFA and DDSi acidolysis of peptide(4)-resin*

Time	TFA method[a]	DDSi method[b]	
		0.5M	1.0M
20 min	69.1%	74.1%	97.2%
40 min	---	93.5%	97.0%
60 min	96.0%	95.8%	97.0%

[a] TFA/phenol/EDT/thioanisole/H2O (85:6:3:3:3, v/v).
[b] DDDSi (0.5 or 1M)/phenol (1 or 3M)/3% EDT/3% thioanisole/DCM-HOAc (1:1)

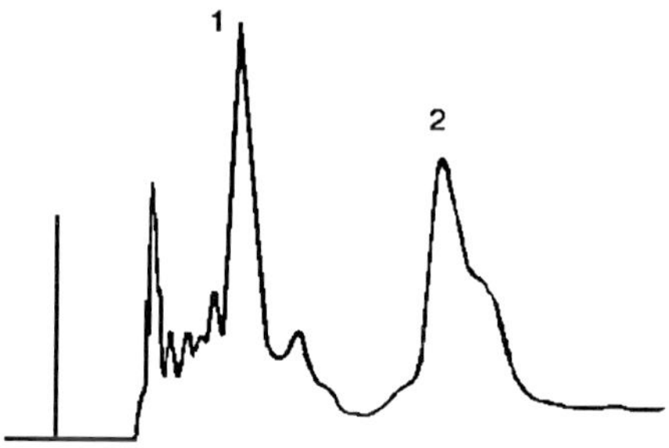

Fig. 1. The profile of HPLC analysis of crude peptide 1 and 2 after DDSi cleavage.

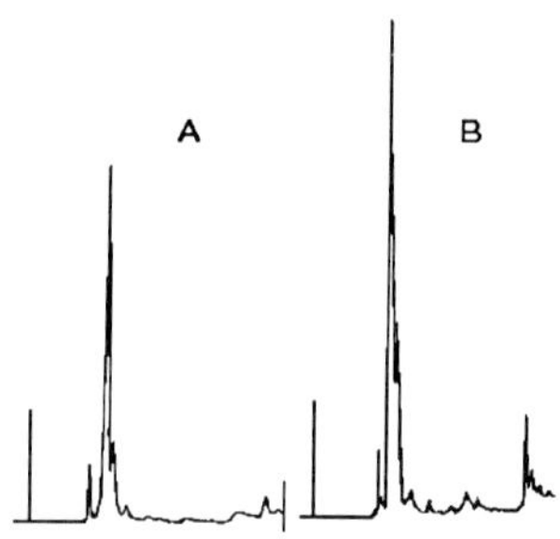

Fig. 2. The comparison of HPLC feature of peptide 4 between A TFA cleavage and B DDSi cleavage.

All crude products were checked by HPLC (Figs. 1 and 2), and amino acid analysis (Table 2). Moreover, peptides 3 and 4 were confirmed by FAB-MS.

From the RP-HPLC analysis of peptide 4, there was no conspicuous difference between the two products obtained by means of TFA and DDSi cleavage. Perfect correspondence of amino acid composition between the observed and calculated data was also found (Table 2).

The DDSi procedure therefore is not only an additional method for the cleavage of peptidyl Wang-resin linkage, but also has the advantage of eliminating the necessity of using TFA, which is expensive and corrosive solvent-TFA.

Table 2 *Amino acid composition of product 1-4*

Residue	Peptide(1)	Peptide(2)	Peptide(3)	Peptide(4)
Asp	2.90(3)	3.30(3)	1.00(1)	1.02(1)
Thr	2.60(3)	2.09(2)	2.95(3)	---(0)
Ser	1.00(1)	1.00(1)	---(0)	---(0)
Tyr	---(0)	---(0)	0.99(1)	---(0)
Gln	0.93(1)	1.19(1)	---(0)	---(0)
Val	---(0)	0.84(1)	---(0)	---(0)
Gly	---(0)	---(0)	---(0)	1.00(1)
Leu	0.85(1)	1.00(1)	---(0)	---(0)
Lys	---(0)	---(0)	---(0)	0.95(1)
Phe	1.00(1)	0.98(1)	---(0)	---(0)
Arg	---(0)	---(0)	---(0)	0.95(1)
His	0.91(1)	1.11(1)	---(0)	---(0)
Pro	0.98(1)	1.05(1)	---(0)	---(0)

Acknowledgements

This study constitutes a portion of the project " Studies on structural modification and bioactivity of hF-GRP and its analogs ", which was supported by the National Natural Science Foundation of China, No. 3907951.

References

1. Wang, D.X. and Lu, G.S., Chin. J. Med. Chem., 2(1991)60.
2. Wang, S.S., J. Am. Chem. Soc., 95(1973)1328.

Application of a new type of coupling agent, containing phosphorous, in liquid - phase synthesis of polypeptide

Jun Wu, Yan-Mei He, Yao Lin and Qing-Zhong Zhou
Department of Chemistry, Peking University, Beijing 100871, China

Introduction

A number of scientists have focused on peptide synthesis since the technique was developed. They have tried to find and improve methods of peptide preparation. On one hand, they looked for convenient ways to protect the peptide and to cleave the protecting group. On the other hand, they have attempted new approaches for peptide bond formation. Organic phosphorous compounds, used in both protecting and coupling peptide, have been extensively studied. 3-diethylphosphoryloxy-1,2,3-benzotriazin-4(3H)-one (DEPBT, Fig. 1.), which has been used by us in liquid-phase synthesis of pentapeptide, is a new type of coupling agent.

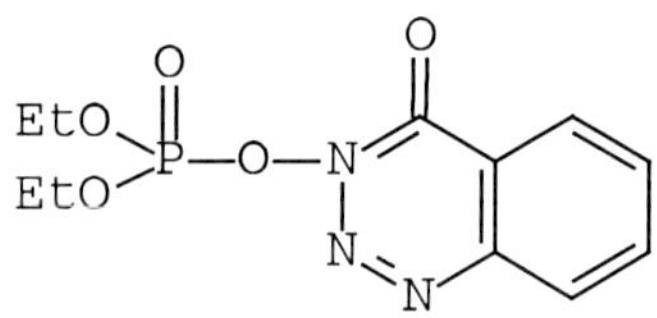

Fig. 1.

We designed and synthesized the pentapeptide Boc-Ile-Asn-Met-Trp(CHO)-Gly-OMe (I) which exists in human immunodeficiency virus-1 (HIV-1). Major envelope protein (GP120) on the surface of HIV-1 has high affinity with the CD_4 molecule on the surface of T-lymphocytes. After GP120 bonds to CD_4, T-lymphocytes are infected and AIDS is caused. The investigation showed that although HIV changes rapidly, there still is a relatively stable region by which GP120 combines with CD_4. Its sequence is Gln-Phe-Ile-Asn-Met-Trp-Gln-Glu-Val-Gly-Lys-Ala-Met-Tyr-Ala-Pro. If we can cleave the peptide bond in this region specifically, the incidence of AIDS may be lowered.

Accordingly, we used the strategy of catalytic antibody to obtain a molecule with proteinase function. We designed an appropriate haptin molecule containing the pentapeptide. We first successfully used DEPBT in liquid-phase synthesis of the pentapeptide and compared its properties with DCC. DEPBT is a new type of coupling agent which has been used in liquid-phase dipeptide synthesis and solid-phase polypeptide synthesis. Compared with DCC, DEPBT has some obvious advantages:

(1) Low-yielding D-isomer. According to the racemization study by Young Test during peptide bond formation by DEPBT, D-isomer was less than 1%. It is lower than of DCC in the same condition.

(2) Few side reactions. Hydrolysis reaction of the side chain of Gln, Asn was not observed when DEPBT was used.

(3) Convenient post-treatment. It is easy to purify the product because the byproduct of the reaction can be removed thoroughly by conventional means.
(4) Easy storage. DEPBT is a white solid, not easy to be deliquesced and decomposed.

There are few reports on DEPBT being applied in liquid-phase synthesis of smallpeptides, even though it has many good properties. We therefore investigated further its application in liquid-phase polypeptide synthesis and obtained pentapeptide(I).

Results and Discussion

We synthesized the pentapeptide by liquid-phase method from the carboxyl terminal. The procedure is shown in Scheme 1.

Scheme 1 *Synthesis of pentapeptide (1)*[a]

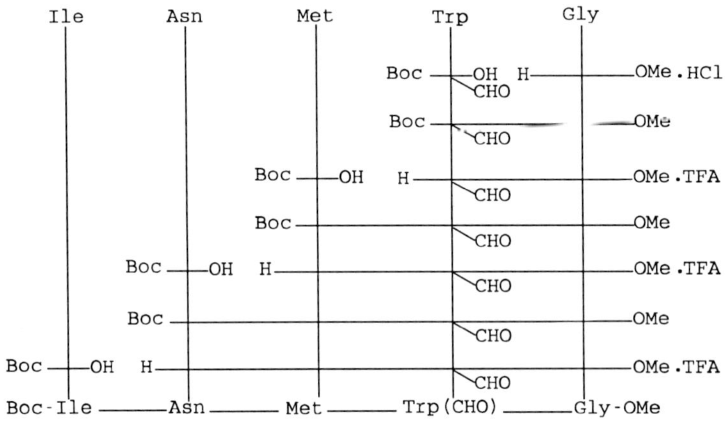

[a] DEPBT was used as coupling agent of peptide bond formation. N-Boc was removed by TFA.

The general procedure is as the follows:

To a stirred mixture of the carboxyl component (1 mmol) and the hydrochloride of the amino component (1 mmol) in DMF was added Et_3N (2 mmol) and DEPBT (1.1 mmol). The reaction time was 2 hr. The reaction mixture then was diluted with saturated aqueous sodium chloride and extractedby ethyl acetate 3 times. The product was washed with 5% aqueous citric acid, 5% aqueous sodium carbonate, water and saturated aqueous sodium chloride. Drying over magnesium sulfate followed by evaporation *in vacuo* yielded the crude peptide derivative, which was recrystallized to gain pure product.

In Scheme 2 the melting points and yields of the thus synthesized peptide derivatives are listed.

Scheme 2 *Peptide derivatives Boc-R-OMe prepared by DEPBT method*

Compound R	m.p.	Yield(%)
Trp(CHO)-Gly	121-4°C	64.5[a]
Met-Trp(CHO)-Gly	95-7°C	98.4[b]
Asn-Met-Trp(CHO)-Gly	202-8°C	67.3[b]
Ile-Asn-Met-Trp(CHO)-Gly	232-7°C	47.0[b]

[a] Yield of peptide bond formation.
[b] Yield of deprotection and peptide bond formation.

We also used different methods, such as solid-phase, DCC-HOBt and active ester (DCC), to prepare the pentapeptide.

From the experiment, we found that a 2 hr. reaction was sufficient for preparing dipeptide. For larger peptide derivative, reaction time should be longer.

Application of DEPBT for synthesis of linear and cyclic heptapeptide from a *Caryophyllaceae* plant

Yun-Hua Ye[a], Hai-Bo Xie[a], Gui-Ling Tian[a], Chong-Xu Fan[a], Ying Liu[a], Yong-Jun Lu[b] and Gui-Yang Xie[a]
[a]Department of Chemistry and [b]Department of Biology, Peking University, Beijing 100871, China

Introduction

Two novel cyclic organophosphorus compounds DOPBO and DOPBT as coupling reagents for peptide synthesis were reported by us previously [1]. In order to improve the solubility of DOPBT, another oranophosphorus compound, 3-(diethyloxyphosphoryl-oxy)-1,2,3-benzotriazin-4(3H)-one was studied recently.

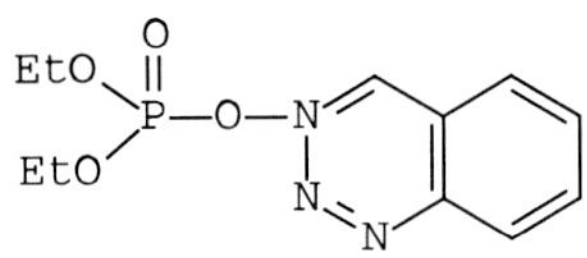

DEPBT

DEPBT is a stable colourless crystal and is easily to be prepared. The solubility of DEPBT is much better than DOPBO and DOPBT. It can be soluble in common organic solvents such as THF,CH_2Cl_2 etc.

DEPBT is a very efficient coupling reagent for peptide synthesis either by solution method or by solid-phase method. It can also be used for synthesis of cyclic peptides.

Results and Discussion

1. The preparation of DEPBT was accomplished by the reaction of diethy1 phosphory1 chloride and 3-hydroxy-1,2,3-benzotriazin-4(3H)-one in THF and Et_3N at room temperature.

The product DEPBT was confirmed by MS,NMR,IR and elemental analysis. MS (EI): 299 (M+), ^{31}PNMR ($CDCl_3$) δ (ppm): -32, ^{1}HNMR ($CDCl_3$) δ (ppm): 1.36-1.61 (6H,m,-OCH_2CH_3), 4.43-4.64 (4H,m,-OCH_2CH_3), 7.84-8.40 (4H,m,ArH)

2. DEPBT is a very efficient reagent for peptide synthesis in one-pot method. N-protected amino acid can be readily condensed with another amino component by equivalent DEPBT in organic solvent (THF,CH_2Cl_2 etc.) at room temperature to give the expected peptide. In using DEPBT for peptide synthesis, the hydroxy group in the amino component need not be protected and dehydration of Asn was not observed. The peptides are easily purified and racemization is less than 1% detected by Young test (Table 1).

3. The cyclic heptapeptide cyclo (Gly•Tyr•Gly•Gly•Pro•Phe•Pro) is one of cyclic peptides isolated from a kind of Chinese medicinal herbs, Caryopy llaceae plant-Stellaria yunnanensis Franch (M) [2]. DEPBT was used as coupling reagent for synthesis linear intermediate HGLY•Tyr•Gly•Gly•Pro•Phe•ProOH by solution and solid-phase method respectively. The linear heptapeptide was cyclized successfully by DEPBT in dilute solution (10^{-3}M)for 24h to obtain 61% cyclic product. In the same reaction condition, the cyclic product was obtained by BOP [3] in 57% and BBC [4] in 51%, these yields were based on the peak areas of HPLC. The cyclic product was identified by MS (FAB)$(M+H)^+$: 676 and amino acid analysis: Gly (2.72), Tyr (1.08), Pro (1.92), Phe (1.00).

Table 1 *The physical constants and yield of some oligopeptides synthesized by DEPBT*[a]

Peptides	m.p. (°C)[b]	MS (FAB) (M+H)+	$[\alpha]_D^{25}$ (C,sol)[c]	Yield(%)
Z-Ala•PheOMe	101.0~102.5	/	-11.5 (1,EtOH)	94
Boc-Ile•TyrOMe	144~146	/	-17.0 (1, EtOH)	91
Z-Asn•PheOMe	199~201	/	-3.0 (1, DMF)	73
Z-Ala•SerOMe	137~138	/	/	66
Z-Gly•TyrOMe	119~121	387	+20.5 (1, EtOH)	83
Boc-Phe•ProOBz	110~112	453	-44.5 (1, EtOH)	75
Boc-Pro•PheOMe	65~67	377	/	85
Z-Gly•Tyr•GlyOMe	110~113	430	-6.0 (1, EtOH)	50
Boc-Pro•Phe•ProOBz	Oil	550	/	64
Z-Gly•Tyr•Gly•GlyOEt	128~130	515	+1.3 (1, EtOH)	52

[a] All amino acids used in the present studies belonged to the L-series.
[b] m.p. were measured by Yanaco melting point apparatus and the thermometer was not corrected.
[c] $[\alpha]_D$ were measured by Perkin-Elmer 241-MC. $[\alpha]_D^{25}$ and m.p. of known peptides were in agreement with reported values.

Acknowledgements

The authors thank the National Natural Science Foundation of China for financial support, Dr. B. Castro for BOP reagent, Professor Jie-cheng Xu for BBC reagent and Professor Hou-ming Wu for the authetic sample of cycloheptapeptide.

References

1. Fan, C.X., Hao, X.L., and Ye, Y.H., In Du, Y.-C., Tam, J.P. and Zhang, Y.-S. (Eds.) Peptides: Biology and Chemistry (Proceedings of the 1992 Chinese Peptide Symposium), ESCOM, Leiden, 1993, pp. 297–298.
2. Zhao, Y.R., Zhou, J., Huang, X.L., and Wu, H.M., Proceedings of the 5th National Natural Organic Chemistry Symposium p.76 (November 16-19, 1993, Guangzhou, China).

3. Castro, B., Dormy, J.R., Tetrahedron Lett., (1972)4747.
4. Chen, S.Q. and Xu, J.C., In Du, Y.-C., Tam, J.P. and Zhang, Y.-S. (Eds.) Peptides: Biology and Chemistry (Proceedings of the 1992 Chinese Peptide Symposium), ESCOM, Leiden, 1993, pp. 311–313.

Session II
Biomedical peptides

Chairs: Pravin T.P. Kaumaya
The Ohio State University
Columbus, Ohio, U.S.A.

Charles M. Deber
University of Toronto
Toronto, Ontario, Canada

and

Yun-Hua Ye
Peking University
Beijing, China

Table 1 *Biologically active peptides*

Peptide	# of Residues	Application/Biological Activity
Captopril	2	Hypertension
Thyrotropin Releasing Hormone (TRH)	3	Release TSH; Thyroid stimulation
Enkephalin	5	Analgesia; Diarrhea; Opiate activity
Cholecystokinin (CCK)	8	Ulcers; Appetite suppression
Angiotensin II	8	Hypertension; Cardiac failure
Oxytocin	9	Induction of labor; Lactation
Vasopressin	9	Diabetes Insipidus; Antidiuretic
Bradykinin	9	Dilation of peripheral blood vessels
Leutenizing Hormone-Releasing Hormone (LHRH)	10	Control of ovulation
Substance P	11	Pain; Inflammation
Neurotensin	13	Hypertension; Pain
α-Melanocyte Stimulating Hormone (MSH)	13	Fever
Somatostatin	14	Diabetes Mellitus; Bleeding Ulcers
Bombesin	14	Hyperthermia; Peptic Ulcers
Motillin	22	Stimulation of gastric motility
Secretin	27	Stimulation of exocrine pancreatic secretion
Vasoactive Intestinal Peptide (VIP)	28	Relax smooth muscle from bronchi
Glucagon	29	Hypoglycemia; Mobilization of liver glycogen
Calcitonin	32	Osteoporosis; Regulation of Ca++ in blood
Adrenocorticotrophic Hormone (ACTH)	39	Severe inflammation; Amnesia
Growth Hormone Releasing Factor (GHRH)	44	Growth hormone deficiency
Insulin	51	Insulin-dependent diabetes mellitus
Parathyroid Hormone (PTH)	84	Osteoporosis; Hypertension

characterized and shown to have important biological activities. Some of the more widely known biologically active peptides are summarized in Table 1. Most of these peptides have been subjected to extensive structure-activity studies. In many cases optimized analogs have been developed which have greater potencies, longer duration of activity or better biological profiles than the naturally occurring, parent peptide. Important breakthroughs in *(i)* classical solution synthesis *(ii)* solid phase peptide synthesis *(iii)* enzymatic synthesis of peptides and *(iv)* recombinant synthesis of peptides have provided opportunities for scale-up synthesis and the commercial development of biologically active peptides. In addition, the availability of high resolution techniques, e.g. reverse-phase high-performance liquid chromatography, capillary zone electrophoresis, fast-atom bombardment mass spectrometry and high-field NMR spectrometry have provided valuable tools for the rapid analysis and evaluation of synthetic peptides.

In recent years research on biologically active peptides has included analogs with antibacterial effects (cecropin A, maganin, polymixin, ranalexin); hormonal activities (growth hormone releasing factor, neuropeptide Y, corticotropin releasing factor, amylin, antagonists of LHRH and GnRH); inhibitors of platelet aggregation (RGD-peptides); neuropeptides (β-amyloid peptides, deltorphin, β-casmorphin and conantokin analogs) and peptide inhibitors (inhibitors of interleukin-1 and interleukin-2, HIV-1 protease, endothelin and thrombin).

Of course it is not possible to predict whether the modifications that are summarized in this brief review apply to other families of biologically active peptides. Each system requires thorough evaluation of the various principles that have been described. Moreover, other modifications of biologically active peptides including the incorporation of unnatural amino acids, modified peptide bonds and other structural changes designed to allow the peptide to "fit" into a receptor pocket should be incorporated into any serious effort to design a novel peptide therapeutic. The opportunities for the development of biologically active peptides into therapeutics for the treatment of a variety of ailments are excellent. However, the continued advancement of synthetic methodology, biomolecular structural analysis, drug delivery systems and the development of rapid and multiple bioassay procedures are essential components for biologically active peptides to attain their full potential.

References

1. Rivier, J., Spiess, J., Thorner, M.O. and Vale, W., Nature, 300(1982)276.
2. Guillemin, R., Brazeau, P., Bühlen, P., Esch, F., Ling, N. and Wehrenberg, W., Science, 218(1982)585.
3. Ling, N., Esch, F., Bühlen, P., Brazeau, P., Wehrenberg, W.B. and Guillemin, R., Proc. Natl. Acad. Sci. USA, 81(1984)4302.
4. Madison, V., Berkovitch-Yellin, Z., Fry, D., Greeley, D. and Toome, V. In Tam, J. and Kaiser, E.T. (Eds.), Synthetic Peptides: Approaches to Biological Problems, Alan R. Liss, NY, 1989, p. 109.
5. Madison, V.S., Fry, D.C., Wegrzynski, B.B., Williamson, M.P., Campbell, R.M., Danho, W., Heimer, E.P. and Felix, A.M. In Renugopalakrishnan, V., Carey, P.R., Smith, I.C.P., Huang, S.G. and Storer, A.C. (Eds.) Proteins: Structure, Dynamics and Design, ESCOM, Leiden, 1991, pp. 234–239.
6. Felix, A.M., Heimer, E.P., Mowles, T.F., Eisenbeis, H., Leung, P., Lambros, T.J., Ahmad, M., Wang, C.T. and Brazeau, P. In Theodoropoulus, D. (Ed.), Peptides 1986, Walter de Gruyter & Co., NY, 1987, p. 481.

7. Ling, N., Baird, A., Wehrenberg, W.B., Ueno, N., Munegumi, T., Chiang, T.-C., Regno, M. and Brazeau, P., Biochem. Biophys. Res. Commun., 122(1984)304.
8. Coy, D.H., Hocart, S.J. and Murphy, W.A., Eur. J. Pharmacol., 204(1991)179.
9. Frohman, L.A., Downs, T.R., Williams, T.C., Heimer, E.P., Pan, Y.-C. and Felix, A.M., J. Clin, Invest., 78(1986)906.
10. Felix, A.M., Wang, C.-T., Heimer, E., Fournier, A., Bolin, D.R., Ahmad, M., Lambros, T., Mowles, T. and Miller, L., In Marshall, G.R. (Ed.) Peptides: Chemistry and Biology (Proceedings of the 10th American Peptide Symposium), ESCOM, Leiden, 1988, pp. 465–467.
11. Felix, A.M., Heimer, E.P., Wang, C.-T., Lambros, T.J., Fournier, A., Maines, S., Campbell, R.M., Wegrzynski, B.B., Toome, V., Fry, D. and Madison, V.S., Int. J. Pept. Prot. Res., 32(1988)441.
12. Campbell, R.M., Lee, Y., Mowles, T., McIntyre, K.W., Ahmad, M., Felix, A. and Heimer, E., Peptides, 13(1992)787.
13. Felix, A.M., Wang, C.-T., Campbell, R.M., Toome, V., Fry, D.C. and Madison, V.S., In Smith, J.A. and Rivier, J.E. (Eds.) Peptides: Chemistry and Biology (Proceedings of the 12th American Peptide Symposium), ESCOM, Leiden, 1992, pp. 77–79.
14. Zhao, Z. and Felix, A.M., Peptide Research, 7(1994)218.
15. Felix, A.M., Zhao, Z.C., Lee, Y. and Campbell, R.M., In Du, Y.-C., Tam, J.P. and Zhang, Y.-S. (Eds.) Peptides: Biology and Chemistry (Proceedings of the 1992 Chinese Peptide Symposium), ESCOM, Leiden, 1993, pp. 255–258.
16. Campbell, R.M., Bongers, J., Felix, A.M., Peptide Science, in press.
17. Nucci, M.L., Shorr, R. and Abuchowski, A., Advanced Drug Delivery Reviews, 6(1991)133.
18. Lu, Y.-A. and Felix, A.M., Int. J. Pept. Prot. Res., 43(1994)127.
19. Felix, A.M., Lu, Y.-A., Lee, Y. and Campbell, R.M., In Lu, G.-S., Tam, J.P. and Du, Y.-C. (Eds.) Peptides: Biology and Chemistry (Proceedings of the 1994 Chinese Peptide Symposium), ESCOM, Leiden, 1995, pp. 79–83.

Synthesis of pegylated growth hormone-releasing factor: Effect of site of pegylation on biological activity

Arthur M. Felix, Yi-An Lu, Yung Lee and Robert M. Campbell
Roche Research Center, Hoffmann-La Roche Inc., Nutley, NJ 07110, U.S.A.

Introduction

Monomethoxypoly (ethylene glycol) [PEG] modified proteins have been reported to have enhanced stability, increased resistance to proteolytic inactivation, improved plasma half-life, increased solubility and decreased antigenicity and immunogenicity [1]. A number of proteins have been pegylated and are currently being evaluated as potential therapeutics since their *in vivo* potencies are considerably greater than the parent protein. Unfortunately, conditions commonly used for protein pegylation [2,3] result in heterogeneous mixtures of pegylated proteins since neither the site nor extent of pegylation can be readily controlled. Our interest in developing peptides with longer half-life, which may eventually be used as therapeutics, led us to develop conditions for the site-specific pegylation of peptides. PEG side-chain protected amino acids have been successfully used for liquid-phase synthesis of hydrophobic peptides[4]. These earlier studies encouraged us to develop general methods for the solid phase synthesis of pegylated peptides in which pegylation is specifically achieved at the amino-, carboxy- or side-chain positions using the Fmoc/tBu strategy [5,6]. The pegylated peptides were purified by preparative HPLC and structures confirmed by ^{1}H-NMR spectroscopy, amino acid analysis, and laser desorption mass spectrometry. We have now extended this work and studied the effect of pegylation of a potent analog of Growth Hormone-Releasing Factor, [Ala^{15}]-GRF(1-29)-NH_2 [7], at various positions in the peptide. Studies were also carried out using PEG with varying molecular weights.

Results and Discussion

NH_2-Terminal pegylation of [Ala^{15}]-GRF(1-29)-NH_2 was achieved by solid phase synthesis by assembling the peptide on the Rink-MBHA-resin using the Fmoc/tBu strategy. The N-terminal Fmoc-protection group was removed and the peptide-resin was coupled with PEG_n-O-CH_2-CO-Nle-OH (PEG_{2000} and PEG_{5000}) using BOP reagent as previously described [6]. Side-chain pegylation of [Ala^{15}]-GRF(1-29)-NH_2 was achieved by one of two pathways (Fig. 1). Direct side-chain pegylation of Asp^8 or Asp^{25} was carried out by coupling with Fmoc-Asp(Nle-NH-CH_2CH_2-O-PEG_n)-OH during the Fmoc/tBu solid phase assemblage. An alternative post-pegylation procedure was developed in which Asp^8 or Asp^{25} was protected as the allyl-ester and selectivily removed by palladium-catalyzed deprotection after assemblage of the peptide-resin, followed by pegylation with H-Nle-NH-CH_2CH_2-O-PEG_n. Pegylation of the side-chain of Lys^{12} or Lys^{21} was achieved analogously by either direct side-chain pegylation by solid phase coupling with Fmoc-Lys(PEG_n-CH_2-CO-Nle)-OH or *via* incorporation of

Lys(Alloc), selective palladium-catalyzed deprotection and post-pegylation with PEG_n-O-CH_2-CO-Nle-OH. COOH-Termina; pegylation of [Ala^{15}]-GRF(1-29)-NH_2 was accomplished by Fmoc/tBu solid phase assemblage. Cleavage and deprotection of the pegylated peptides was accomplished with 95%TFA (Reagent K) and purification was achieved by the combined processes of dialysis and preparative HPLC. Alternatively, COOH-terminal pegylation was achieved in solution by the reaction of C-terminal cysteinyl peptide with dithiopyridyl-activated PEG (Fig. 2). This latter product was prepared by the reaction of 3-(2-pyridyldithio)propionic acid N-hydroxysuccinimide ester (SPDP) with H-Nle-NH-CH_2-CH_2-O-PEG_n. This pegylated GRF analogs were shown to be homogeneous by analytical HPLC. Structures were confirmed by amino acid analysis, ^{1}H-NMR spectroscopy and laser desorption mass spectrometry.

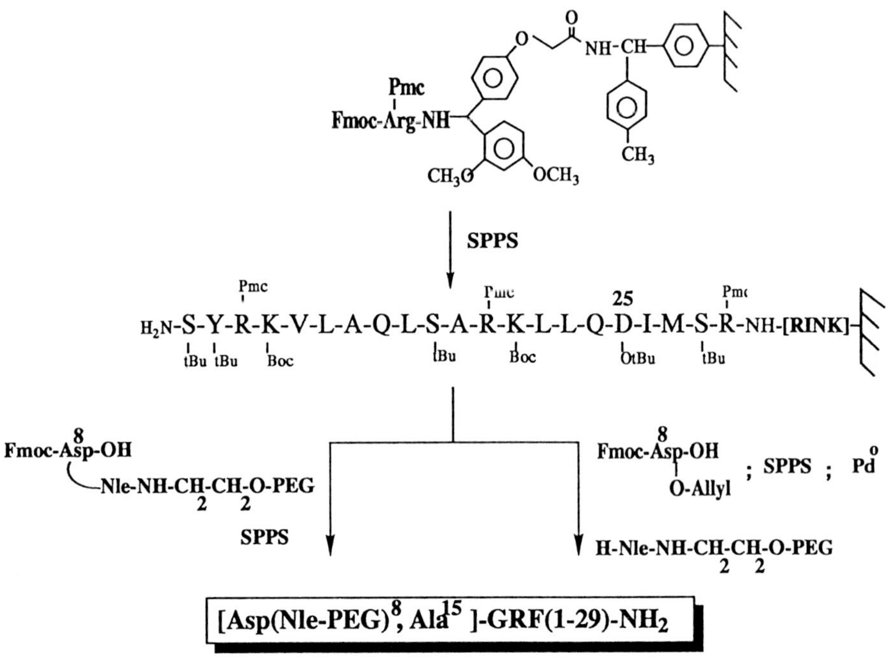

Fig. 1. Internal Asp^8-side-chain pegylation of [Ala^{15}]-GRF(1-29)-NH_2 by (left) direct pegylation via Fmoc-Asp-(Nle-CH_2CH_2-O-PEG_n)-OH during solid phase synthesis and (right) post-pegylation via Fmoc-Asp(OAllyl)-OH. Similar procedures are used for internal Asp^{25}-side-chain pegylation.

In vitro biological studies (rat pituitary bioassay) were carried out on all the pegylated GRF derivatives and correlations were made on the relationship between site and degree of pegylation on biological activity and are summarized in Table 1. Pegylation at position 1, 12, 21 and at the C-terminus did not alter the *in vitro* potencies relative to the parent GRF analog. However, pegylation at positions 8 andd 12 resulted in a decrease in potency which became more pronounced with the degree of pegylation (i.e. PEG_{5000} was less active than PEG_{2000}). *In vitro* studies in mice and pigs were carried out with COOH-terminal pegylated [desNH$_2$Tyr^1, D-Ala^2, Ala^{15}]-GRF(1-29)-Gly-Gly-Cys-NH$_2$ (Table 2). In both species the pegylated analog was more active since it stimulated the release of more growth hormone than the non-pegylated peptide. COOH-Terminal pegylated [desNH$_2$Tyr^1, D-Ala^2, Ala^{15}]-GRF(1-29)-NH$_2$ has the highest in vitro potencies that we have observed (30-50 times that of GRF(1-44)-NH$_2$). Circular dichroism studddies of the pegylated GRF analogs revealed that the PEG-function had no effect on conformation which is substantially (~85%) α-helical in 75% methanol (Fig. 3). CD spectra of the pegylated and non-pegylated peptides were also similar in aqueous solvent.

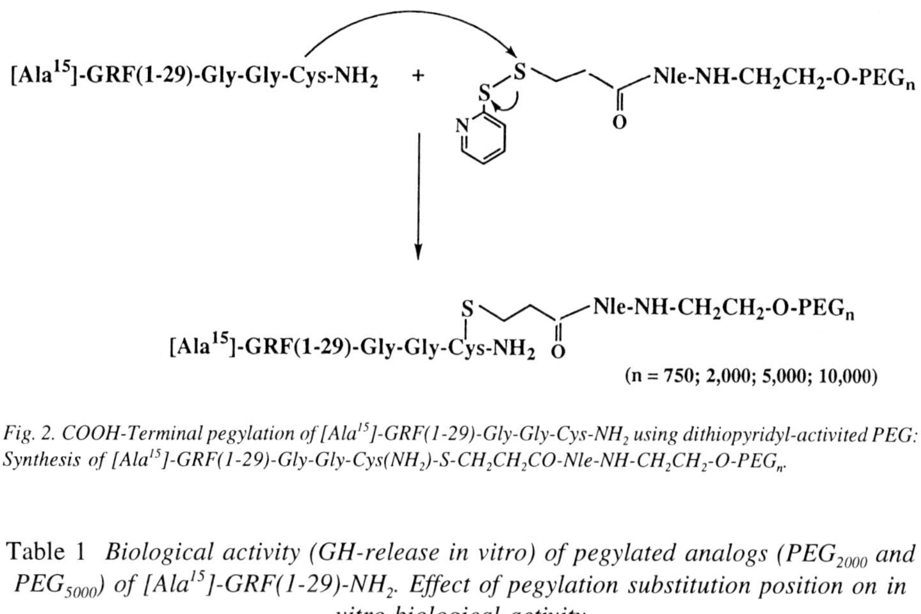

Fig. 2. COOH-Terminal pegylation of [Ala^{15}]-GRF(1-29)-Gly-Gly-Cys-NH$_2$ using dithiopyridyl-activited PEG: Synthesis of [Ala^{15}]-GRF(1-29)-Gly-Gly-Cys(NH$_2$)-S-CH$_2$CH$_2$CO-Nle-NH-CH$_2$CH$_2$-O-PEG$_n$.

Table 1 *Biological activity (GH-release in vitro) of pegylated analogs (PEG$_{2000}$ and PEG$_{5000}$) of [Ala^{15}]-GRF(1-29)-NH$_2$. Effect of pegylation substitution position on in vitro biological activity*

GRF analog	Relative potency[a] substitution position					
	1	8	12	21	25	30
PARENT[b]	1.02	4.68	0.77	2.86	3.20	4.36
PEG2000	1.04	1.59	0.34	3.21	3.09	4.63
PEG5000	1.28	0.17	0.03	3.24	3.27	4.83

[a] In vitro rat pituitary bioassay (concentration: 3.1-200 pM); potencies relative to GRF(1-44)-NH$_2$.
[b] NH$_2$- or -COOH positions of parent peptides are protected by N$^\alpha$-Ac and -CONHEt, respectively.

Table 2 *Effect of COOH-terminal pegylation on biological activity (GH-release in vivo in mice and pigs)*

GRF analog	Relative potency[a]	
	mice[b]	pigs[c]
GRF(1-44)-NH_2	1.00	1.00
[desNH$_2$Tyr^1, D-Ala^2, Ala^{15}]-GRF(1-29)-Gly-Gly-Cys-NH_2	9.81	10.5
[desNH$_2$Tyr^1, D-Ala^2, Ala^{15}]-GRF(1-29)-Gly-Gly-Cys(NH_2)-S-Nle-PEG$_{5000}$	33.6	54.9

[a] Potencies calculated on a molar basis and are relative to GRF(1-44)-NH_2.
[b] *In vivo* in mice (concentration: 0-30 μg/kg, iv); Growth hormone determined by AUC for 60 min.
[c] *In vivo* in pigs in a Latin Squares Design (doses: 0.3, 1.0, 3.0, 6.0 and 10μg/kg,sc); Growth hormone determined by area under the curve (AUC) for 4 hours.

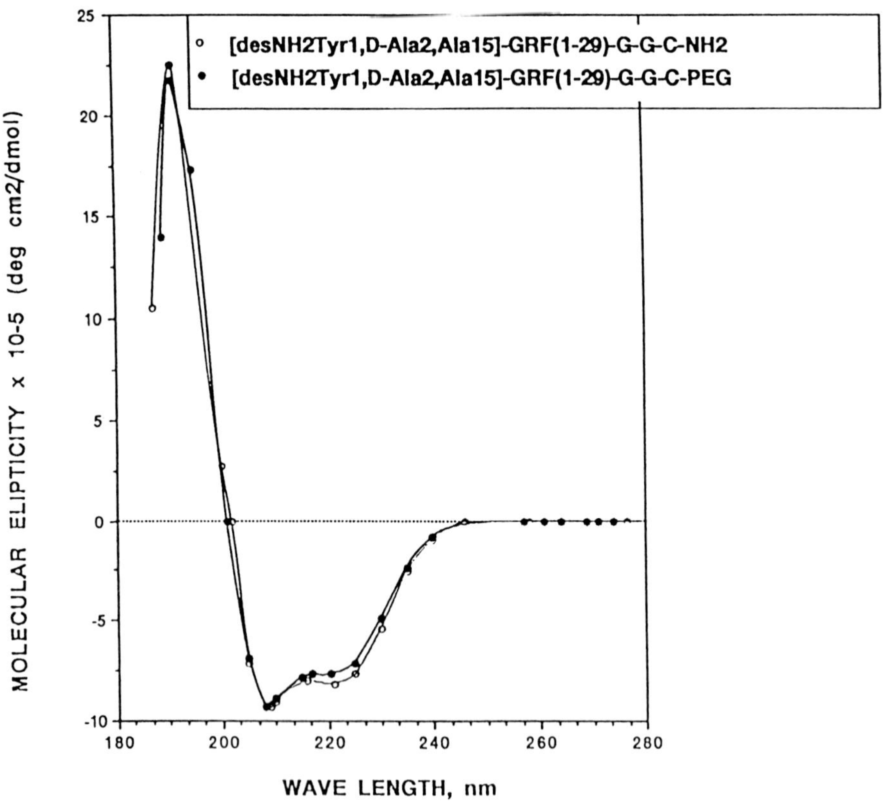

Fig. 3. Circular dichroism of pegylaed vs nnon-pegylated analogs of [desNH$_2$Tyr^1, D-Ala^2, Ala^{15}]-GRF(1-29)-Gly-Gly-Cys-NH$_2$ in 75% MeOH-H$_2$O (conc: 2×10^{-5} M).

Conclusions

Conditions have been developed for the site-specific pegylation of a biologically active peptide, Growth Hormone-Releasing Factor. These pegylated peptides were fully characterized and were evaluated for their in vitro and biological activity. It was determined that pegylation at the N-terminus, C-terminus and at positions 21 and 25 did not reduce potency. Pegylation at positions 8 and 12 caused a decrease in potency which decreased further with higher molecular weight poly(ethyleneglycol). The COOH-terminal pegylated analog was found to have extremely high biological potency in vitro. The increase potency of the pegylated GRF analogs is not due to any conformational effect and may be attributed to longer half-life in the peripheral circulation i.e. decreased metabolic clearance.

References

1. Nucci, M.L., Shorr, R.and Abuchowski, A., Adv. Drug. Del. Res., 6(1991)133.
2. Abuchowski, A., NcCoy, J.R., Palczuk, N.C., Van Es, T. and Davis, F.F., J.Biol. Chem., 252(1977)3582.
3. Beauchamp, C.O., Gonia, S.L., Menapace, D.P. and Pizzo, S.V., Anal.Biochem., 131(1983)25.
4. Mutter, M., Oppliger, H. and Zier, A., Markromol. Chem. Rapid Commmun., 13(1992)151.
5. Lu, Y.-A. and Felix, A.M.,Peptide Res., 6(1993)140.
6. Lu, Y.-A. and Felix, A.M., Int. J. Pept.Prot. Res ., 43(1994)127.
7. Felix, A.M., Heimer, E.P., Mowles, T.F., Eisenbeis, H., leung, P., Lambros, T., Ahmad, M., Wang, C.-T. and Brazeau, P., in Proc. 19th European Peptide Symposium, Ed. by Theodoropouos, D., W. deGruyter and Co., Berlin, 1987, pp. 481.

Design of a potent thrombin receptor radioligand

Dong-Mei Feng[a], D.F. Veber[a], T.M. Connolly[b] and R.F. Nutt[a]
[a]Department of Medicinal Chemistry and [b]Department of Biological Chemistry, Merck Research Laboratories, West Point, PA 19486, U.S.A.

Introduction

Thrombin-induced platelet activation plays a pivotal role in arterial thrombosis. According to a novel mechanism of receptor activation, thrombin cleaves its receptor to create a new amino terminus, which can then act as a tethered ligand to activate the receptor [1]. A 14-amino acid prptide, Ser-Phe-Leu-Leu-Arg-Asn-Pro-Asn-Asp-Lys-Tyr-Glu-Pro-Phe-OH, derived from the new amino terminus of the receptor, is also able to fully activate human platelets and cause aggregation in the presence of thrombin with an EC_{50} of 10,000 nM. Carboxy-truncated analogs of this 14-residue peptide have also been found to have full receptor activating potencies [2-9]. In order to develop a high affinity ligand, position modifications were carried out enhancing agonist potency. SAR studies resulted in the discovery of potent thrombin receptor-activating peptides. Potent peptides suitable for radiolabelling were also found.

Results and Discussion

Peptide analogs of Ser-Phe-Leu-Leu-Arg-Asn-Pro-Asn-Asp-Lys-Tyr-Glu-Pro-Phe-OH were synthesized incorporating modifications which would help elucidate structural features critical for thrombin receptor activation. Structure modifications involved: a) peptides of shorter chain-length and b) side-chain and backbone modifications. Chemical syntheses were carried out by the solid phase method using an ABI 430 (t-Boc based chemistry) instrument, HF deprotection, and purification by HPLC. Products were characterized by HPLC (>99% purity), amino acid analysis and FABMS. Biological potencies of analogs were measured as peptide effects on the extent of human blood platelet aggregation [9].

As shown in Table 1, N-terminal pentapeptide amide analog **2** shows a 10-fold increase in potency (EC_{50}=1,000 nM) when compared with the 14-residue prototype peptide, analog **1** (Table 4), and was used as a reference structure for further structure-activity relationship (SAR) studies. Methylation of Ala^1(analog **9**), Leu^3(analog **11**), or Leu^4(analog **12**) lowers potency 600-, 350- or 260-fold, respectively. The (N-Me)Phe^2(analog **10**), and (N-Me)Arg^5(analog **13**) are both inactive at 800 μM. The results of a D-amino acid scan are also shown in Table 1. Substitution with the corresponding D-amino acid at positions 1, 2, or 3 (analogs **3**, **4**, or **5**) results in inactive analogs at 800 μM.

A series of side-chain modifications of position-1 were carried out and the bioassay results are shown in Table 2. Removal of the hydroxyl group of Ser^1, as in the Ala analog **8**, does not alter potency. On the other hand, replacement of Ser^1 with Asp (analog **22**) or Arg (analog **21**) results in no agonist activity at >800 μM. Substitution of Ser[1] with the neutral, aromatic residue Tyr (analog **15**) or His (analog **16**) reduces activity 300-400 fold. However, a change to Gly[1] (analog **17**) maintains the same activity at 1,000 nM(EC_{50}). Replacement of Ala[1] with γ-NH_2-Butyric acid (analog **19**) has a 900-fold potency lowering effect. Removal of Ser[1] or acetylation of Phe (analog **23**, **20**) results in reduced agonist activity of >800 μM, or 205 μM, respectively. These results show that a neutral, small or no side-chain in position-1 is essential for agonist activity.

Table 1 *D-AA and N-Methyl Scans*

No.	Structure						Platelet Aggregation[b] EC_{50} (μM)
2	Ser	Phe	Leu	Leu	Arg	$-NH_2$	1
3	DSer	DPhe[a]					800
4							>800
5							>800
6			DLeu[a]				500
7					DArg		110
8	Ala	Phe	Leu	Leu	Arg	$-NH_2$	0.8
9	N-Me-Ala						600
10		N-Me-Phe					>800
11			N-Me-Leu				350
12				N-Me-Leu			260
13					N-Me-Arg		>800

[a] Ala[1] in analogs **4**, **5**.
[b] EC_{50} for stimulation of aggregation of human platelets.

A number of Phe[2] replacements were carried out (see Table 2). The replacement of Phe[2] with Ala causes a total loss in activity [2-3]. Analogs featuring Phe(p-Cl) (analog **32**) retain agonist activity at 700 nM/EC_{50}. Replacement of Phe[2] with Phe(3,4-Cl_2) (analog **33**) or Phe(p-F) analog **34**) has a potency increasing effect of two or four fold, whereas incorporation of Homo-Phe (analog **25**) or PhGly (analog **24**) shows no activity at 800 μM. Analogs featuring Tyr(Me) (analog **31**) or Cha (analog **26**) have a 20 or 30-fold potency lowering effect. An analog featuring 3-Qal (analog **27**), a large aromatic residue, was inactive at 800 μM. The exchange of Phe[2] with APhe (analog **29**), or Amf (analog **30**) has a potency lowering effect of 300 fold or abolishes activity at 800 μM, respectively. Incorporation of F_5-Phe containing the strongly electron deficient aromatic pentaflourophenyl group (analog **28**) causes a decrease from 0.8 to 600 μM(EC_{50}). The exchange of Phe^2 with Phe(D-3,4-Cl_2)(analog **35**), or Phe(D-p-F) (analog **36**) totally abolishes activity at 800 μM. It seems that L-chirality and a specific hydrophobic side-chain are critical for effective binding and/or activation.

Table 2 *Effect of modifications in position 1 or 2 on Agonist potencies*

No.	Structure						Platelet Aggregation
	Xaa^1	Phe	Leu	Leu	Arg	$-NH_2$	
14	α-Me-Ala						415
15	Tyr						400
16	His						310
17	Gly						1
18	β-Ala						35
19	γ-Abu						730
20	Acetyl						200
21	Arg						>1200
22	Asp						>800
23	---						>800
	Ala	Yaa^2	Leu	Leu	Arg	$-NH_2$	
24		PheGly					>800
25		HPhe					>800
26		Cha					28
27		3-Qal					>800
28		F_5-Phe					600
29		APhe					235
30		Amf					>800
31		Tyr(Me)					21
32		Phe(p-Cl)					0.5
33		Phe(3,4-Cl_2)					0.7
34		Phe(p-F)					0.2
35		D-Phe(3,4-Cl_2)					>800
36		D-Phe(p-F)					>800

In position 3, 4 and 5, combinations of positional modifications Ala-Phe-Leu-Leu-Arg- NH_2 (analog **8**) have been carried out (Table 3, and 4). Replacement of Leu^3 in analog **46** with the Tyr(3,5-I_2) (analog **47**) shows a decrease in agonist activity from 300 to 1,500 nM(EC_{50}). An $arginine^3$ (analog **39**) retains activity and gives added solubility, and Homo-Arg in position 5 increases potency slightly (analog **41**). Leu^4 can be replaced with the more hydrophobic Cyclohexylalanine (Cha) (analog **37**) with a two-fold increase in potency. However, a change to Arg^4 (analog **40**) reduces potency.

Table 3 *Effect of modifications in Position 3, 4 and 5 on Agonist Potencies*

No.	Structure							Platelet Aggregation EC_{50} (nM)
37	Ala	Phe	Leu	Cha	Arg	$-NH_2$		400
38	Ala	Phe(p-F)	Leu	Cha	Arg			140
39			Arg					130
40				Arg				40,000
41					HArg			120
42			Arg		Arg	$-NHEt_2$		160
43			Arg		HArg	$-NH_2$		270
44			Arg		HArg	Tyr	$-NH_2$	10

The potency enhancing modifications of Phe(p-F)[2] and Cha[4] were combined and resulted in Ala-Phe(p-F)-Leu-Cha-Arg-NH_2 (analog **38**) with an EC_{50} of 140 nM. Replacement of the C-terminal amide with an ethyl-amide retains agonist actiities (analog **42**). The C-terminally extended hexapeptide was shown to be more potent than the pentapeptide. Thus, to further increase potency and make available analogs for potential radiolabelling, the potent pentapeptide (analog **43**) was elongated with a tyrosine amide. This modification resulted in Ala-Phe(p-F)-Arg-Cha-HArg-Tyr-NH_2 (analog **44**) with an EC_{50} of 10 nM, a 27-fold potency enhancement over the corresponding pentapeptide (analog **43**). Analog **44** is the most potent agonist reported

Table 4 *Activity of Radioligand Peptides*

No.	Structure	Platelet Aggregation EC_{50}(nM)
1	Ser-Phe-Leu-Leu-Arg-Asn-Pro-Asn-Asp-Lys-Tyr-Glu-Pro-Phe-OH	10,000
45	Ser-Phe-Leu-Leu-Arg-Asn-Pro-Asn-Asp-Lys-Tyr(3,5-I_2)-Glu-Pro-Phe-OH	12,500
46	Ala-Phe(3,4-Cl_2)-Leu-Cha-Arg-NH_2	300
47	Ala-Phe(3,4-Cl_2)-Tyr(3,5-I_2)Cha-Arg-NH_2	150
48	Ala-Phe(p-F)-Arg-Cha-HArg-Tyr-NH_2	10
49	Ala-Phe(p-F)-Arg-Cha-HArg-Tyr(3-I)-NH_2	30

to date. The replacement of Tyr with Tyr(3-I) resulted in Ala-Phe(p-F)-Arg-Cha-HArg-Tyr(3-I)-NH_2 (analog **48**), which shows an EC_{50} of 30 nM. This level of potency is suitable for potential use as a radioligand in receptor binding assays.

Structural features of hexapeptide agonists of the thrombin receptor have been elucidated. A small neutral side-chain is preferred in position-1. Small, electron withdrawing groups on aromatic rings in position-2 increase potency. Neutral or basic side-chains are preferred in position-3, aliphatic side-chains in position 4, and a cationic site is necessary in position-5. Basic side-chain in position-3, combined with Tyr in position-6, are beneficial for potency and solubility. All L-stereochemistry is necessary for agonist activities.

The agonist potency of shortened thrombin receptor peptides have been enhanced up to 1000 fold. These peptides provide powerful tools for the development of a receptor radioligand and the potential establishment of a binding assay. It may provide the basis for a new class of antithrombotic drugs.

Abbreviations

β-Ala=β-Amino-Ala	γ-Abu=4-Amino-Butyric acid	PheGly=Phenyl-Gly
3-Qal=3-(3-Quinolyl)Ala	APhe=p-Amino-Phe	Tyr(Me)=0-Methyl-Tyr
Amf=p-Aminomethyl-Phe	Cha=Cyclohexyl-Ala	Tyr(3,5-I_2)=3,5-Diiodo-Tyr
Tyr(3-I)=3-Iodo-Tyr	HPhe=Homo-Phe	HArg=Homo-Arg

References

1. Vu, T.-K. H., Hung, D.T., Wheaton, V.I. and Coughlin, S.R., Cell, 64(1991)1057.
2. Scarborough, R.M., Naughton, M.A., Teng, W., Hung, D.T., Rose, J., Vu, T.-K.H., Wheaton, V.I., Turck, C.W. and Coughlin, S.R., J. Biol. Chem., 267(1992)13146.
3. Feng, D.-M., Veber, D.F., Connolly, T. and Nutt, R.F., In Du, Y.-C., Tam, J.P. and Zhang, Y.-S. (Eds.) Peptides: Biology and Chemistry (Proceedings of the 1992 Chinese Peptide Symposium), ESCOM, Leiden, 1993, pp. 141–142.
4. Vassallo, R.R., Kieber-Emmons, T., Cichowski, K. and Brass, L.F., J. Biol. Chem., 267(1992)6081.
5. Hui, K.Y., Jakubowski, J.A., Wyss, V.L. and Angleton, E.L., Biochem. Biophys. Res. Comm., 184(1992)790.
6. Chao, B.H., Kalkunte, S., Maraganore, J.M. and Stone, S.R. Biochem., 31(1992)6175.
7. Feng, D.-M., Veber, D.F., Connolly, T.M. and Nutt, R.F., In Hodges, R.S. and Smith, J.A. (Eds.) Peptides: Chemistry, Structure and Biology (Proceedings of the 13th American Peptide Symposium), ESCOM, Leiden, 1994, pp. 387–389.
8. Reilly, C.F., Connolly, T.M., Feng, D.-M., Nutt, R.F. and Mayer, E., J. Biochem. Biophys. Res. Comm., 190(1993)1001.
9. Connolly, T.M., Condra, C., Feng, D.-M., Nutt, R.F. and Gould, R.J., Thromb. Haemostas., 69(1993)787.

Membrane partitioning and helix formation studied with designed hydrophobic peptides

Shun-Cheng Li and Charles M. Deber
Division of Biochemistry Research, Research Institute, The Hospital for Sick Children, Toronto M5G 1X8; and Department of Biochemistry, University of Toronto, Toronto, M5S 1A8, Ontario, Canada

Introduction

Transmembrane (TM) domains of integral membrane proteins are structured either in the α-helical conformation or occasionally in β-sheet barrels. For single TM segments, however, only the α-helix is energetically permissible. Although available structural data for membrane proteins determined by X-ray [1] and electron crystallography [2] suggest that TM helices are virtually indistinguishable in structure from α-helices in globular proteins, the lipid bilayer imposes a different environment for membrane protein folding from that in aqueous media. For instance, hydrophobic interactions between the peptide and the lipid bilayer are considered important for the initiation and stabilization of TM helices [3,4; and references therein]. We report here the use of a series of model peptides with membrane-interactive potential for a conformational study in the membrane-mimetic environment of lipid micelles. Results suggest that hydrophobic partitioning of the peptide into the matrix of the lipid is both necessary and sufficient for helix formation.

Results and Discussion

The rationale for the design of these peptides and the methods for their synthesis and purification are reported elsewhere [3]. Peptides employed in the present study have a prototypical sequence of H_2N-(Ser-Lys)$_2$-<u>Ala-**X**-Ala-Ala-**X**-Ala-Trp-Ala-**X**-Ala</u>-(Lys-Ser)$_3$-OH, where X = Ile, Leu, Val, Ala, Gly, Ser, Thr, or Phe. These peptides are designated A**X**A to reflect the triad sequences in their mid-hydrophobic segments (underlined above) which are expected to partition into the hydrophobic matrix of the lipid.

Circular dichroism (CD) spectra of the peptides were examined in micellar solutions of lyso-phosphatidylcholine (LPC). Lipid micelles are physically more fluid than lipid bilayers, and therefore are more amenable to interactions with the peptides. As shown in Fig. 1, peptides with relatively greater hydrophobicity, *e.g.*, AIA, ALA, AVA, and AFA, adopt largely α-helical conformations in 10 mM LPC micelles, while peptides that are less hydrophobic, *e.g.*, ASA, ATA, and AGA, exhibit little secondary structure under the same conditions. Interestingly, peptide AAA, which should, in principle, be a good helix-former [3,4], is found to be only partially α-helical in the environment of LPC micelles.

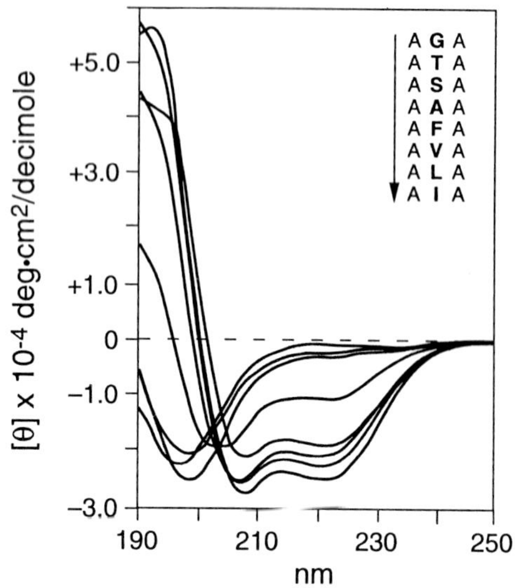

Fig. 1. Circular dichroism spectra of AXA peptides in 10 mM LPC micellar solutions, pH 7.0. CD spectra were recorded on a Jasco J-720 spectrometer from freshly prepared peptide samples (30 mM). Spectra shown were averaged over 4 consecutive scans with solvent background absorbance subtracted. Curves are as labelled in the diagram from top to bottom.

A nearly linear correlation is obtained when the observed peptide ellipticities at 222 nm are plotted against the corresponding peptide segmental hydrophobicities (Fig. 2). It can be rationalized that peptide helicity in LPC micelles is determined almost exclusively by peptide hydrophobicity, rather than the precise chemical natures of its constituent amino acid residues [4]. This notion is strongly supported by the observation that peptide AIA, the most hydrophobic in the group, displays the greatest α-helicity despite the fact that it may be intrinsically less favorable than peptides AAA and ALA in forming the α-helix. As all of these peptides are largely unstructured in aqueous buffer [4], formation of significant secondary structure in some of the peptides in LPC necessarily reflects their stable association with the lipid micelles. As well, since LPC micelles have a net neutral surface, electrostatic interactions are unlikely to be major contributors to this peptide-micelle association. Thus, partitioning of the peptide mid-sequence into the hydrophobic matrix of LPC micelles is most likely driving peptide-lipid interaction [4]. Hydrophobic partitioning is also verified by the fact that relatively large blue shifts of Trp fluorescence in LPC can be detected in those peptides with greater hydrophobicity [data not shown].

Stable interactions between model peptides and lipid micelles/vesicles have been demonstrated to be sufficient to promote helix-formation in the corresponding peptides [3,4]. Data presented here thus suggest that there exists a threshold hydrophobicity for stable partitioning of a peptide segment into the lipid matrix. It is possible that this threshold value corresponds to the approximate hydropathy of a stretch of alanines.

Segments that are less hydrophobic - such as those in peptides ASA, ATA, and AGA - will tend to partition into the aqueous phase instead, as manifested by their inability to form secondary structure in LPC micelles (Fig. 1).

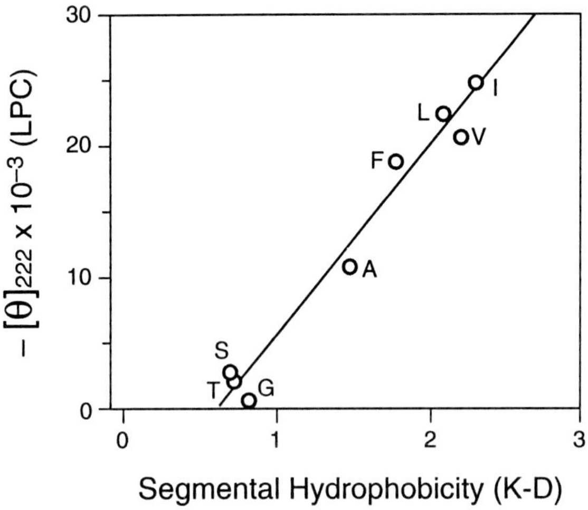

Fig. 2. Correlation of peptide helicity with segmental hydropathy. Helicity of the peptides is represented by their corresponding ellipticities at 222 nm. The average hydrophobicity of segment -AXAAXAWAXA- in each peptide was calculated from the Kyte-Doolittle scale [5]. Points corresponding to peptides AXA are identified as single letters (X) for clarity. The regression line was drawn by a computer program.

Acknowledgements

This work was supported, in part, by grants to C.M.D. from the Natural Sciences and Engineering Research Council of Canada (NSERC) and the Medical Research Council of Canada (MRC). S.-C.L. holds a University of Toronto Open Studentship.

References

1. Deisenhofer, J., Epp, O., Miki, K., Huber, R. and Michel, H., Nature, 315(1985)618.
2. Kuhlbrandt, W., Wand, D. D. and Fujiyoshi, Y., Nature, 367(1994)614.
3. Li, S.-C. and Deber, C. M., J. Biol. Chem., 268(1993)22975.
4. Li, S.-C. and Deber, C. M., Nature Struct. Biol., 1(1994)368.
5. Kyte, J. and Doolittle, R. F., J. Mol. Biol., 157(1982)105.

Structural investigation of cysteine-rich chelating peptides by 1D- and 2D-NMR spectroscopy

Bahram Hemmasi, Weiguang Zeng, Klaus Albert and Ernst Bayer
Institut für Organische Chemie, Universität Tübingen,
72076 Tübingen, Germany

Introduction

Many organisms have developed mechanisms for resisting the cytotoxic effects of transition metal ions. Two classes of cysteine-rich peptides function in the detoxification of these ions. One class is glutathione-related isopeptides phytochelatins ($[\gamma EC]_nG$ or PCs) and homophytochelatins (h-PCs) with the general structure H-[γ-Glu-Cys]$_n$-AA-OH (n = 2-11; AA = Gly, PCs; AA = β-Ala, h-PCs), isolated from several fungi and plants [1]. Exposure of these organisms to certain metal ions leads to the biosynthesis of $[\gamma EC]_nG$. Phytochelatins bind these ions by building metal-cysteinyl clusters. The second class is the metallothionein family of proteins [2].

NMR techniques are widely used for acid-base study of peptides. Sensitivity of the chemical shift and spin-spin coupling constants of non-labile carbon-bound protons to the ioniziation of nearby functional groups makes NMR spectroscopy a direct method for studying acid-base chemistry at the molecular level [3].

Except for our present studies, so far only 1D- and 2D-NMR acid-base studies on the smallest PC member, i.e. GSH, have been reported [4]. Because there have been no systematic NMR studies on PCs besides the assignment of ^{13}C-NMR signals of some PCs (n=2,3, and 4) which were recently synthesized [5-7] and reported by us [5], we carried out more detailed NMR studies on a PC (n = 2) reported here.

Results and Discussion

Assignment of ^1H resonances

The ^{1}H signals of $[\gamma EC]_2G$ were assigned from the ^{1}H-NMR spectrum taken in H_2O:D_2O (9:1) at pH 1.9. The assignment of the amino acids was accomplished using 2D-TOCSY. The five AAs showed the expected five spin systems in the TOCSY spectrum. The resonances of the two Glu residues showed different chemical shifts and coupling patterns. $Glu^1_{\alpha H}$ appeared at 3.91 ppm; $Glu^3_{\alpha H}$ at 4.44 ppm, the β-protons of Glu^3 are non-equivalent and appear as two isolated peaks. By contrast, the two spin systems of Cys residues show very similar chemical shifts, and resonances of the methylene groups overlapped.

The sequence assignment was completed by using 2D-ROESY [8] shown in Fig. 1. In isopeptides such as PCs the sequential resonance assignment [9] has been established via NH(i)-CHγ(i-1)-NOE contacts (Fig. 1).

The 2D-NOESY [10] was unsuccessful due to the weakness of the signal intensities, which was probably due to the unfavorable molecular correlation times of such a small peptide.

Assignment of ^{13}C resonances

The ^{13}C-NMR spectrum of [γEC]$_2$G was taken in H_2O:D_2O (9:1) at pH 1.9. The assignment of the proton-bound carbons was accomplished by using heteronuclear multiple-quantum correlation (^{1}H,^{13}C-HMQC) experiment in D_2O [11,12], and HMQC-TOCSY transfer experiment in H_2O:D_2O (9:1) [13].

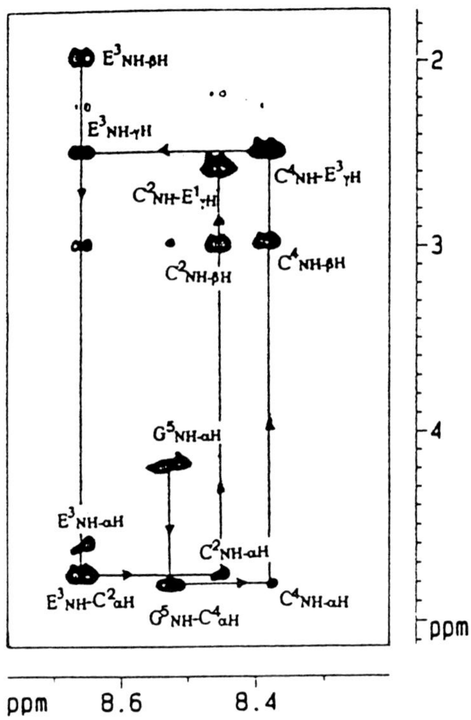

Fig. 1. Part of the 2D-ROESY spectrum of [γEC]$_2$G in H_2O:D_2O (9:1) at pH 1.9.

The seven carbonyl signals (Table 1) were identified by using inverse-detected ^{1}H-^{13}C long-range correlation spectroscopy (heteronuclear multiple-bond correlation, HMBC) experiments [14] shown in Fig. 2.

The pH-dependencies of resonances

The pH-dependency of the chemical shifts is due to the conformational changes and changes of the electronic environments of the carbon-bound protons accompanying titration of carboxyl, ammonium and SH groups.

Table 1 *C-H correlation for the C=O groups of* $[\gamma EC]_2G$ *at pD 1.9*

$^{13}C{=}O$	^{1}H
Glu^{1}_{α} sees	$Glu^{1}_{\alpha H}$ and $Glu^{1}_{\beta H}$
Glu^{1}_{γ} sees	Cys^{2}_{NH}, $Cys^{2}_{\alpha H}$, $Glu^{1}_{\gamma H}$ and $Glu^{1}_{\beta H}$
Cys^{2} sees	Glu^{3}_{NH}, $Cys^{2}_{\alpha H}$ and $Cys^{2}_{\beta H}$
Glu^{3}_{α} sees	$Glu^{3}_{\alpha H}$ and $Glu^{3}_{\beta H}$
Glu^{3}_{γ} sees	Cys^{4}_{NH}, $Cys^{4}_{\alpha H}$, $Glu^{3}_{\gamma H}$ and $Glu^{3}_{\beta H}$
Cys^{4} sees	Gly^{5}_{NH}, $Cys^{4}_{\alpha H}$ and $Cys^{4}_{\beta H}$
Gly^{5} sees	$Gly_{\alpha H}$

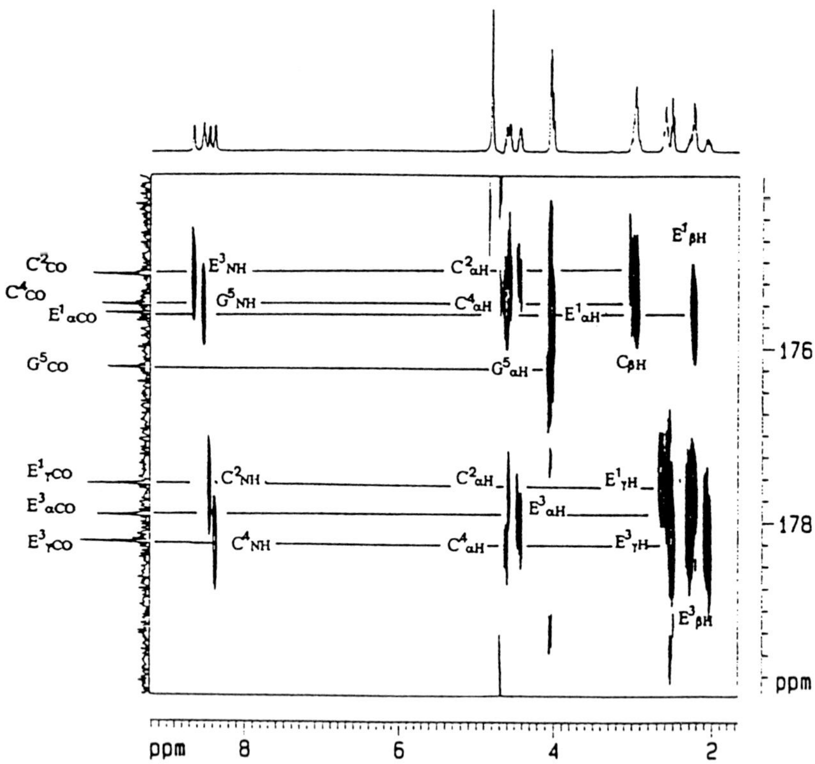

Fig. 2. Part of the inverse-detected ^{1}H-^{13}C-long-range COSY-HMBC (heteronuclear multiple-bond correlation) of $[\gamma EC]_2G$ in H_2O:D_2O at pH 1.9.

Over the pD range 1-6 the three carboxyl groups of $[\gamma EC]_2G$ are titrated and chemical shifts of the resonances for the adjacent protons, i.e., $Gly^{5}_{\alpha H}$, $Glu^{3}_{\alpha H}$ and $Glu^{1}_{\alpha H}$ are the most affected. The chemical shifts of these resonances change over this pD range, indicating simultaneous titration of the three carboxyl groups. The chemical shifts of the resonances of $Glu^3_{\beta H}$, $Glu^1_{\beta H}$, $Glu^3_{\gamma H}$ and $Glu^1_{\gamma H}$ protons also change as a result of the titration of glutamyl carboxyl groups. Titration of the carboxyl groups of $[\gamma EC]_2G$ has little effect on the average chemical shifts of the cysteinyl protons ($Cys^2_{\alpha H}$, $Cys^4_{\alpha H}$, $Cys^2_{\beta H}$ and $Cys^4_{\beta H}$).

Compared with the spectrum at pD 6.0, at pD 7.8 there are much bigger changes of the chemical shifts of $Cys^2_{\alpha H}$, $Cys^4_{\alpha H}$ and $Glu^1_{\alpha H}$ than those of $Gly^5_{\alpha H}$, $Glu^3_{\alpha H}$. Moreover, the non-equivalence of the cysteinyl methylene groups, as inferred from the peak width, increased and the $Cys_{\alpha H}$-$Cys_{\beta H}$ coupling constant changed, indicating a change of the conformation at the cysteinyl residues.

At pD 10.2 the $Glu^1_{\alpha H}$ resonance shifted from 3.7 ppm at pD 7.8 to 3.4 ppm. The high field shifts of $Cys^2_{\alpha H}$ and $Cys^4_{\alpha H}$ are in the same scale as that of $Glu^1_{\alpha H}$ and overlapped with $Glu^3_{\alpha H}$.

The pD-dependency of the chemical shifts of resonances $Cys^2_{\alpha H}$, $Cys^4_{\alpha H}$ and $Glu^1_{\alpha H}$ indicates simultaneous titration of the SH groups and ammonium group of Glu^1 over the pD range 8-10. There is also a large change in the non-equivalence of the methylene protons ($Cys^2_{\beta H}$ and $Cys^4_{\beta H}$),which now appear as two isolated multiplets.

The ^{13}C chemical shifts of the α carbons of $[\gamma EC]_2G$ also change as a function of pD indicating the influence of titration of functional groups of the peptide on the ^{13}C chemical shifts. Their behavior is very similar to that observed for the α-protons in the ^1H-NMR spectrum. For example, the change of the chemical shift of $Glu^1_{\alpha C}$ shows the influence of the progressive dissociation of carboxyl groups of the peptide between pD 1.8-4.5 and the ammonium group between 9.3-11.8.

Experimental

The NMR measurements were performed on a 400 MHz spectrometer (AMX 400, Bruker). For 1D and 2D ^1H-NMR measurements 5.4 mg of peptide was dissolved in D_2O or H_2O:D_2O (9:1, 0.5 mL). For ^{13} C-NMR measurements 16.2 mg of sample was used. ^1H and ^{13}C chemical shifts are reported relative to an internal dioxane reference. TOCSY, ROESY, heteronuclear multiple-quantum correlation (HMQC) and TOCSY-transfer spectra were recorded in the phase-sensitive mode using the time-proportional phase incrementation (TPPI) method [15].

Acknowledgement

We gratefully acknowledge Norbert Zimmermann for helpful discussion.

References

1. Grill, E., Löffler, S., Winnacker, E.L. and Zenk, M.H., Proc. Natl. Acad. Sci. USA, 86(1989)6838.
2. Kägi, J.H.R. and Schäffer, A., Biochemistry, 27(1988)8509.
3. Nuclear Magnetic Resonance, Part A, Spectral Techniques and Dynamics; Part B, Structure and Mechanism: in Methods in Enzymology, (Oppenheimer, N.J. and James, T.L., Eds.), Academic Press,

Inc., San Diego, New York, Berkeley, Boston, London, Sydney, Tokyo and Toronto, Vols. 176 and 177, 1989.
4. Huckerby, T.N. and Tudor, A.J., J. Chem. Soc., Perkin Trans. II, (1985)759.
5. Zeng, W. and Hemmasi, B., Liebigs Ann. Chem., (1992)315.
6. Chen, Z. and Hemmasi, B., Biol. Chem. Hoppe-Seyler, 374(1993)1057.
7. Hemmasi, B., Chen, Z. and Bayer, E., In Schneider, C.H. and Eberle, A.N. (Eds.) Peptides 1992 (Proceedings of the 22nd European Peptide Symposium), ESCOM, Leiden, 1993, pp. 205–206.
8. Bax, A. and Davis, D.G., J. Magn. Reson., 63(1985)207.
9. Jeener, J., Meier, B.H., Bachmann, P. and Ernst, R.R., J. Chem. Phys., 71(1979)4546.
10. Wüthrich, K. in: NMR of proteins and nucleic acids 1986, John Wiley & Sons, New York, Chichster, Brisbane, Toronto and Singapore.
11. Bax, A., Griffy, R.H. and Hawkins, B.L., J. Magn. Reson., 55(1983)301.
12. Bax, A. and Subramanian, S., J. Magn. Reson., 67(1986)565.
13. Lerner, L. and Bax, A., J. Magn. Reson., 69(1986)375.
14. Bax, A. and Summers, M.F., J. Am. Chem. Soc., 108(1986)2093.
15. Marion, D. and Wüthrich, K., Biochem. Biophys. Res. Commun.,113(1983)967.

Design and synthesis of alpha-helix peptide and DNA-binding domain

Xiao-Jie Xu, Yu Luo, Ruo-Heng Zhang, Zhen-Wei Miao and Yun-Hua Ye
College of Chemistry and Molecular Engineering, Peking University, Beijing 100871, China

Introduction

The problem of protein folding problem is very important, and also very difficult. De novo protein design provides an inverse approach to the protein folding problem. This approach seems more promising for understanding the stabilizing factors of protein structure and the relationship of protein structure and function.

Our interest is to develop a systematic approach that includes a computational assistance system aimed at de novo design, synthesis and structure study of designed proteins.

Results and Discussion

Approach for de novo design of protein

The approach for de novo protein design is divided into three stages: knowledge abstraction, modelling and stability evaluation. First, a molecular viewing program (MVP) has been made to assist novo protein design. MVP is a sophisticated computer graphics program. It bears the general features of visual computation and is easy to use and connect with other program packages.

Secondly, a dynamic programming algorithm of structure comparision with the advantages of automation and independence of sequence homology has been developed. We have developed a full spectrum of algorithms for model building.

Thirdly, we have developed an evaluation algorithm of structure stability. Statistical work has been done on globular proteins to search for common properties of correctly folded proteins. The program attempts to evaluate the rationality of model structure by considering average residue molecular weight, maximum polar fraction, average residue maximum accessible surface and so on.

Design and synthesis for alpha-helix peptides

In order to study the factors influencing formation and stability of structure for helical peptides, we designed and synthesized 10 peptide fragments with 8 to 23 amino acid residues. The sequences for 10 peptide fragments are list in Fig. 1.

CD spectroscopic analysis was made for synthesized peptide fragments. The results show that conformation stability of peptide fragments is influenced by peptide length, side-chain ion interactions, hydrophobic interaction, ion strength, and pH value.

Protein unfolding studied by FTIR spectroscopy

Hou-Li Jiang[a], Bo Jiang[b], Zhan-Jun Song[a], Zhen-Hua Deng[a], Song-Cheng Yang[a] and De-Xu Zhu[b]

[a]*National Center of Biomedical Analysis, Academy of Military Medical Sciences, Beijing 100850, China*

[b]*National Laboratory of Biomedical Technology, Department of Biochemistry, Nanjing University, Nanjing 210008, China*

Introduction

Bovine pancreatic trypsin inhibitor (BPTI) is a small globular protein which inhibits the activity of trypsin and other proteases by forming inert complexes with the enzymes [1]. BPTI has 58 amino acid residues with a molar weight of 6500 and it has long been the subject of a large number of experimental and theoretical studies on protein conformation and protein unfolding. A refined crystal structure at 1.5 Å has been reported [2] and systematic NMR studies on BPTI solution conformation have been carried out [3]. Urotrypsin inhibitor (UTI) is a newly isolated peptide consisting of two domains each of which is homologous to BPTI. UTI is very effective in treating thymitis and pneumonectasis. Its conformation is still unknown. Using Fourier-transform infrared (FTIR) spectroscopy, we have compared their solution structures and studied some conformational changes of BPTI and UTI with respect to temperature and disulfide reduction.

In the infrared spectra of proteins, the secondary structure is most clearly reflected by the amide I region, which is from 1620 cm^{-1} to 1700 cm^{-1} and is primarily due to the stretching vibrations of peptide carbonyl groups. Specific conformational types give rise to discrete bands in the amide I region. With the development of Fourier-transform technologies, infrared spectroscopy has become recognized as a valuable tool for probing details of the secondary structure of proteins in solution. Peak characteristics of various secondary structures can be identified by derivation and Fourier-self-deconvolution. Using deconvoluted spectra and curve-fitting techniques, Byler and Susi have developed FTIR methods for quantitative analysis of secondary structures of proteins in D_2O [4]. Here we focus mainly on the infrared analysis of protein unfolding caused by disulfide reduction.

Results and Discussion

Secondary Structure of BPTI and UTI

The deconvoluted amide I' regions of BPTI and UTI together with their fitted bands are shown in Fig. 1. Deconvolution was performed using a Lorentzian half-width of 18 cm^{-1} and a resolution enhancement factor (k) of 2.4 in Bio-Rad Data Station 3200. Curve-fitting was carried out using the Bio-Rad BANDFIT and CURVEFIT, both iterative Gauss-Newton nonlinear regression programs. Curve-fitting is extremely useful for determining the exact peak positions, widths, heights and areas of a set of overlapping peaks. However, curve-fitting is one of the most frequently misused and misunderstood mathematical processing techniques and the results should be carefully scrutinized. This is no fault of the method itself, but rather that of the personnel employing the technique. Three aspects are essential for performing curve-fit: 1. The exact number of peaks; 2. The true peak shapes (Lorentzian, Gaussian, etc.); 3. An approximate estimate of the peak parameters (widths, height, etc.). Since the deconvolution procedure removes the Lorentzian contribution to the bandwidth, the deconvoluted spectra were fitted with Gaussian line shapes in our experiment. Using two different manipulating programs, we have obtained consistent fitting results for each protein spectruxWe found that the amide I' bands of BPTI and UTI are very similar except that BPTI has an extra peak at 1669 cm^{-1}. This indicates that the two peptides bear some analogy to each other.

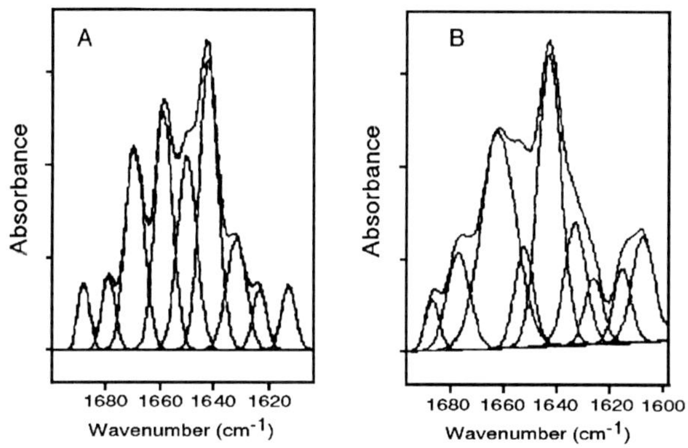

Fig. 1. Deconvoluted amide I' region and fitted bands of (A) BPTI; (B) UTI.

Correlation between specific backbone folding types and individual component bands in the infrared spectrum is the foundation of the application of FTIR in examining protein conformation. Theoretical studies of model peptide systems [5] and previous FTIR studies of many proteins [4, 6] have provided guidelines for the interpretation of protein infrared spectra. However, the inherent complexity, size and versatility in globular proteins often make direct interpretation of these infrared bands questionable. Good structure-spectra correlations require more infrared protein studies. Based on previous studies of proteins by FTIR spectroscopy, preliminary assignment of amide I' bands of BPTI and UTI are listed in Table 1.

Our results showed that BPTI is composed of 17.2% α-helix, 35.1% β-sheets, 23.2% turns and 24.4% unordered structure, while UTI contains 8.2% α-helix, 26.6% β-sheets, 35.9% turns and 29.3% unordered structure. The infrared result of BPTI is very similar to the X-ray crystal and NMR solution structure. Our study showed that UTI had fewer sheet and helical structures than BPTI, while their general topologies seemed were similar.

Table 1 *Deconvolved amide I' frequencies, fractional areas and preliminary assignments[a] of BPTI and UTI*

BPTI			UTI		
cm^{-1}	Area%	Assignment	cm^{-1}	Area%	Assignment
1688	3.7	Turns	1685	3.1	Turns
1679	4.5	Extended Chains	1675	9.7	Extended Chains
1669	16.9	Extended Chains			
1659	19.5	Turns	1660	32.8	Turns
1650	17.2	Helix	1650	8.2	Helix
1642	24.4	Unordered	1641	29.3	Unordered
1632	9.9	Extended Chains	1631	11.6	Extened Chains
1623[b]	3.8	Extended Chains	1624	5.3	Extened Chains
1612[b]		Side Chains	1613[b]		Side Chains
			1605[b]		Side Chains

[a] Assignments are based on results of Byler and Susi[4].
[b] Side chain absorbances.

Thermal Effects on Protein Conformation

Conformational changes caused by temperature variation can be readily seen by analysis of their corresponding amide I' bands. Fig. 2. shows some deconvoluted amide I' bands of UTI at different temperatures. From 20 to 60°C, there were only slight changes of the amide I' band which indicated little denaturation. Significant spectral changes occurred between 60 and 80°C, demonstrating major thermal denaturation. Several component absorbances diminished and the main peak position shifted to 1645 cm^{-1} during the thermal denaturation process.

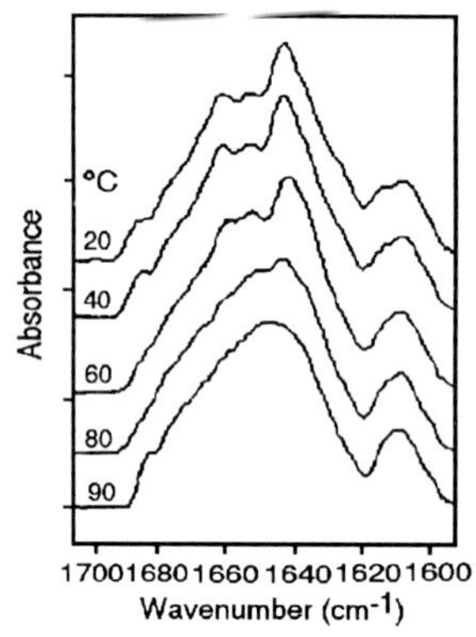

Fig. 2. Deconvoluted amide I' bands of UTI at 20, 40, 60, 80 and 90°C.

The bottom spectrum in Fig. 2. represents the final stage of the thermal denaturation process. It becomes almost featureless. This shows that no regular secondary structures can be detected in the thermal denatured state of UTI. The main band centered at 1645 cm^{-1}, which ischaracteristic of an unordered structure [4, 6].

Protein Unfolding Caused by Disulfide Reduction

It is well known that disulfides play an important role in the folding of protein conformation. In a liquid cell containing a 0.050 ml solution, 0.62 mg BPTI (1.24%)

was mixed with 0.15 mg DTT (20mM). The first FTIR spectrum was taken within 10 minutes of the reaction. The result is shown in Fig. 3. Although some component features of the deconvoluted amide I' band still exist in the first amide I' band, it is quite different from the spectrum of natural BPTI (Fig. 1). This shows that significant changes in the conformation of BPTI set in once the disulfides are disturbed. Under these conditions, the reduction process takes about 12 hours to reach the final stage.

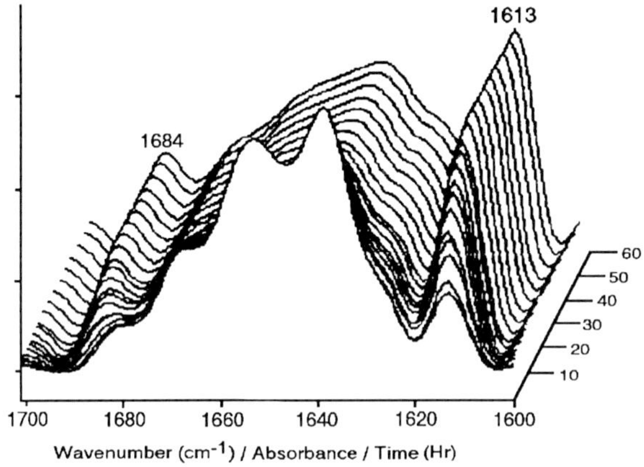

Fig. 3. Deconvoluted amide I' bands of BPTI during reduction of disulfides with DTT. The concentration of BPTI is 1.24%, DTT 20mM, pD=8.5 (without deuterium correction).

Characteristics of reduction of disulfides as learned from FTIR spectra are as follows. First, the main region of the amide I' band, from 1620 cm^{-1} to 1680 cm^{-1}, gradually becomes featureless, demonstrating the unfolding process. Second, there is a considerable time-dependent increase of two peaks, 1684 cm^{-1} and 1613 cm^{-1}. The peak at 1613 cm^{-1} was previously considered due to C=C stretching vibrations in aromatic side chain residues [7]. However, the infrared absorbance of the hydrophobic groups should be weak in natural proteins since they are usually folded in the inner part of the protein molecule. Peaks around 1684 cm^{-1} are assigned to turns in most published papers, although uncertainty and questions it have been expressed in some of the literatures [6, 8]. In this reduction experiment, since this band arises concurrently with the 1613 cm^{-1} band. This fact indicated that they most likely arise from the inter-molecular β-sheet [9].

Disulfide reduction of UTI and other peptides (results not shown) show the same result as that of BPTI. Repeated experiments with different concentrations of protein and DTT showed different reduction speeds. The results indicate that formation of disulfide bonds is essential for stabilization of folded protein conformation.

The final spectra of disulfide reduction and of thermal denaturation are different, as shown in Fig. 4. The final spectrum of thermal denaturation peaked at 1645 cm^{-1}, which represents unordered or irregular conformations having no regular structures. The final spectrum of disulfide-reduced proteins had a top absorbance in the main amide I' region at around 1639 cm^{-1}, which is often associated with extended chains [4, 6]. In addition, the thermal denaturation evolved slowly until the temperature reached a certain degree, while reduction of disulfides caused an abrupt loosening of the folded conformation followed by a gradual folding into inter-molecular β-sheet.

We have studied the solution conformation of BPTI and UTI using FTIR method. The conformational changes caused by thermal denaturation and by reduction of disulfides have been investigated. In particular, we have shown that the rising bands at 1613 cm^{-1} and 1m^{-1} indicate the the formation of folding aggregation in the disulfide reduction process. Our study showed that FTIR spectroscopy, combined with resolution-enhancement and curve-fitting techniques, is a very sensitive method for probing protein conformation in solution.

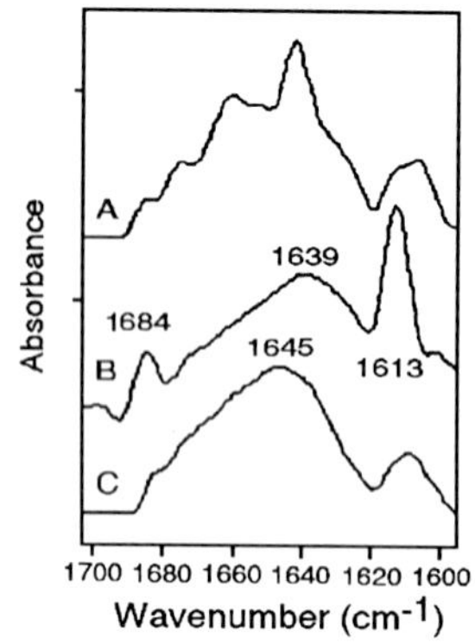

Fig. 4. Deconvoluted amide I' bands of (A) natural UTI ; (B) Thermal denatured UTI; (C) Disulfide reduced UTI.

Acknowledgements

Thanks to our engineer Ms. Jin An for designing the infrared cell allowing temperature perturbation, and the Bio-Rad Company Ltd. China, for providing the spectral analysis program.

References

1. Tschesche, H., Angew. Chemie, Int. Ed. Engl., 13(1974)10.
2. Deisenhofer, J. and Steigemann, W., Acta Crystallogr. sect. B, 31(1975)238.
3. Wuthrich, K., in NMR of proteins and Nucleic Acids, NY: Wiley, 1986, pp176-199.
4. Byler, D.M. and Susi, H., Biopolymers, 25(1986)469.
5. Krimm, S. and Bendear, J., Adv. Protein Chem., 38(1986)181.
6. Prestrelski, S., Byler, D.M. and Liebman, M.N., Biopolymers, 30(1991)133.
7. Chirgadze, Y.N., Fedorov, O.V. and Trushina, N.P., Biopolymers, 14(1975)679.
8. Surewicz, W.K., Szabo, A.G. and Mantsch, H.H., Eur. J. Biochem., 167(1987)519.
9. Venyaminor, S.y. and Kalnin, N.N., Biopolymers, 30(1990)1259.

The glucagon signaling system: The ligand's view of the receptor

C.G. Unson
The Rockefeller University, New York, NY 10021, U.S.A.

Introduction

The glucagon sensitive adenylate cyclase system has for many years been one of the most prolific models in the study of hormone signal transduction. One of a vast number of biologically active chemical compounds that are secreted into the blood, the polypeptide hormone glucagon triggers the production of glucose within its target cell to maintain an internal milieu critical to the survival of the organism. Glucagon conveys its message or signal through a highly specific cell-surface receptor whose primary structure is now known. A dual approach combining peptide synthesis of glucagon analogs and site-directed oligonucleotide replacement in a synthetic receptor gene should provide greater insight into the mechanism of glucagon signal transduction.

Results and Discussion

For several years, we have been investigating the relationship of glucagon structure to its function. Since it is widely accepted that recognition and activation are separate events, our efforts were directed towards discovering which amino acid residues are important for receptor recognition and binding, and which are responsible for the biological response. Early structure/function reports have demonstrated that virtually the entire sequence is required for binding and expression of the full response, although there were indications that the more hydrophobic C-terminus is primarily important in binding while the N-terminus mediates transduction.

Our synthetic studies showed that Asp^9 provides a major functional requirement for agonist activity and is not so much involved in binding to the receptor [1]. Aspartic acid is critical at this position and cannot be substituted even by the closely related glutamic acid. Position 9 of the hormone is thus a point of uncoupling of the binding function from activation. It also became evident that His^1 provides a functional group partially capable of transducing the hormonal signal and could be involved in an electrostatic interaction with Asp^9 [2].

In the absence of Asp^9, His^1-containing analogs retain low agonist activity. We later showed [3] that aspartic acid residues at positions 15 and 21 behave as surrogates to Asp^9. In addition, the carboxyl group of Asp^{15} contributes to the potency of the binding function.

Since a histidine and an aspartic acid were involved in activation, we speculated that a serine residue might cooperate with His^1 and Asp^9 to complete an active site triad reminiscent of the active residues of serine proteases. The idea warranted an analysis of synthetic replacement analogs of all four serine residues in glucagon. Ser^2 and Ser^{11}

are not important for activity, while Ser^{8} enhances binding. In contrast, Ser^{16} seemed to contribute more towards activity than binding and could be the third residue in our active triad [4].

We postulated that upon binding, the glucagon-receptor complex mediates proteolysis of the receptor to a smaller protein. The truncated protein might be the active form of the receptor that interacts with Gs to activate cyclase. There were precedents to our thinking.

In photoaffinity labeling experiments of the glucagon receptor in liver membranes, the appearance of the expected molecular weight receptor of roughly 60k, but also a smaller 33k protein that retains its binding affinity for ligand as well as a sensitivity to GTP [5]. In exploratory experiments, we have also shown that we can inhibit the production of cAMP with the addition of the serine protease inhibitor p-aminophenylsulfonylchloride and to a lesser extent, difluorophosphorylfluoride [4]. The thrombin signaling system is an interesting case. Upon binding to its receptor, thrombin cleaves its receptor at the sequence LDPR/S in the amino terminus of the extracellular domain to unmask a new N-terminus. This newly exposed amino terminus functions as a tethered ligand that activates the receptor [6].

It is unlikely that glucagon itself has enzymatic properties. Preincubation of the hormone with serine protease inhibitors did not result in inhibition of its ability to activate adenylate cyclase. Likewise, if the membranes are preincubated with inhibitors and then exposed to glucagon, no attenuation of the cyclase response occurred. Loss of cyclase activity was observed only when inhibitors were added to both glucagon and receptor [4].

That His^{1}, Asp^{9}, and Ser^{16} are involved in transduction was borne out by the fact that replacement of these putative active residues resulted in potent antagonists of the hormone [7]. Structure activity data from synthetic studies of glucagon analogs no doubt has had an important impact on the design of clinically useful antagonists. However, the mechanism by which the hormone signal is transmitted upon binding from receptor to the effectors on the intracellular side of the cell remains unknown. Our theory that proteolysis of the receptor occurs as part of the triggering mechanism remains to be proven by studying the receptor protein that furnishes the complementary site of the interaction.

The glucagon receptor has only recently been identified electrophoretically, when it was cloned from a rat liver library and expressed in COS cells [8]. Its primary structure determined from the cDNA sequence confirmed that it belongs to a subgroup of the G protein-coupled receptor superfamily, characterized by a seven transmembrane helix motif that is a recurring structural feature in all of the G-protein coupled receptors known to date. It is highly homologous to receptors of the subfamily which includes secretin, glp1, vip, calcitonin, and pth receptors. It has a number of conserved cysteines, four potential glycosylation sites, and the RLAR sequence in the third intracellular loop believed to interact with Gs. It has a relatively long extracellular amino terminus which contains at least part of the ligand binding site. It is likely that glucagon interacts with some of the extracellular loops and transmembrane sequences (Fig. 1).

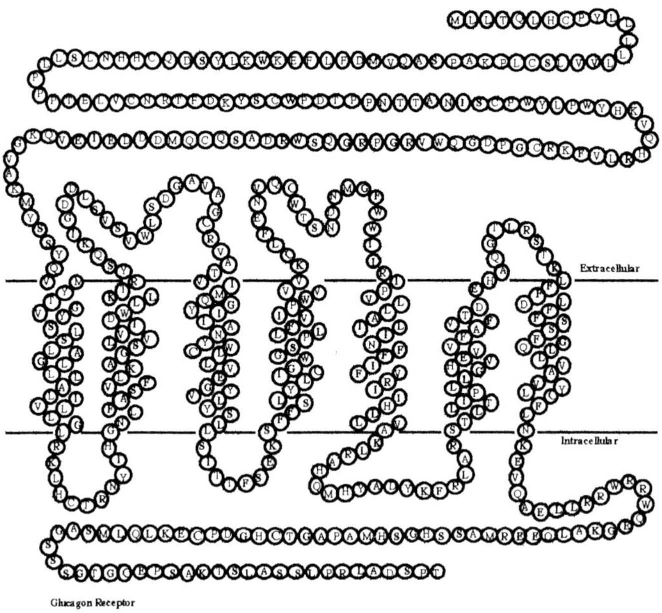

Fig. 1. The rat glucagon receptor.

To further our studies on the mechanism of glucagon action, we have synthesized the gene that codes for the glucagon receptor [9]. This work was done in collaboration with Tom Sakmar's laboratory at the Howard Hughes Institute at the Rockefeller University. The synthetic gene was designed to contain 90 unique restriction sites to facilitate restriction fragment replacement for mutagenesis studies.

The synthetic gene was transiently expressed in COS cells and identified on immunoblots with a specific antibody raised against the extreme C-terminus of the receptor [9]. The expressed receptor in COS cells bound glucagon and transduced the signal to stimulate adenylate cyclase.

Four Asp^{64} glucagon receptor mutants (D64K, D64G, D64,N, and D64E) were prepared and characterized to assess the role of Asp^{64} of the extracellular region of the receptor in glucagon binding. An Asp to Gly mutation of the analogous aspartic acid in the growth hormone releasing factor receptor had been shown to be responsible for the (lit) mouse phenotype [10]. It was postulated that the mutation interfered with stimulation of adenylate cyclase. Consistent with the molecular defect of the GRF receptor, all of the D64 mutants failed to recognize glucagon (Fig. 2).

The results suggest that there might be a direct interaction between Asp64 of the receptor and a corresponding positively charged site in glucagon. A structure/function study of glucagon analogs to identify the complementary site in the hormone is underway. A combination of chemical synthesis and molecular biology should help us understand the mechanism of glucagon-mediated signal transduction at the molecular level.

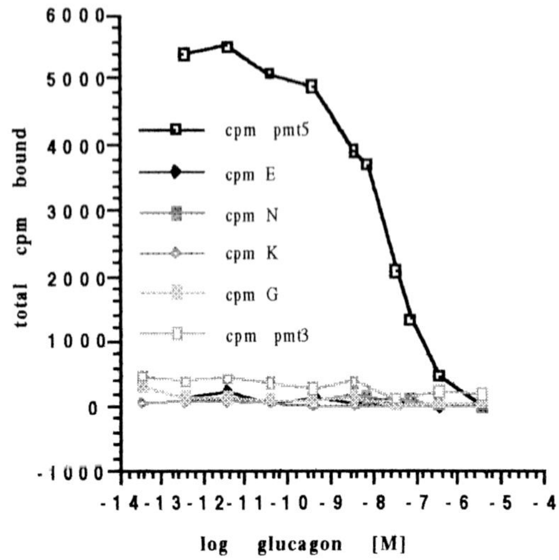

Fig. 2. Competitive binding of wild type glucagon receptor and D64 receptor mutants.

References

1. Unson, C.G., Macdonald, D., Ray, K., Durrah, T.L. and Merrifield, R. B., J. Biol. Chem., 266(1991)2763.
2. Unson, C.G., Macdonald, D. and Merrifield, R.B., Arch. Biochem. Biophys., 300(1993)747.
3. Unson, C.G., Wu, C.R. and Merrifield, R.B., Biochemistry, 33(1994)6884.
4. Unson, C.G. and Merrifield, R.B., Proc. Natl. Acad. Sci. USA., 91(1994)454.
5. Iyengar, R. and Herberg, J.T., J. Biol. Chem., 259(1984)5222.
6. Vu, T.K.H., Hung, D.T., Wheaton, V.I. and Coughlin, S.R., Cell, 64(1991)1057.
7. Unson, C.G., Wu, C.R., Fitzpatrick, K. and Merrifield, R.B., J. Biol. Chem., 269(1994)12548.
8. Jelinek, L.J., Lok, S., Rosenberg, G.B., Smith, R.A., Grant, F.J., Biggs, S., Bensch, P.A., Kuifper, J.L., Sheppard, P.O., Sprecher, C.A., O'Hara, P.J., Foster, D., Walker, K.M., Chen, L.H.J., Mckernan, P.A. and Kindsvogel, W., Science, 259(1993)1614.
9. Carruthers, C.J.L., Unson, C.G., Kim, H.N. and Sakmar, T.P., (1994) Synthesis and expression of a gene for the rat glucagon receptor. Replacement of an aspartic acid in the extracellular domain prevents glucagon binding, submitted to J. Biol. Chem.

Synthesis of the DNA-binding domain of human thyroid hormone receptor segments

Hong Ji, Hong Wang, Ruo-Heng Zhang and Mei-Chang Shao
Chemistry Department, Peking University, Beijing 100871, China

Introduction

Thyroid hormone receptor is a zinc-finger motif of DNA-binding protein that contains two zinc-fingers, with each zinc atom liganded by four cysteines [1,2]. Its sequence is shown in Fig. 1. hTR(140-169)-NH_2 was synthesized by SPPS method for structure study by 2DNMR spectroscopy. hTR(101-111)-S-$CH_2CH_2CONH_2$ and Troch-hTR(112-124)-NH_2 segments were also synthesized [3,4], and will be used for the segment condensation of hTR(102-124)-NH_2.

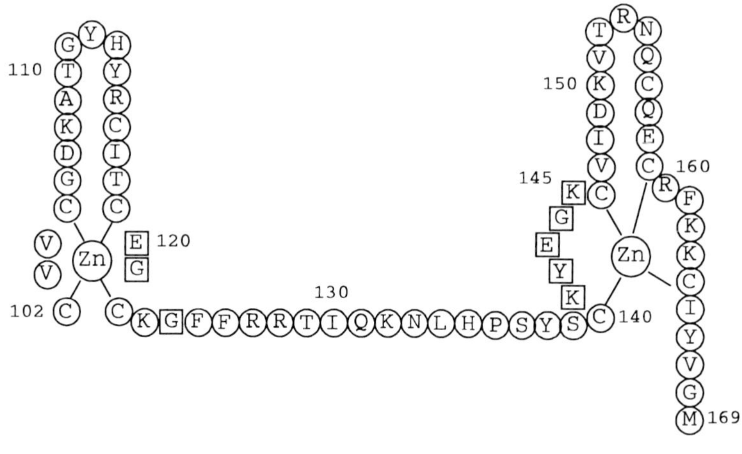

Fig. 1. The sequence of hTR (102-169).

Results and Discussion

In order to synthesize hTR(102-124), the alkyl thioester of a partially protected peptide segment, hTR(102-124)-SR, was used as a building block and the α-amino group of hTR(112-124)-NH_2 segment was protected by 2,2,2-trichloroethoxycarbonyl group. Before the condensation of these two segments, all lysines were protected with Boc-OSU and the Troc group was cleaved with Zn/HOAc. hTR(140-169)-NH_2 was synthesized by standard Boc chemistry. The segments were purified with RP-HPLC and confirmed by AA analysis and FAB-MS. Their HPLC profiles are shown in the Figs. 2-4.

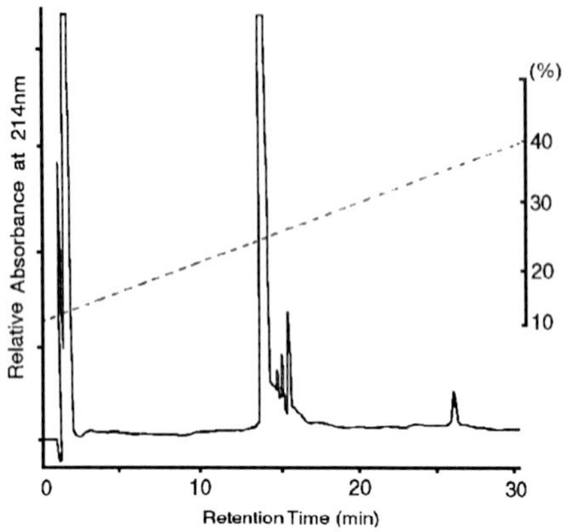

Fig. 2. HPLC profile of hTR(102-111)-SR. Chromatography was carried out on a Nova-Pak C_{18} column at a flow rate of 1ml/min. Broken line indicates the acetonitrile concentration in 0.1% aq TFA solution.

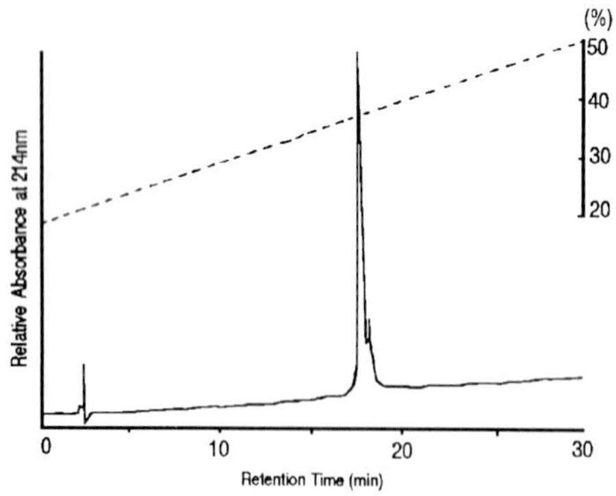

Fig. 3. HPLC profile of Troc-TR(112-124)-NH_2. Chromatography was carried out on a Nova-Pak C_{18} column at a flow rate of 1ml/min. Broken line indicates the acetonitrile concentration in 0.1% aq TFA solution.

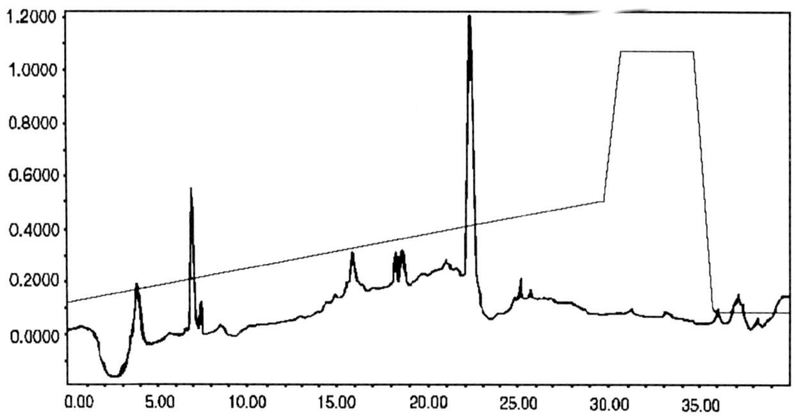

Fig. 4. HPLC profile of hTR(140-169)-NH_2. Conditions: A, 0.1% TFA/H_2O B, 0.1% TFA/CH_3CN. Flow rate: 2ml/min. C18 RP semipreparatiive column.

In the synthesis of hTR(102-111)-SR, we first used α-S-alkyl thioester as an anchoring group to prepare peptide thioester. We found that the thioester bond was quickly cleaved under TFA treatment. This suggested that the α-S-alkyl thioester bond was not an appropriate anchoring group for peptide segment condensation. Using β-S-primary-alkyl thioester, the synthesized peptide was pure but the yield was low.

It was easy to synthesize trichloroethyl chloroformate with phosphene generated from trichloroethanol. Troc group is stable under HF treatment conditions and can be deprotected against Boc groups on the side chain amino groups by zinc dust treatment in 50% acetic acid. Troc group was a convenient protecting group but it accompanies some by-products during the deprotection step. Thus it is necessary to improve this protecting group to obtain pure peptide. To extend our research, we will study interactions with zinc ions using various spectra such as UV,fluorescene,CD and 2D NMR spectroscopy.

References

1. Evans, R.M., Science, 240(1988)889.
2. Glass, C.K. and Holloway, J.M., Biochimica et Biophysica, 1032(1990)157.
3. Kwon, Y.D., Zhang, R.H., Bemquerer, M.P., Tominaga, M., Hojo, H. and Aimota, S., Chem. Lett., 1993, 881.
4. Hojo, H. and Aimoto, S., Bull. Chem. Soc. Jpn., 64(1991)111.

Crystal structure of an acidic neurotoxin from East Asia scorpion *Buthus martensii* Karsch

Hong-Min Li, Lei Jin, Zong-Hao Zeng, Miao Wang and Da-Cheng Wang
Institute of Biophysics, Chinese Academy of Sciences, Beijing 100101, China

Introduction

The East Asia scorpion *Buthus martensii* Karsch (BmK) is widely distributed in China. The whole body of BmK is a rare traditional Chinese medicine. Because scorpion neurotoxins are potential ion channel blockers, the study of their 3-dimension structure is significant for both structure-function analysis and practical applications.

The scorpion neurotoxins are mainly basic polypeptides. Here we report the crystal structure of an acidic scorpion neurotoxin BmK8, a component from the venom of the East Asia scorpion *Buthus martensii* Karsch. This is also the first X-ray structure of the East Asia scorpion neurotoxin.

Results and Discussion

Crystallization

The crystallization of BmK8 was succeeded in a solution system containing NaH_2PO_4 with low concentration by means of sitting drop method on a microscale. Preliminary X-ray analysis indicates that the crystals belong to monoclinic form with space group $P2_1$ with one molecule in an asymmetric unit. The cell parameters are: a=23.42Å b=38.53Å c=29.17Å and β=107.38Å.

Structure analysis

BmK8, whose isoelectric point is 5.3, is the first acidic neurotoxin found in BmK. It is weakly toxic to scorpion to date. The X-ray structure of BmK8 was determined at 1.85Å resolution using molecular replacement method with the structure of AaH II at 1.8Å resolution as the initial model. The parameters were refined by X-PLOR with the final R-value of 0.192. 12 unknown residues at the C-terminal have been determined by high quality electron density maps.

Features of tertiary structure

The structure looks like a flat fist. It contains a dense core of secondary structure elements in which two and a half α-helix are crosslinkedthrough two disulfide bridges to the middle strand of a three-segment of antiparallel β-sheet, and several loops project from this core. Some conserved amino acids with several hydrophobic residues form a conservative-hydrophobic surface.

The most interesting finding of BmK8 structure is the concentrated distribution of hydrophobic and charged residues. The charged amino acids, especially the basic residues, are obviously distributed in one side of the molecule, while the discharged residues are concentrated in the other side. The particular distribution in 3-dimensional structure may be related to the binding properties of the molecule to sodium channel.

Comparison with AaH II and CsE V3

Comparing the crystal structure of BmK8 with that of two other scorpion toxins, AaH II from a native of North Africa [1], and CsE V3 from a native of North America [2], the structure of BmK8 is close to that of AaH II (Fig. 1a), in keeping with the fact that both are α type toxins. The structure of BmK8 is markedly different from that of CsE V3, a β type toxin, in the loops protruding from the dense core (Fig. 1b). There are three loops clearly differ in the structures of BmK8 and CsE V3. (1) Their tails are oriented, in almost the opposite direction; (2) In BmK8, there is a loop composed of residues 38 to 44 that are not present in CsE V3; (3) The loop composed of residues 31 to 36 in CsE V3 does not exist in BmK8. It is reasonable to suppose that these three different loops may display functional roles in the action of scorpion neurotoxins.

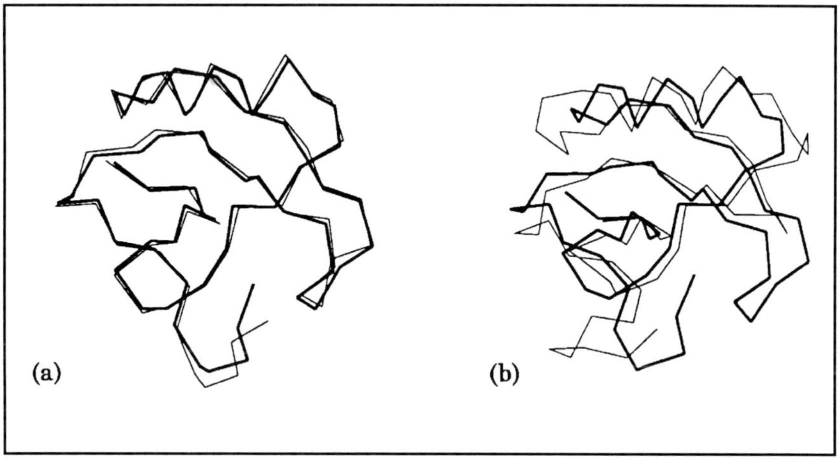

Fig. 1. Comparison of the α-carbon backbones of BmK8(thick lines) with AaH II(a) and CsE V3(b). Common Structure Motif (αβββ motif).

To date, with the exception of the crystal structures of CsE V3, AaH II and Bmk8, there are five scorpion toxins whose solution structures have been determined by 2D-NMR technique. The targets of these toxins, mainly the ion channels, are different, and the sequences of these toxins vary in length from 35 to 70. In their structures, however, they all contain two and a half α-helixes and β-sheet three segments even though their sequences display very low homology. Thus it can be supposed that there is a common motif in the 3-D structure of scorpion neurotoxins. The characteristics of the motif is that one segment of α-helix is crosslinked with three-strand of anti-parallel β-sheet by two disulfide bridges and they intersect about 30 degree. Therefore we designate it as αβββ motif (Fig. 2). This common structure motif may be the skeletonal core for all scorpion neurotoxins.

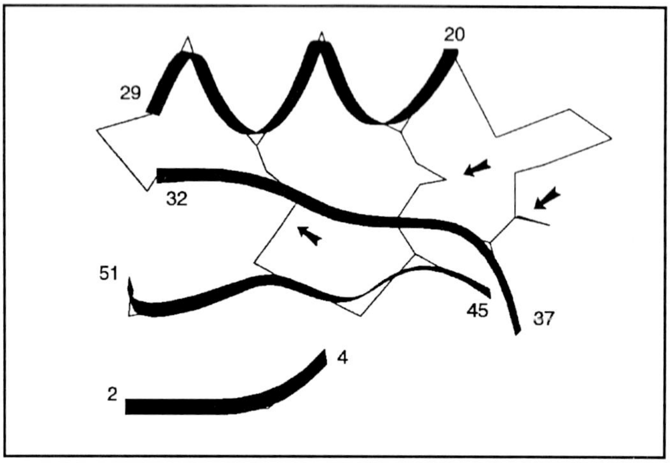

Fig. 2. Common structure motif(αβββ motif) shown by ribbon found in structures of scorpion neurotoxins. Arrow: Disulfide bridges.

Residues important to binding activity

BmK8, in which residues 28, 32, 50, 53 and 59 that are near the conservative-hydrophobic surface activity-important residue Lys58, greatly differs from AaH II. These differences may be related to the rather low toxicity of BmK8 and be functionally important for binding to sodium channel.

Possible binding model

Though α and β scorpion toxins have different binding sites on sodium channel, they all have a common conservative-hydrophobic surface and a common structure motif. It is therefore conceivable that the binding sites for α and β scorpion toxins are structurally related and mainly involved with residues composed of conserved residues 2 to 5, 45 to 48, 50 to 53, Lys58 and Glu59. Alternatively, the common conservative-hydrophobic surface could be important in the preliminary binding of α and β toxins to a general site, from which they could move to more specific binding sites.

The sequence of one sodium channel protein has been determined [3]. It contains 1820 residues and is composed of four repeated homologue units. On basis of this, a channel model was proposed. One of the features of this model is that there are two non-polarity areas and two negative charge areas on the outside of membrane. It has also been proven that the scorpion neurotoxins bind to sodium channel on its repeated unit I. Considering the main data evaluated here, we proposed a model for the binding of scorpion neurotoxins to the sodium channel proteins as shown in Fig. 3. In other words, from the polarity distribution of the molecule, the preliminary binding may also be divided into two aspect, one the hydrophobic interaction and the other the charge interaction.

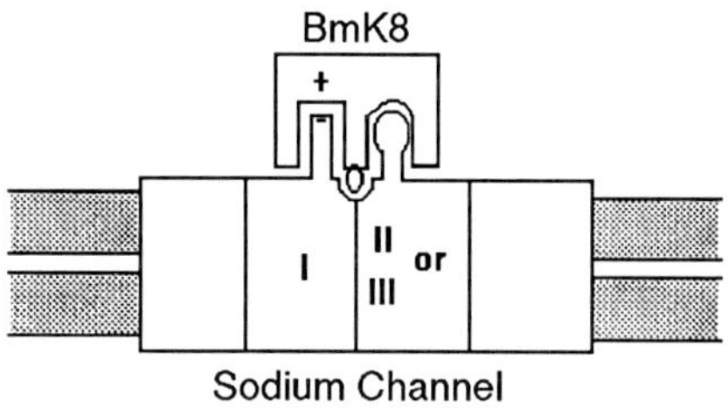

Fig. 3. Schematic representation of the hypothetical mode of binding of BmK8 to the sodium channel.

Acknowledgements

The authors thank Professor Fontecilla-Camps, J. C. for sending the coordinates of model AaH II. This work was supported by National Foundation of Nature Science (39370153).

References

1. Fontecilla-Camps, J. C., Rochat, C. H. and Rochat H., Proc. Natl. Acad. Sci. USA, 85(1988)7443.
2. Fontecilla-Camps, J. C., Almassy, R. J. Suddath, F. L. and Bugg, C.E., Toxicon, 20(1982)1.
3. Noda, M., Shimizu, S., Tanabe, T., Takai, T., Kayano, T., Ikeda, T., Takahashi, H., Nakayama, H., Kanaoka, Y., Nature, 312(1984)121.

Synthesis and studies on properties of water soluble Chinese rape pollen dodecapeptide

Kai-Ke Wan, Yan-Ling Song, Ai-Xue Ji and Sheng Jin
College of Chemistry and Molecular Engineering, Peking University, Beijing 100871, China

Introduction

Using solid-phase synthesis method we have prepared a dodecapeptide RPP I which was first extracted from chinese rape pollen [1]. From the sequence of RPP I we know that its most dominant characteristic is that the N-terminal is hydrophobic and the C-end hydrophilic, similar to the structure of melittin [2]. In order to study the structure-function relationship of the dodecapeptide RPPI, seven peptide segments RPP II - RPP VIII were synthesized. According to Chou-Fasman's conformation prediction method [3], $Ala^1 \rightarrow Val^5$ and $Gln^8 \rightarrow Gln^{11}$ in RPP I can form two β-sheets which are connected by a β-turn formed by $Val^5 \rightarrow Gln^8$. Based on this conformation, we designed and synthesized one anologue (RPP IX) with α-helix confomation and other anologue (RPP X) with β-sheet conformation. In addition, to determine the influence of Ser^7, Asn^{10} and charge on the activity of RPP I, we designed and synthesized three anologues (RPP XI-RPP XIII). Their sequences are shown in Table 1.

Table 1 *List of synthetic peptides*

A I G L V P S Q T N Q N	RPPI
A I G L V P	RPPII
S Q T N Q N	RPPIII
I G L V P S Q T N Q N	RPPIV
A I G L V P S Q T N Q	RPPV
A I G L V P S Q T N	RPPVI
L V P S Q T N Q N	RPPVII
A I G L V P S Q	RPPVIII
A I A L V A S Q T N Q N	RPPIX
A I G L V S Q T N Q N	RPPX
A I G L V P G Q T N Q N	RPPXI
A I G L V P S Q T D Q N	RPPXII
A I G L V P S E T D E D	RPPXIII

We have initially discussed their transitions of secondary structures in different solvents using CD spectra. In addition, their immunopromotive activities have been observed by E-rosette measurement *in vitro* [4], tumor necrosis factor (TNF) measurement [5] and T-lymphocyte transformation.

Results and Discussion

(1) The amino acid sequences of the peptides are shown in Table 1. Each purified peptide showed only one peak on RP-HPLC. Amino acid analysis of the purified peptides revealed residue ratios.

(2) Immunopromotive activiities of RPP I - RPP XIII were determined by E-rosette test. The results in Table 2 show that Ala^1, Ile^2,Gly^3 and Asn^{12} have no effect on the activity of RPPl, Ser^7 has some influence while Asn^{10},Gln^{11} and β-Sheet conformation play important roles. β-sheet conformation (RPP X, ΔERFC%=20.3) is more favorable activity of RPPI than α-helix conformation (RPPIX, ΔERFC%=10.8). For the above result, RPPIV, RPPV and RPPVII have similar activity to RPPI in promoting maturation of pig thymocyte (PTC) *in vitro*. Their ΔERFC% are 22.0, 21.1, 22.8 and 24.4, respectively. RPPIX, XII, XIII maintain only half of the activity. These results may indicate that the dodecapeptide RPPI and some of its segments can cause the maturation of pig thymocytes *in vitro*, and thus have immunopromotive activity. The result of tumor necrosis factor (TNF) in Table 3 shows that RPPI is able to enhance the activity of TNF on murine tumor cells, but the T-lymphocyte transformation test indicat that RPPI can inhibit the proliferation of T-lymphocytes stimulated by PHA.

Table 2 *The maximum activities segments and analogues of RPPI to promote the maturation of pig thymocytes*

Peptides	I	II	III	IV	V	VI	VII
$\Delta ERFC\%_{max}$	24.4	-1.0	-1.8	22.0	21.1	11.3	22.8
C_{max}(mg/ml)	0.1	0.0001	0.0012	0.00011	0.001	0.012	0.001
Peptides		VIII	IX	X	XI	XII	XIII
$\Delta ERFC\%_{max}$		8.1	10.8	20.3	18.9	10.9	12.8
C_{max}(mg/ml)		0.1	0.14	0.11	0.001	0.0011	0.13

Table 3 *TNF activity of RPP I and the effects of RPP I on the proliferation of T-lymphocytcs*

Concen-tration (mg/ml)	blank	only TNF	TNF+RPP I				Concen-traion (mg/ml)	blank	only PHA	PHA+RPP I			
			10^{-1}	10^{-2}	10^{-3}	10^{-4}				10^{-1}	10^{-2}	10^{-3}	10^{-4}
OD	1.20	0.86	0.84	0.63	0.65	0.48	CPM[b]	1.72	37381	34161	32569	31642	31226
ΔOD%[a]		0	-2.3	-26.7	-24.2	-44.4	ΔCPM%[c]	-	0	-8.8	-12.9	-15.4	-16.5

[a] $\Delta OD\%=[OD_{TNF+sample}-OD_{TNF}]\div OD_{TNF}\times 100$ [b] CPM: count per minute

[c] $\Delta CPM\%=[CPM_{(PHA+sample)}-CPM_{(PHA)}]\div CPM_{(PHA)}\times 100$

(3) The CD spectra of RPP1 indicated solvent inducing transitions of secondary structures. It can undergo random coil to β-sheet conformation transition induced by 95% trifluorethanol (TFE) and to α-helix conformation transition induced by 95% 2-

chloroethanol. A segment as small as RPPII also contains secondary structure in some organic solvents. RPPII contains some β-sheet in 96% TFE or in 96% 2-chloroethanol. It is possible that this is related to the high β-sheet (P_{β}=1.24) formation tendency of Ala^{1}-Val^{5} according to the Chou-Fasman's prediction method. The CD spectra of RPPIX showing the presence of α-helix conformation in 96% 1,4-dioxane, agrees with Chou-Fasman's prediction, but the results of CD spectra and prediction of RPPXI do not.

Acknowledgements

This work was supported by the National Natural Science Foundation of China.

References

1. Zhang, Y.L., Jin, S., Chinese Chem. Lett., 1(1990)85.
2. Drawson, C.R., Drake, A.F., Biochem. Biophys. Acta., 510(1978)75.
3. Chou, P.Y., Fasman, G.D., Biochemistry., 13(1974)222.
4. Wybran, J., Clin. Immunol. Immunopathol., 1(1973)408.
5. Sherwood, E.R., J. Biol. Response Mod., 9(1990)44.

Studies on synthesis and bioassay of hF-GRP analogs

De-Xin Wang, Qing-Chai Xu, Gui-Shen Lu, Ying Sun, Nai-Gong Wang and Mu-Zhen Guan

Institute of Materia Medica, Chinese Academy of Medical Sciences, Beijing 100050, China

Introduction

Human follicular gonadotropin releasing peptide (hF-GRP), a 14-amino acid peptide, was isolated from human follicular fluid several years ago [1]. It is interesting to note that hF-GRP is located in residues 11-24 of the primary structure of human α_1-antitrypsin [2], which does not exhibit gonadotropin releasing activity. However hF-GRP with an amino acid sequence completely different from that of LHRH was found to have LHRH-like activity in vitro [1].

Six hF-GRP analogs have been synthesized and evaluated for stimulating pituitary LH/FSH secretion in vitro by Li's group [3]. Of those analogs, [Tyr^4]-hF-GRP had a higher LH releasing activity (2.4 fold) and the C-terminal decapeptide, hF-GRP(5-14), had a slightly higher activity (1.5 fold) in the RIA than that of hF-GRP.

In order to understand the structure-function relationship, we have designed, synthesized and tested the natural sequence of hF-GRP and twenty of its analogs.

Table 1 *Structure of hF-GRP and its analogs*

No.	Sequence													
hF-GRP	Thr	Asp	Thr	Ser	His	His	Asp	Gln	Asp	His	Pro	Thr	Phe	Asn
1	—	—	—	—	—	—	—	—	—	—	—	—	—	— (NH_2)
2	—	—	—	—	—	—	—	—	—	—	—	Tyr	—	—
3					—	—	—	—	—	—	—	Tyr	—	—
4							Lys	—	Glu	Tyr	—	—	—	Tyr
5						Lys	Glu	—	Glu	Lys	—	—	—	Glu
6									—	Lys	—	—	—	—
7							Glu	—	Glu	Lys	—	—	—	—
8	—	—	—	—	—	—	—	—	—	—	—	—	—	Phe
9					—	—	—	—	—	—	—	—	—	Phe
10			—	Tyr	—	—	—	—	—	—	—	—	—	
11			Tyr	—	—	—	—	—	—	—	—	—	—	
12									—	—	—	—	—	Lys
13	—	—	—	Leu	—	Lys	—	—	—	—	—	—	—	—
14	—	—	—	Leu	—	Leu	—	—	—	—	—	—	—	—
15	—	—	—	(D)[a]	—	Lys	—	—	—	—	—	—	—	—
16	—	—	—	(D)	—	Leu	—	—	—	—	—	—	—	—
17	—	—	Val	—	- - -	- - -	(D)	—	—	—	—	—	—	—
18	—	—	(D)	—	- - -	- - -	(D)	—	—	—	—	—	—	—
19	—	—	(D)	—	- - -	- - -	Leu	—	—	—	—	—	—	—
20	—	—	Val	—	- - -	- - -	Leu	—	—	—	—	—	—	—

a (D): the D form residue substituted for the natural one

Results and Discussion

Synthesis of hF-GRP and its analogs (Table 1) was performed by the solid-phase method. New protocols, such as KI/Cs_2CO_3 (1:10) to induce the esterification of Boc-Asn onto chloromethyl resin[4], the use of DDSi-phenol instead of TFA to remove Nα-Boc group [5], and simultaneous multiple peptide synthesis (SMPS) [6] for analogs 13-16 and 17-20, were adopted. All synthetic peptides were purified by reverse-phase HPLC on a C-18 column and verified by amino acid analysis.

After treatment of synthetic peptide in the mouse pituitary incubation system, the in vitro bioasssay data (Table 2) for LH secretion was determined by the use of Van Damme's method [7].

Among all analogs of hF-GRP, two mutants (8 and 9) with a substitution of Phe for Asn14 at the C-terminal had markedly higher LH release in vitro. It was obvious that the configurational variation of native residue to D form at whatever position in the sequence involved in present study (15-19) could abolish the original activity. The C-terminal amide analog (1) retained bioactivity comparable to hF-GRP. As for the other analogs, no structure-function conclusions could be drawn.

Table 2 *Effect of hF-GRP analogs on pituitary LH release in vitro*

Peptide	LH activity(%)
control	100
hF-GRP	115.4[a]
[C-NH_2]-GRP	119.0[a]
[$Lys^{6,10}$, $Glu^{7,9,14}$]-GRP(6-14)	135.0[a]
[Phe^{14}]-GRP(5-14)	162.0[b]
[Phe^{14}]-GRP	179.2[b]

[a] $P<0.05$ [b] $P<0.02$

Acknowledgement

This study was supported by grant from the National Natural Science Foundation of China, No. 39070951.

References

1. Li, C.H., Ramasharma, K., Yamashiro, D and Chung, D., Proc. Natl. Acad. Sci. USA., 84(1987)959.
2. Carrell, R.W., Jeppssons, J.O., Laurell, C.B., Brennan, S.O., Owen, M.C., Vaughan, L. and Boswell, D.K., Nature (London), 298(1982)329.
3. Ramasharma, K. and Li, C.H., Int. J. Pept. Prot. Res., 32(1988)419.
4. Wang, D.X., Lu, G.S. and Guo, M.T., In Du, Y.-C., Tam, J.P. and Zhang, Y.-S. (Eds.) Peptides: Biology and Chemistry (Proceedings of the 1992 Chinese Peptide Symposium), ESCOM, Leiden, 1993, pp. 295–296.
5. Wang, D.X. and Lu, G.S., Chin. J. Med. Chem., 1(2)(1991)60.
6. Tjoeng, F.S., Towery, D.S., Bulock, J.W., Whipple, D.E., Fok, K.F., Willoams, M.H., Zupec, M.E. and Adams, S.P., Int. J. Pept. Prot. Res., 35(1990)141.
7. Van Damme, M.P., Acta Endocrin. Logoca, 77(1974)655.

Study on local angiotensin in the pericardium

Shao-Jun Wen[a], Wei-Jun Zhang[b], Shi-Da He[a], Yu Liu[a] and Bao-Tian Chen[b]
[a]*Department of Cytogenetics, Xuan Wu Hospital, Capital Institute of Medicine, Beijing 100053, China*
[b]*An Zhen Hospital, Beijing 100029, China*

Introduction

The pericardium arises from the mesodermal layer and consists of mesothelial cells and connective tissue. It is divided into the pericardium fibrosum and serosum. For hundreds of years it has been known that the function of the pericardium is to protect the heart and diminish friction. Components of the local renin-angiotensin system, including angiotensinogen, renin, angiotensin converting enzyme, Ang I, Ang II and Ang receptor, have been found in heart and adrenals tissues [1,2,3]. The purpose of this study was to determine the presence and distribution in human pericardium of immuno-competent peptide hormone,i.e., angiotensin-like materials, which are associated with renin activity and to measure the levels of Ang-like materials in the pericardium of patients with chronic rheumatic heart disease (RHD).

Results and Discussion

Using the radioimmunoassay method, Ang I and Ang II-like materials were found in pericardia of normal individuals and in patients with chronic rheumatic heart disease (Table 1). Ang II-like materials were not significantly different among upper, middle and lower parts of the pericardia as determined by analysis of variance. The concentration of Ang I was significantly higher than Ang II in the middle part of the pericardium. ($P<0.05$). There was renin activity in human pericardium. Compared with normal persons, patients with chronic rheumatic heart disease had a much higher concentration of Ang II-like materials ($P<0.01$).

Adopting a method combining radioimmunoassay with HPLC, immunoreaction of Ang II-like materials was found (Fig. 1). Ang II-like materials were found in human pericardium under the optical microscope. The immuno-reactive granules were mainly located in the cytoplasm of mesothelial cells of human pericardium and rat epicardium. There were no significant linear correlations between pericardial Ang II and the peptide in plasma ($P>0.05$).

Recent studies have demonstrated the presence of a local renin-angiotensin system in target organs. The local renin-angiotensin system plays an important role in regulating local physiological functions and maintaining homeostasis [4,5]. We postulate that the Ang"r found in plasma are derived from the circulating renin-angiotensin system and that there may be a local renin-angiotensin system producing the peptide materials in the pericardium.

In the present study, most cardiac patients had severe heart dysfunction with marked cardiac enlargment. Pericardial tension increased considerably in these patients as the result of compensatory mechanisms. The increase of local Ang II in the pericardium of the patients with RHD may serve to enhance the contractility of the pericardium and prevent the heart from over-expansion.

Table 1 *Concentrations of angiotensin I and angiotensin II(ng/g wet weight) in human pericardium*

Locations in pericardium	Ang I		Ang II	
	Mean	SD	Mean	SD
	Normal[a] (n=5)			
Upper			1.42	0.84
Middle	11.64	7.63	1.06	0.63
Lower			1.35	0.53
	Rheumatic Heart Disease (n=11)			
Middle			4.34	2.62

[a] Those who died from accident.

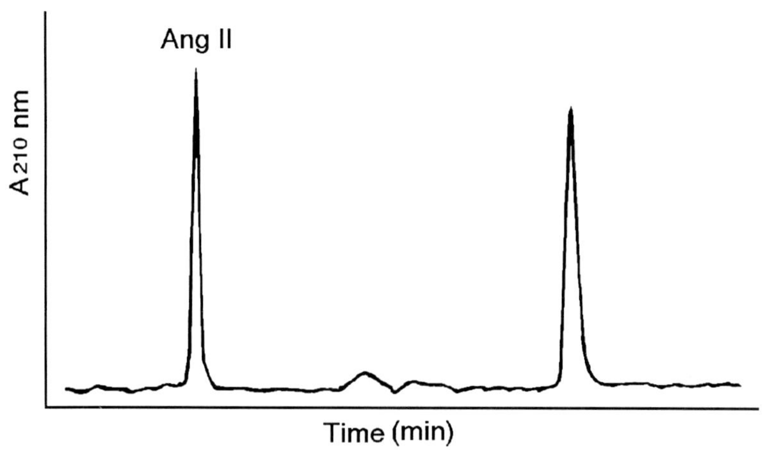

Fig. 1. Reverse-phase HPLC-RIA of angiotensin II.

The data presented here are the first to suggest that the human pericardium may have compensatory endocrine functions which operate under normal and abnormal heart conditions.

References

1. Dzau, V.J., Circulation (suppl I), 77(1988)I4.
2. Sarzani, R., Fallo, F., Dessi-Fulgheri, P., Pistorello, M., Lanari, A., Paci, V.M., Mantero, F., Rappelli, A., Hypertension, 19(1992)702.
3. Wen, S.J., Wang, J.R., He, S.D., Chinese J. Cardiol., 18(1990)91.
4. Richard, R.E., Am. J. Cardiol., 59(1987)56A.
5. Yamada, H., Fabris, B., Allen, A.M., Jackson, B., Johnston, C.I., Mendelsohn, F.A.O., Circulation Research, 68(1991)141.

Synthesis of some new analogs and segments of papaver somniferum pollen peptides

Jia-Xi Xu[a], Yuan Ma[b], Meng-shen Cai[a] and Sheng Jin[c]
[a]School of Pharmaceutical Sciences, Beijing Medical University, Beijing 100083, China
[b]Department of Chemistry, Tsinghua University, Beijing 100084, China
[c]College of Chemistry and Molecular Engineering, Peking University, Beijing 100871, China and National Laboratory of Applied Organic Chemistry at Lanzhou University, Lanzhou 730000, China

Introduction

Pollen grain is the plant male reproductive product, which not only contains rich nutritive constituents, such as amino acids, proteins, carbohydrates, nucleic acids, enzymes and cofactors, vitamins, lipids, antibiotics, minerals and trace active substances, but also possesses message substances connected with life science [1]. From our investigation, we found that pollen has many physiological activities, for example, promoting immunity and restraining cancer cell division [2-7]. We therefore investigated peptides in pollen, whose chemical compositions may lead to pharmaceuticals with significant physiological actions.

Results and Discussion

The peptide PSPP3 (Papaver Somniferum Pollen Peptide 3) was obtained from Papaver somniferum pollen by Xu and Jin. This peptide showed strong immunopromotive activities according to the methods of counting ERFC and *in vitro* T-lymphocyte transformation tests [1,2,7].

In order to find the active segment of PSPP3, two new analogs and three new segments were designed and synthesized by using Merrifield solid phase peptide synthesis method. Their T-lymphocyte transformation tests and the relation between their activities and structure are under study. The sequences of analogs and segments are as follows.

```
H-Asp-Glu-Asp-Gly-Ser-Asp-Pro-Lys-Thr-Val-Lys-Glu-Ala-OH
H-Asn-Gln-Asn-Gly-Ser-Asp-Pro-Lys-Thr-Val-Lys-Glu-Ala-OH
H-Gly-Ser-Asp-Pro-Lys-Thr-Val-Lys-Glu-Ala-OH
H-Asn-Gln-Asn-Gly-Ser-Asn-Pro-OH
H-Asp-Glu-Asp-Gly-Ser-Asn-Pro-OH
```

References

1. Jin, S., Xu, J.X. and Miao, P., Youji Huaxue, 13(1993)202.
2. Xu, J.X. and Jin, S., Chin. Chem. Lett., 4(1993)213.
3. Zhang, Y.L. and Jin, S., Chin. Chem. Lett., 1(1990)85.

4. Wan, K.K., Ji, A.X., Wang, Z., Qiu, W.G. and Jin, S., Chin. Chem. Lett., 4(1993)97.
5. Xu, J.X., Cai, M.S. and Jin, S., The 5th National Symposium on Organic Synthesis, pp.244, Huangyan, Zhejian, China, 1993, 10.
6. Xu, J.X., Cai, M.S. and Jin, S., The 5th National Symposium on Organic Synthesis, pp.245, Huangyan, Zhejian, China, 1993, 10.
7. Xu, J.X. and Jin S., Huaxue Xuebao, in press, 1994.

Microcomputer modeling structures of HIV-1 protease with inhibitors in the binding sites

Han-Lin Chi
Department of CADD, Institute of Materia Medica, Chinese Academy of Medical Science, Beijing 100050, China

Introduction

The interactions of drug molecules with enzymes or transport proteins have received particular attention [1]. The studies on docking and research docking tools for fitting ligand/receptor is a very important in CADD (Computer-Aided Drug Design). The microcomputer programs of protein-anatomy PDBPC and ligand-docking DOCK we developed were used to study the interactions of HIV-1 protease with inhibitors in the binding sites. The docking algorithm utilized in our DOCK program is unique in moving and rotating the ligand into the cavity of the receptor by automatically searching for the position of the minimal energy by the interactions. There are no other the similar reports in literature.

Results and Discussion

Publications of the x-ray structures of the complex of HIV-1 protease with inhibitors promoe de novo drug design. The complex contains 1,644 atoms (Fig. 1). Indeed, knowledge and insight gained by modeling the complex and studies on physicochemical requirements would be a great help in designing the new anti-AIDS drugs [2]. The operation of ligating cutting the remote residues down provided by the program depends on the user-given threshold contact between the ligand and receptor. The binding pocket of HIV-1 protease (Fig. 2) was formed by removing the residue whose contact distances are more than 5 Å.

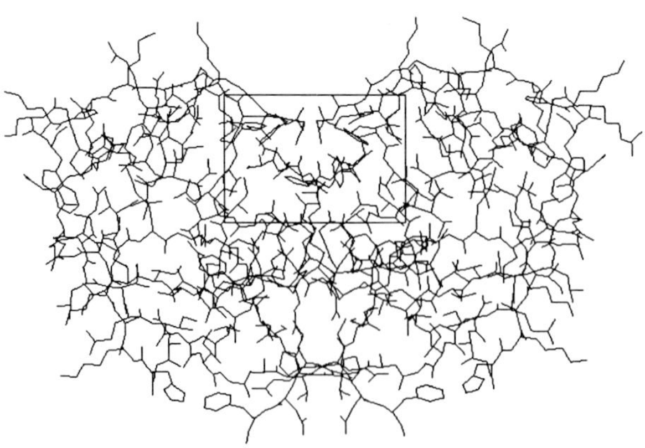

Fig. 1. The view of homodimeric structure of HIV-1 protease with the inhibitor acetyl pepstatin was shown by PDBPC on IBM/PC computer. The window on the screen represents the substrate binding cleft vertical to the plane of the page.

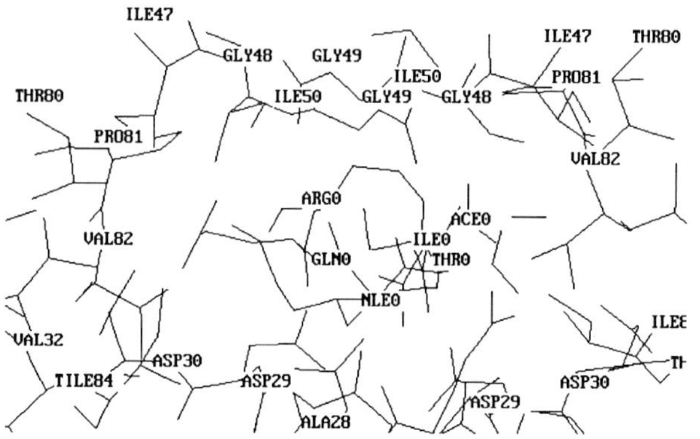

Fig. 2. The cavity of the inhibitor binding sites was separated by the anatomy program. The residues around the inhibitor acetyl pepstatin that formed the wall of the cavity are non-sequential but close. It contains Arg8 Leu23 Asp25 Gly27 Ala28 Asp29 Asp30 Thr31 Val32 Ile47 Gly48 Gly49 Ile50 Leu76 Thr80 Pro81 Val82 Ile84 in the monomer. Most of them are associated with dimer stabilization and confirmed that Asp25 and Gly27 are closed to activities.

The docking procedure is very complicate and deficient in the study. Using DOCK program, the ligand molecule moves along the axial or middle-axial space on the screen trying to find the position of minimum energy. The rotations of the ligand are performed in two different ways: around the cavity of the receptor or around the ligand molecule. The interaction energies can be divided into three parts, vdW interaction, electrostatic or hydrophobic components. In the fitting procedure, these components play different roles in the different phases. There are at least two phases, "the ship-lying phase" and "the casting anchor phase". It is supposed that vdW interaction would be more important in the ship-lying phase, because vdw interaction is the short-range force in the binding sites. The ship-lying phase is concerned with initiation of the docking procedure. The descriptor for this phase could be vdW interaction energy relative to molecular geometry. The second phase, the so-called anchor-casting phase, is the refining procedure, in which the hydrophobic, H-bonding or other small components were refine and direct the ligand.Hydrophobic interactions have very a important influence on substracte binding as we know. The hydrophobic energies are calculated from the determination of atomic types. Hydrophobic parameters then are put into atoms in the file of both molecules according to Viswanadhan's table [3]. The simplified new Yeti equation [4] was selected for calculation of the hydrophobic energy.

We considered that the stepwise docking in the different phases of interactions should be developed as a better strategy in computer-aided drug design.

The activities of HIV-1 protease closed to residues Asp25-Thr26-Gly27. However, most of the residues in the cavity would be related to the stabilization. We have tried to research a series of compounds, which

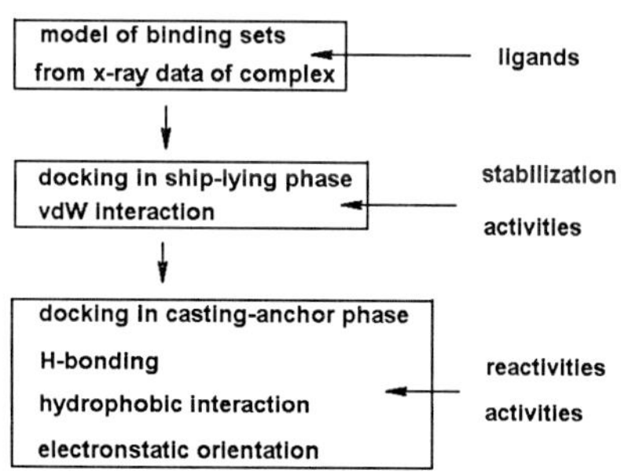

Schema of stepwise dock

can mask off Asp25 and Gly27 using intra-molecular interactions. On an account of the status stability reactive to vdW interactions, we used the stepwise docking method to investigate one of inhibitors of HIV-1 protease [5] (Fig. 3.).

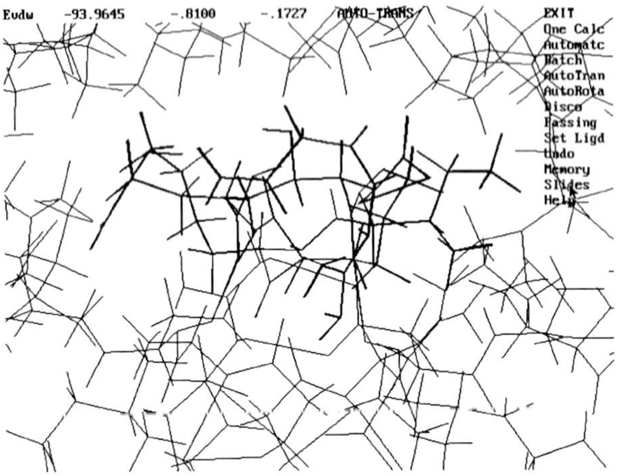

Fig. 3. Acetyl Pepstatin (the center) was docking into the binding cavity of HIV-1 protease. The traces on the screen were left by moving and rotating the ligand by the DOCK program. The number (-93.96 kcal/mol) in the upper part of the screen is the vdW interaction energy. It is demonstrated that vdw interactions is the major descriptor in the ship-lying phase, because the position of the ligand is not much changed in using the data of the fitted complex from the PDB database.

It was not too difficult to dock the very flexible structures into the longer and very narrow pocket. It demonstrated that DOCK and PDBPC programs are more effective and reliable than others to attempt in drug design on the common micro-computer systems. The results of study will be published in sequence.

References

1. Weinsten, H. and Osman, R., in Computer-aided molecular design, IBC Technical Service Ltd., 1989, pp. 199–128.
2. Huff, J. R., J. Am. Chem., 34(1991)2306.
3. Viswanadhan, V.N., Ghose, A.K., Revankar, G.R., Robins, R.K., J. Chem. Inf. Comput. Sci., 29(1989)163.
4. Vedani, A. and Huhta, D.M., J. Am. Chem. Soc., 113(1991)5860.
5. Martin, Y.C., J. Am. Chem., 35(1992)2145.

Preparation and analysis of lactosyl poly(L-lysine) as ligand for the hepatic galactose receptor

Shou-Ming Wen, Pei Wang, Tao Jin, Xiao-Li Ji and Xiao-Yuan Du
Department of Clinical Pharmacology, General Hospital of Air Force, PLA, Beijing 100036, China

Introduction

Galactose-terminal plasma glycoproteins and glycopeptides are specifically recognized and bound by the mammalian hepatic asialoglycoprotein receptor (galactose receptor), and have been successfully used as carriers to deliver drugs to hepatic cells for targeting chemotherapy in hepatomas and hepatitis [1,2]. In an effort to search for the synthetic ligands specific to the galactose receptor, we prepared a series of lactosylated poly (L-lysine) (Lm-PLL) peptides and evaluated their binding affinity and selectivity to the galactose receptor using fluorescent histochemistry approach.

Results and Discussion

1. Preparation of Lm-PLL

Lactose was covalently attached to the primary amino groups of poly (L-lysine) (PLL.HBr, Mr6,000) by reductive amination with sodium borohydride [1]. The resulting stable alkylamines (Lm-PLL) were isolated from unreacted sugar and reductant on a Sephadex G-25 column, and fractions containing Lm-PLL were lyophilized to amorphous powder. The molar ratios of lactose linked to PLL, as determined by the phenol-sulfuric acid method for galactose residues and 2,4,6-trinitrobenzensulfonic acid method for the primary amino groups of PLL [2], increased with the amounts of lactose and $NaBH_4$ added to the reacting mixture untill all the amino groups (28) in PLL were glycosylated.

2. Histochemical Analysis

L_{14}-PLL, in which half of the amino groups in PLL lactosylated,was chosen as a model molecule for histochemical studies of rat tissue sections. L_{14}-PLL was labeled with fluorescein isothiocyanate (FITC) in carbonate buffer (pH 9.0) at 4°C for 16 hr, in 1:1.5~2.0 molar ratio. Frozen tissue sections (4~5μm) of normal adult rat liver, spleen, heart, kidney and lung were mounted and fixed on slides and incubated with the L_{14}-PLL-FITC (10^{-5}~10^{-6} Mol/L in PBS) for 30 min at 30°C in the dark. The fluorescent signal was observed under fluorescent microscope (Olympus model BH2) after PBS washes, compared with three controls by incubating slide-mounted tissues with 1) buffer alone to measure the extent of tissue autofluorescence, 2) excess of L_{14}-PLL before addition of L_{14}-PLL-FITC to ascertain the extent of nonspecific binding, 3) free FITC (10^{-6} Mol/L). Incubation of rat tissue sections with L_{14}-PLL-FITC revealed high fluorescent staining (++~+++) in the liver and considerably less (±~+) in the spleen, heart, kidney and lung. Most of the fluorescence in liver sections was ligand-induced, dose-dependent and blocked with excess concentrations of unlabeled L_{14}-PLL. All the tissue sections showed sparse autofluorescence after incubation with PBS alone but widespread strong staining (+++) with free FITC.

Table 2 *Histochemical analysis of fluorescein labeled L_{14}-PLL on rat tissue sections*

rat tissue sections	compounds	conc. (mol/L)	extent of fluorescence staining
liver	L_{14}-PLL-FITC	10^{-6}	+++
	L_{14}-PLL-FITC	10^{-6}	++
	L_{14}-PLL+L_{14}-PLL-FITC		+
	FITC		++
	PBS		-
spleen	L_{14}-PLL-FITC	10^{5}	+
	L_{14}-PLL-FITC	10^{-6}	±
	FITC		++
heart	L_{14}-PLL-FITC	10^{-5}	+
	L_{14}-PLL-FITC	10^{-6}	±
	FITC		++
kindney	L_{14} PLL-FITC	10^{-5}	+
	L_{14}-PLL-FITC	10^{-6}	±
	FITC		++
lung	L_{14}-PLL-FITC	10^{-5}	+
	L_{14}-PLL-FITC	10^{-6}	±
	FITC		++

Lactosyl poly (L-lysine) can be easily prepared by reductive amination or scale amidation large and is available in a wide range of molecular sizes. The results in this report showed that lactosyl poly (L-lysine) with appropriate galactosyl density had much higher binding affinity and was more selective to liver sections than to other tissues and could be used as a ligand specific to the galactose receptor as well as to hepatotropic drug carriers. Additional studies are needed to confirm the receptor-selective binding of the ligand (L-PLL), using a conventional radiolabeled ligand assay.

References

1. Follon, R.J., Schwartz, A.L., Adv. Drug. Delivery Rev, 4(1989)49.
2. Meijer, D.K.F., Sluijs, P.Van der., Pharmaceut. Res., 6(1989)105.

Session III
Neuropeptides

Chairs: Arthur M. Felix
Hoffmann-La Roche Inc.
Nutley, New Jersey, U.S.A.

and

Yu-Cang Du
Shanghai Institute of Biochemistry
Shanghai, China

Opioid peptide analogs as pharmacological tools and as potential drugs

P.W. Schiller
Laboratory of Chemical Biology and Peptide Research, Clinical Research Institute of Montreal, 110 Pine Avenue West, Montreal (Que.) Canada H2W 1R7

Introduction

Since the discovery of the enkephalins in 1975 [1] a large number of endogenous opioid peptides have been detected in mammals and at present three distinct families of opioid peptides are known (for a review, see ref. [2]). These are 1) the enkephalins, 2) the endorphins (α-, β- and γ-) and 3) the dynorphins and neoendorphins. Peptides with opioid activity have also been isolated from tryptic digests of milk casein (β-casomorphins) and from frog skin (dermorphins and deltorphins). Like the morphine-related opiates, opioid peptides produce a large spectrum of central and peripheral effects which, aside from analgesia, include tolerance and physical dependence, respiratory depression, euphoria, dysphoria and hallucinations, sedation, feeding and other behavioral effects, hypothermia/hyperthermia, miosis, effects on tumor growth, control of release of several peptide hormones and catecholamines, effects on transit in the gut, various cardiovascular effects, etc. All these effects are elicited through interaction with specific opioid receptors. The existence of at least three different classes of opioid receptors (μ, δ, κ) was proposed in the mid-seventies on the basis of pharmacological evidence [3] and has been definitely established through the recent cloning of δ [4,5], κ [6] and μ [7] receptors. There is increasing evidence for the existence of receptor subtypes in the case of the μ receptor (μ_1, μ_2) [8], δ receptor (δ_1, δ_2) [9] and κ receptor (κ_1, κ_2, κ_3) [8]; however, the concept of opioid receptor subtypes has not yet found general acceptance and needs to be corroborated by further receptor cloning and expression experiments.

It has not yet been possible to establish clear-cut relationships linking specific opioid receptor types to distinct opioid effects because, until recently, potent, stable agonists and antagonists with high specificity for the various receptor types and subtypes have not been available. Most of the naturally occurring opioid peptides show limited receptor selectivity, the exception being the highly δ receptor-selective deltorphins [10]. Thus, the development of receptor-specific opioid agonists and antagonists represents a challenge for peptide chemists and medicinal chemists alike. Progress made in the development of receptor-specific non-peptide opiates has been excellent (for a review, see ref. [11]), an example being the highly selective μ agonist ohmefentanyl [12]. The design of receptor-specific opioid agonists and antagonists through structural modification of naturally occurring opioid peptides has also been successful and will be described in some detail in this review. Design strategies used in the development of receptor-specific opioid peptide analogs and peptidomimetics included (i) the substitution, deletion or addition of natural or artificial (non-proteinogenic) amino acids;

(ii) the design of bivalent ligands; (iii) the concept of conformational restriction; and (iv) peptide bond replacements. Very recently, the use of synthetic peptide combinatorial libraries has led to the discovery of novel peptide ligands with opioid agonist or antagonist properties. These peptides also represent excellent starting points for the development of highly specific opioid receptor ligands through appropriate structural modifications. Highly specific opioid agonists and antagonists are not only of interest as pharmacological tools but also as potential therapeutic agents with fewer side effects as compared to non-selective opioid receptor ligands. Recently, it has been demonstrated that pretreatment of mice with the non-peptide δ opioid antagonist naltrindole attenuated the development of morphine tolerance and dependence [13]. This interesting observation suggested that a single compound with mixed μ agonist/δ antagonist properties might have therapeutic potential. Initial success in the development of compounds with this profile will be discussed in this article.

An issue of particular concern to the development of opioid peptide-derived drugs is the blood-brain barrier (BBB). In efforts to design opioid compounds with either primarily central or exclusively peripheral action, the development of opioid peptide analogs with greatly enhanced ability to cross the BBB or totally unable to penetrate into the central nervous system (CNS) is currently being attempted. Progress made along these lines will be discussed in the final section of this article.

Receptor-selective opioid peptide agonists

Agonists with selectivity for the various opioid receptor types have been obtained by using various design strategies (for reviews, see refs. [14,15]). The development of potent and highly selective *μ agonists* has been particularly successful. The classical approach based on amino acid substitutions and deletions produced the enkephalin analogs H-Tyr-D-Ala-Gly-Phe(NMe)-Gly-ol (DAMGO) and H-Tyr-D-Met(O)-Gly-Phe(NMe)-ol (syndyphalin), the β-casomorphin analogs H-Tyr-Pro-Phe-Pro-NH_2 (morphiceptin) and H-Tyr-Pro-Phe(NMe)-D-Pro-NH_2 (PLO17), and the dermorphin analogs H-Tyr-D-Arg-Phe-Lys-NH_2 (DALDA) and H-Tyr-D-Ala-Phe-Phe-NH_2 (TAPP), all of which are highly selective μ agonists. (H-Tyr-D-Ala-Gly-NH-$CH_2)_2$ is an example of a dimeric peptide ligand capable of simultaneously interacting with two μ receptor binding sites and, therefore, displaying high μ selectivity for entropic reasons. The most μ-selective cyclic opioid peptide analog with μ agonist properties reported to date is the cyclic dermorphin tetrapeptide analog H-Tyr-D-Orn-Phe-Asp-NH_2. The development of specific agonists for either the μ_1 or the μ_2 receptor has not yet been achieved and will represent an important future challenge in case the existence of these receptor subtypes can be corroborated. Conformational studies have led to models of the μ receptor-bound conformations of the cyclic peptide H-Tyr-D-Orn-Phe-Asp-NH_2 [16] and of the morphiceptin analog PLO17 [17]. However, there is some ambiguity about these models because the Tyr^1 residue and the Phe^3 side chain in these peptides still enjoy considerable orientational freedom. In the case of the cyclic peptide further structural rigidification was achieved through synthesis of the side chain-constrained analog H-Hat-D-Orn-Aic-Glu-NH_2 (Hat = 6-hydroxy-2-aminotetralin-2-carboxylic acid; Aic = 2-

aminoindan-2-carboxylic acid) which retained good μ agonist potency and represents the most rigid opioid peptide analog reported to date. Conformational analysis revealed that this analog is compatible with the originally proposed model of the bioactive conformation but, of course, this compound still contains two freely rotatable bonds and further conformational constraints will have to be imposed in order to unambiguously determine the μ receptor-bound conformation.

Excellent progress has also been made in the development of highly selective *δ agonists*. Peptide analogs with highest δ selectivity are the linear enkephalin analog H-Tyr-D-Cys(StBu)-Gly-Phe-Leu-Thr(OtBu)-OH (BUBUC) and the cyclic peptides H-Tyr-D-Pen-Gly-Phe-D-Pen-OH (DPDPE), H-Tyr-D-Pen-Gly-Phe(p-Cl)-D-Pen-OH and H-Tyr-D-Cys-Phe-D-Pen-OH (JOM-13). The naturally occurring hexapeptides H-Tyr-D-Met-Phe-His-Leu-Met-Asp-NH_2 (dermenkephalin), H-Tyr-D-Ala-Phe-Asp-Val-Val-Gly-NH_2 (deltorphin I) and H-Tyr-D-Ala-Phe-Glu-Val-Val-Gly-NH_2 (deltorphin II) also are highly selective δ agonists. Evidence recently obtained from pharmacological studies *in vivo* suggested that DPDPE and the deltorphins may represent selective agonists for the δ_1 and δ_2 receptor subtypes, respectively [9]. Extensive conformational studies of DPDPE [18] and JOM-13 [19] have been performed but again have not resulted in definitive models of the bioactive conformations due to the structural flexibility of the exocyclic Tyr^1 residue and the Phe side chain.

Attempts to develop selective *κ agonists* through structural modification of dynorphin A so far have had little success. The 11-peptide analog [D-Pro^{10}]dynorphin A-(1-11) showed somewhat improved κ selectivity but further efforts clearly are necessary to obtain highly κ-selective dynorphin analogs. The most selective currently available κ agonists are non-peptides.

Receptor-selective opioid peptide antagonists

In the case of μ and δ opioid antagonists, recently obtained peptide analogs are far more selective than non-peptide antagonists. On the other hand, dynorphin-derived κ antagonists are much less selective than currently available non-peptide κ antagonists. For reviews, see refs. [14,15].

The most potent and selective *μ antagonists* known to date are not derived from opioid peptides but from somatostatin. The Sandoz somatostatin analog H-D-Phe-Cys-Phe-D-Trp-Lys-Thr-Cys-Thr-ol (SMS-201995) was first shown to be a potent opioid antagonist with high preference for μ receptors over δ receptors. Structural modification of SMS-201995 led to compounds with further improved μ antagonist properties. The most selective μ antagonist developed so far is the octapeptide H-D-Tic-Cys-Tyr-D-Trp-Orn-Thr-Pen-Thr-NH_2 (TCTOP).

Substantial progress has been made in the development of highly selective peptide *δ antagonists*. The recently discovered prototype tetrapeptide H-Tyr-Tic-Phe-Phe-OH (TIPP; Tic = tetrahydroisoquinoline-3-carboxylic acid) showed high antagonist potency against DPDPE and deltorphin I in the MVD assay, high δ receptor affinity, excellent δ receptor selectivity and, unlike the non-peptide δ antagonist naltrindole, no μ or κ antagonist properties in the GPI assay [20]. Unprecedented δ antagonist potency (K_e =

0.152 nM) was displayed by the TIPP analog H-Dmt-Tic-Phe-Phe-OH (Dmt = 2',6'-dimethyltyrosine) [21] and the potent pseudopeptide δ antagonist H-Tyr-TicΨ[CH_2-NH]Phe-Phe-OH (TIPP[Ψ]) not only was totally stable against chemical and enzymatic degradation but also turned out to be the most selective δ receptor ligand reported to date [22]. Plausible models of the δ receptor-bound conformation of the tripeptide δ antagonist H-Tyr-Tic-Phe-OH (TIP) were obtained by theoretical conformational analysis and superimposition of low energy conformers with naltrindole [23].

Mixed agonist/antagonists

Opioid compounds that act as agonists at one receptor type and as antagonists at another may be of therapeutic interest. For example, it has recently been suggested that *mixed μ agonist/δ antagonists* may have potential as analgesics with a low propensity for producing tolerance and dependence [13]. Two prototype peptide analogs with this profile have recently been developed. The first, H-Tyr-Tic-Phe-Phe-NH_2 (TIPP-NH_2), showed moderate μ agonist potency and high δ antagonist activity [20], whereas the second, the cyclic dermorphin analog H-Tyr-c[-D-Orn-2-Nal-D-Pro-Gly-], displayed somewhat higher potency but relatively weak δ antagonist activity [24]. Substitution of 2',6'-dimethyltyrosine (Dmt) for Tyr^1 in TIPP-NH_2 led to a compound, H-Dmt-Tic-Phe-Phe-NH_2 (DIPP-NH_2), which showed greatly enhanced μ agonist activity as well as extraordinary δ antagonist potency [23]. The development of further analogs with mixed μ agonist/δ antagonist properties is required to determine whether a particular μ agonist/δ antagonist potency ratio may be optimal for potent analgesia and attenuated development of tolerance and dependence. Another interesting example of a mixed agonist/antagonist is the enkephalin analog H-Tyr-D-Ala-Gly-NH-$(CH_2)_2CH(CH_3)_2$ (TRIMU-5) which displays mixed μ_2 agonist/μ_1 antagonist properties [25].

Discovery of novel opioid agonists and antagonists through the use of synthetic peptide combinatorial libraries

Recently, the construction of a synthetic combinatorial library composed of over 52 million N-acetylated hexapeptide amides and its screening based on iterative selection led to the discovery of three novel opioid antagonists, Ac-Arg-Phe-Met-Trp-Met-Thr-NH_2, Ac-Arg-Phe-Met-Trp-Met-Lys-NH_2 and Ac-Arg-Phe-Met-Trp-Met-Arg-NH_2 [26]. These so-called acetalins displayed high μ and κ_3 receptor affinities, high antagonist potency against μ agonists in the GPI assay and weak antagonist activity against κ agonists in the MVD assay. They represent totally novel peptides, since they are not listed in the Registry produced by Chemical Abstract Service obtained through The Scientific and Technical Information Network as known peptides or as part of known protein sequences. Most remarkable is the fact that the acetalins require an acetylated N-terminal amino group for optimal receptor binding, whereas acetylation of the N-terminal amino group in all other known opioid peptides drastically reduces opioid receptor affinity. Since interaction of a positively charged nitrogen moiety in opioid compounds with a negatively charged receptor moiety is known to be important for high affinity, it is possible that the positively charged Arg[1] side chain of the acetalins plays a role in opioid receptor binding similar to that of the N-terminal amino group in "classical" opioid peptides.

More recently, examination of an analogous synthetic combinatorial library composed of N-acetylated hexapeptides containing only D-amino acids resulted in the discovery of a novel opioid peptide, H-D-Arg-D-Phe-D-Trp-D-Ile-D-Asn-D-Lys-NH_2, which also showed high affinity and selectivity for μ opioid receptors but, unlike the acetalins, turned out to be a μ agonist in the GPI assay [27]. This represents a most important result, since for the first time screening of a library based on a receptor binding assay led to the discovery of an *agonist*. Obviously, the chances of obtaining an antagonist by this approach are much higher, since antagonists only have to bind to the receptor in one mode or another, whereas agonists also have to be able to produce the specific change in receptor conformation required for signal transduction. In the mouse warm-water tail flick assay i.c.v. administration of H-D-Arg-D-Phe-D-Trp-D-Ile-D-Asn-D-Lys-NH_2 produced an antinociceptive effect which, compared with morphine, was of similar potency but of longer duration. Obviously, this peptide is highly stable against enzymatic degradation due to its all D-configuration. Both the acetalins and H-D-Arg-D-Phe-D-Trp-D-Ile-D-Asn-D-Lys-NH_2 represent excellent starting points for further structural modifications aimed at developing novel potent agonists and antagonists with high selectivity for the various opioid receptor types and subtypes.

Opioid peptide analogs as potential drugs

More than a decade ago two enkephalin analogs, FK 33824 and metkephamid, underwent quite extensive clinical testing. Initially, these compounds looked promising because they produced a long-lasting analgesic effect after systemic administration. However, both of them showed a number of quite serious side effects and eventually were not further pursued clinically as candidates for drug development. These negative results led to a certain discouragement with regard to the potential of opioid peptide analogs as drug candidates. However, it is important to point out that at least some of these side effects may have been due to the demonstrated lack of receptor selectivity of these early enkephalin analogs. Many of the more recently developed opioid peptide-derived agonists and antagonists show much improved receptor selectivity and may have better potential as therapeutic agents. Of particular current interest are δ agonists and δ antagonists. δ Agonists have been shown to be potent analgesics acting at the supraspinal level and appear to produce somewhat reduced physical dependence as compared to μ agonists [28]. The ability of δ antagonists to attenuate the development of morphine tolerance and dependence [13] has been discussed above. On the basis of these interesting results it seems likely that concomitant administration of a δ opioid antagonist and morphine or other analgesics acting at the μ receptor could be advantageous for pain management. Furthermore, δ antagonists may also have therapeutic potential as immunosuppressants [29]. One of the most exciting recent achievements in the analgesic field is the development of mixed μ agonist/δ antagonists that may turn out to be analgesics with reduced propensity to produce tolerance and dependence, as discussed above and elsewhere in this volume [23].

In order to be able to exert centrally mediated effects after peripheral administration, opioid peptide analogs must be able to cross the blood-brain barrier (BBB). This hurdle can be overcome by using a number of peptide analog design strategies. Thus, highly lipophilic peptides of low molecular weight (< 600) may cross the BBB by passive transmembrane diffusion, whereas peptides that carry a high positive charge or are glycosylated may penetrate into the CNS via adsorptive endocytosis. In an effort to develop peripherally acting analgesics, it is also of interest to develop opioid peptide analogs that are unable to cross the BBB. Peripherally acting analgesics have therapeutic potential for the treatment of painful inflammatory conditions because they will not produce centrally mediated side effects such as respiratory depression. The opioid peptide analogs H-Tyr-D-Arg-Gly-Phe(pNO_2)-Pro-NH_2 (BW443C) and H-Tyr-D-Arg-Phe-Lys-NH_2 (DALDA) are examples of analgesics with predominantly but not exclusively peripheral action. Current efforts in several laboratories are aimed at developing opioid peptide analogs that are totally excluded from the CNS.

Acknowledgement

This work was supported by grants from the MRCC (MT-5655), the Canadian Heart and Stroke Foundation and the U.S. National Institute on Drug Abuse (DA-04443).

References

1. Hughes, J., Smith, T.W., Kosterlitz, H.W., Fothergill, L.A., Morgan, B.A. and Morris, R.H., Nature, 258(1975)577.
2. Höllt, V., Ann. Rev. Pharmacol. Toxicol., 26(1986)59.
3. Lord, J.A.H., Waterfield, A.A., Hughes, J. and Kosterlitz, H.W., Nature, 267(1977)495.
4. Evans, C.J., Keith, Jr., D.E., Morrison, H., Magendzo, K. and Edwards, R.H., Science, 258(1992)1952.
5. Kieffer, B.L., Befort, K., Gaveriaux-Ruff, C. and Hirth, C.G., Proc. Natl. Acad. Sci. USA, 89(1992)12048.
6. Yasuda, K., Raynor, K., Kong, H., Breder, C.D., Takeda, J., Reisine, T. and Bell, G.I., Proc. Natl. Acad. Sci. USA, 90(1993)6736.
7. Chen, Y., Mestek, A., Liu, J., Hurley, J.A. and Yu, L., Mol. Pharmacol., 44(1993)8.
8. Clark, J.A., Liu, L., Hersch, B., Edelson, M. and Pasternak, G.W., J. Pharmacol. Exp. Ther., 251(1989)461.
9. Mattia, A., Farmer, S.C., Takemori, A.E., Sultana, M., Portoghese, P.S., Mosberg, H.I., Bowen, W.D. and Porreca, F., J. Pharmacol. Exp. Ther., 260(1992)518.
10. Erspamer, V., Melchiorri, P., Falconieri-Erspamer, G., Negri, L., Corsi, R., Severini, C., Barra, D., Simmaco, M. and Kreil, G., Proc. Natl. Acad. Sci. USA, 86(1989)5188.
11. Zimmerman, D.M. and Leander, J.D., J. Med. Chem., 33(1990)895.
12. Jin, W., Chen, X. and Chi, Z., Sci. Sin., 30(1987)176.
13. Abdelhamid, E.E., Sultana, M., Portoghese, P.S. and Takemori, A.E., J. Pharmacol. Exp. Ther., 258(1991)299.
14. Hruby, V.J. and Gehrig, C.A., Med. Res. Rev., 9(1989)343.
15. Schiller, P.W. In Ellis, G.P. and West, G.B. (Eds.) Progress in Medicinal Chemistry, Vol. 28, Elsevier, Amsterdam, 1991, p. 301.
16. Wilkes, B.C. and Schiller, P.W., Biopolymers, 29(1990)89.
17. Yamazaki, T., Ro, S., Goodman, M., Chung, N.N. and Schiller, P.W., J. Med. Chem., 36(1993)708.

18. Pettit, B.M., Matsunaga, T., Al-Obeidi, F., Gehrig, C., Hruby, V.J. and Karplus, M., Biophys. J., 60(1991)1540.
19. Lomize, A.L., Flippen-Anderson, J.L., George, C. and Mosberg, H.I., J. Am. Chem. Soc., 116(1994)429.
20. Schiller, P.W., Nguyen, T.M.-D., Weltrowska, G., Wilkes, B.C., Marsden, B.J., Lemieux, C. and Chung, N.N., Proc. Natl. Acad. Sci. USA, 89(1992)11871.
21. Schiller, P.W., Weltrowska, G., Nguyen, T.M.-D., Chung, N.N., Lemieux, C. and Wilkes, B.C., Regulatory Peptides (Suppl. 1), (1994)S63.
22. Schiller, P.W., Weltrowska, G., Nguyen, T.M.-D, Wilkes, B.C., Chung, N.N. and Lemieux, C., J. Med. Chem., 36(1993)3182.
23. Schiller, P.W., Schmidt, R., Wilkes, B.C., Weltrowska, G., Nguyen, T.M.-D., Chung, N.N. and Lemieux, C., In Lu, G.-S., Tam, J.P. and Du, Y.-C., (Eds.) Peptides: Biology and Chemistry (Proceedings of the 1994 Chinese Peptide Symposium), ESCOM, Leiden, 1995, pp. 140–143.
24. Schmidt, R., Vogel, D., Mrestani-Klaus, C., Brandt, W., Neubert, K., Chung, N.N., Lemieux, C. and Schiller, P.W., J. Med. Chem., 37(1994)1136.
25. Tive, L.A., Pick, C.G., Paul, D., Roques, B.P., Gacel, G.A. and Pasternak, G.W., Eur. J. Pharmacol., 216(1992)249.
26. Dooley, C.T., Chung, N.N., Schiller, P.W. and Houghten, R.A., Proc. Natl. Acad. Sci. USA, 90(1993)10811.
27. Dooley, C.T., Chung, N.N., Wilkes, B.C., Schiller, P.W., Bidlack, J.M., Pasternak, G.W. and Houghten, R.A., Science, in press.
28. Cowan, A., Zhu, X.Z., Mosberg, H.I., Omnaas, J.R. and Porreca, F., J. Pharmacol. Exp. Ther., 246(1988)950.
29. Arakawa, K., Akami, T., Okamoto, M., Akioka, K., Nakai, I., Oka, T. and Nagase, H., Transplant. Proc., 25(1993)738.

Development of potent opioid δ antagonists and mixed μ agonist / δ antagonists

P.W. Schiller, R. Schmidt, B.C. Wilkes, G. Weltrowska, T.M.-D. Nguyen, N.N. Chung and C. Lemieux
Laboratory of Chemical Biology and Peptide Research, Clinical Research Institute of Montreal, 110 Pine Avenue West, Montreal (Que.) Canada H2W 1R7

Introduction

Recently, we reported the discovery of a new class of opioid peptide-derived δ antagonists that contain a tetrahydroisoquinoline-3-carboxylic acid (Tic) residue in the 2-position of the peptide sequence [1]. The two prototype antagonists were the tetrapeptide H-Tyr-Tic-Phe-Phe-OH (TIPP) and the tripeptide H-Tyr-Tic-Phe-OH (TIP). TIPP showed high antagonist potency against various δ agonists in the mouse vas deferens (MVD) assay (K_e = 3-5 nM), high δ opioid receptor affinity (K_i^δ = 1.22 nM) and excellent δ selectivity (K_i^μ/K_i^δ = 1410). In contrast to the nonpeptide δ antagonist naltrindole [2], TIPP displayed no μ or κ antagonist properties in the guinea pig ileum (GPI) assay at concentrations as high as 10 μM. The tripeptide TIP also showed good δ antagonist potency and δ selectivity but was somewhat less potent and less selective than TIPP [1]. In comparison with TIPP, its pseudopeptide analog H-Tyr-TicΨ[CH_2-NH]Phe-Phe-OH (TIPP[Ψ]) turned out to be an equally potent but even more selective δ antagonist and was shown to be chemically stable and totally resistant to enzymatic degradation [3]. In the present paper, we describe TIPP analogs with further improved δ antagonist potency and δ selectivity. Futhermore, two models of the bioactive conformation of TIP based on structural comparison with naltrindole are presented.

It has been demonstrated that pretreatment of mice with naltrindole prevented the development of morphine tolerance and dependence [4]. This interesting observation suggested that the development of a single compound with mixed μ agonist/δ antagonist properties may have therapeutic potential. The first known example of a mixed μ agonist/δ antagonist is the recently reported opioid tetrapeptide analog H-Tyr-Tic-Phe-Phe-NH_2 (TIPP-NH_2) which was found to be a moderately potent μ agonist in the GPI assay and a highly potent δ antagonist in the MVD assay [1]. In comparison with TIPP-NH_2, the cyclic β-casomorphin analog H-Tyr-c[-D-Orn-2-Nal-D-Pro-Gly-] turned out to be a 4 times more potent μ agonist and a significantly less potent δ antagonist [5]. Here we report on a TIPP analog which is a mixed μ agonist/δ antagonist with both greatly enhanced μ agonist potency and still very high δ antagonist activity.

Results and Discussion

Substitution of 2',6'-dimethyltyrosine (Dmt) for Tyr^1 in TIPP[Ψ] resulted in an analog, H-Dmt-TicΨ[CH_2-NH]Phe-Phe-OH (DIPP[Ψ]), which displayed the same high δ antagonist potency as H-Dmt-Tic-Phe-Phe-OH (DIPP; K_e = 0.152 nM) [6] but showed somewhat lower δ selectivity (Tables 1 and 2). On the other hand, DIPP[Ψ] can be expected to display higher stability against chemical and enzymatic degradation than DIPP due to the presence of the reduced peptide bond between Tic^2 and Phe^3. Reduction of the peptide bond in the 2-3 position and substitution of homophenylalanine (Hfe) for Phe^3 in Tyr(NMe)-Tic-Phe-Phe-OH [6] led to a stable pseudopeptide analog showing about the same high δ antagonist potency as the parent peptide. Methylation of the secondary amino group contained in the reduced peptide bond of TIPP[Ψ] only slightly decreased δ antagonist potency and δ receptor affinity. Since the drop in μ receptor affinity was somewhat more pronounced, the resulting compound, H-Tyr-TicΨ [CH_2-NCH_3]Phe-Phe-OH, displayed extraordinary δ receptor selectivity (K_i^μ/K_i^δ = 15900) and represents the most selective δ opioid receptor ligand reported to date.

Table 1 *GPI and MVD assay of opioid peptide analogs*

Compound	GPI	MVD
	IC50, nM[a]	K_e, nM[a,b]
H-Tyr-Tic-Phe-Phe-OH	>10000	2.96 ± 0.02
H-Tyr-TicΨ[CH_2-NH]Phe-Phe-OH	>10000	2.58 ± 0.29
H-Dmt-TicΨ[CH_2-NH]Phe-Phe-OH	>10000	0.157 ± 0.028
Tyr(NMe)-TicΨ[CH_2-NH]Hfe-Phe-OH	>10000	0.418 ± 0.098
H-Tyr-TicΨ[CH_2-NCH_3]Phe-Phe-OH	>10000	2.89 ± 0.31
H-Tyr-Tic-NH_2	>10000	5750 ± 450
Naltrindole	antagonist	0.632 ± 0.161
H-Tyr-Tic-Phe-Phe-NH_2	1700 ± 220	18.0 ± 2.2
H-Tyr-c[-D-Orn-2-Nal-D-Pro-Gly-]	384 ± 52	202 ± 24
H-Dmt-Tic-Phe-Phe-NH_2	18.2 ± 1.8	0.260 ± 0.064

[a]Mean of 3 determinations SEM. [b]K_e value determined against [D-Ala^2]deltorphin I.

Table 2 *Binding affinities of opioid peptide analogs at and opioid receptors*

Compound	[^{3}H]DAMGO	[^{3}H]DSLET	
	K_i^μ, nM[a]	K_i^δ, nM[a]	K_i^μ/K_i^δ
H-Tyr-Tic-Phe-Phe-OH	1720 ± 50	1.22 ± 0.07	1410
H-Tyr-TicΨ[CH_2-NH]Phe-Phe-OH	3230 ± 439	0.308 ± 0.060	10500
H-Dmt-TicΨ[CH_2-NH]Phe-Phe-OH	95.5 ± 11.0	1.70 ± 0.40	56.2
H-Tyr-TicΨ[CH_2-NCH_3]Phe-Phe-OH	13400 ± 700	0.842 ± 0.116	15900
Naltrindole	3.86 ± 0.74	0.182 ± 0.024	21.2
H-Tyr-Tic-Phe-Phe-NH2	78.8 ± 7.1	3.00 ± 0.15	26.3
H-Tyr-c[-D-Orn-2-Nal-D-Pro-Gly-]	5.89 ± 0.11	17.2 ± 4.9	0.342
H-Dmt-Tic-Phe-Phe-NH2	1.19 ± 0.11	0.118 ± 0.016	10.1

[a]Mean of 3 determinations ± SEM.

In an effort to examine the importance of the Tic residue for the δ antagonist behavior of TIPP peptides, the dipeptide H-Tyr-Tic-NH_2 was synthesized and tested. Careful analysis revealed that diketopiperazine formation did not occur under the conditions of the performed MVD assay. This compound indeed displayed very weak δ antagonist potency and the observed K_e value (5.75 ± 0.45 M) was considerably higher than that previously reported in the literature (K_e = 1.00 μM) [7]. Nonetheless, this result confirms the earlier observation that the Tic^2 aromatic ring in TIPP peptides plays a crucial role for δ antagonist behavior [8].

A molecular mechanics study (grid search and energy minimization) of the tripeptide δ antagonist TIP resulted in a number of low energy conformers with energies within about 2 kcal/mol of that of the lowest energy structure. Attempts to demonstrate spatial overlap between the pharmacophoric moieties of TIP low energy conformers and the nonpeptide δ antagonist naltrindole were made by superimposing either the Tyr^1 and Tic^2 aromatic rings and the N-terminal amino group of the peptide, or the Tyr^1 and Phe^3 aromatic rings and the N-terminal amino group of the peptide, with the corresponding aromatic rings and nitrogen atom in the alkaloid structure. In the first case best spatial overlap was achieved with an all-*trans* conformer which was 1.2 kcal/mol higher in energy than the lowest energy structure (Fig. 1, left panel). In this conformer the intramolecular distance between the Tyr^1 and Tic^2 aromatic rings (6.5 Å) was close to that observed between the aromatic moieties in naltrindole (6.4 Å) and the spatial overlap of the three proposed functional groups with those in naltrindole was excellent, the rms deviation being 0.6 Å. In the second case, best results were obtained with the lowest energy conformer containing a *cis* peptide bond between the Tyr^1 and Tic^2 residues, which was 2.1 kcal/mol higher than the lowest energy conformer (Fig. 1, right panel). In this conformation the intramolecular distance between the Tyr^1 and Phe^3 aromatic rings was 6.3 Å and the overall spatial overlap between the three proposed functional groups was also excellent (rms deviation of 0.6 Å). Either one of these two TIP conformers may represent the δ receptor-bound conformation, albeit the all-*trans* conformer appears the more likely candidate structure in view of the demonstrated importance of the Tic^2 aromatic ring for δ antagonism (see above).

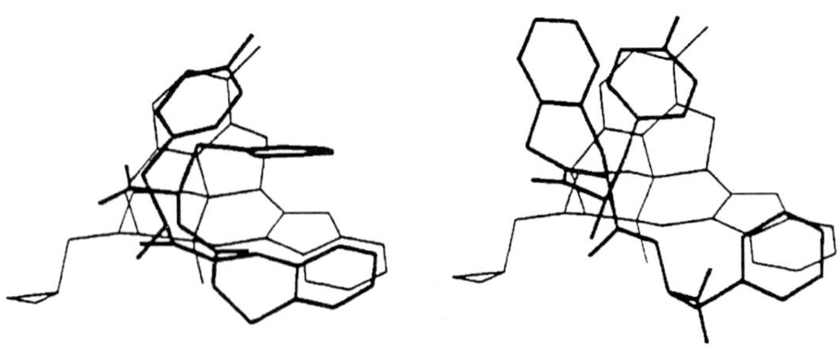

Fig. 1. Superimposition of naltrindole (drawn in light lines) with an all-trans low energy conformer of TIP (left panel) and with a low energy conformer of TIP containing a Tyr^1-Tic^2 cis peptide bond (right panel). Peptide structures are drawn in heavy lines.

In an attempt to strenghten the μ agonist component of TIPP-NH_2, the N-terminal tyrosine residue was substituted with 2',6'-dimethyltyrosine (Dmt), since the analogous substitution in DPDPE has previously been shown to drastically enhance μ receptor affinity and, to a lesser extent, δ receptor affinity [9]. Indeed the resulting compound, H-Dmt-Tic-Phe-Phe-NH_2 (DIPP-NH_2), displayed about 100 times higher μ agonist potency than TIPP-NH_2 in the GPI assay as well as very high δ antagonist potency (Table 1). These results are in agreement with the receptor binding assay data which show a 70-fold increase in μ receptor affinity for DIPP-NH_2 (K_i^{μ}= 1.19 nM) and a significantly reduced preference for δ receptors over μ receptors (Table 2).

Acknowledgement

This work was supported by grants from the MRCC (MT-10131) and NIDA (DA-04443).

References

1. Schiller, P.W., Nguyen, T.M.-D., Weltrowska, G., Wilkes, B.C., Marsden, B.J., Lemieux, C. and Chung, N.N., Proc. Natl. Acad. Sci. USA, 89(1992)11871.
2. Portoghese, P.S., Nagase, H., MaloneyHuss, K.E., Lin, C.-E. and Takemori, A.E., J. Med. Chem., 34(1991)1715.
3. Schiller, P.W., Weltrowska, G., Nguyen, T.M.-D., Wilkes, B.C., Chung, N.N. and Lemieux, C., J. Med. Chem., 36(1993)3182.
4. Abdelhamid, E.E., Sultana, M., Portoghese, P.S. and Takemori, A.E., J. Pharmacol. Exp. Ther., 258(1991)299.
5. Schmidt, R., Vogel, D., Mrestani-Klaus, C., Brandt, W., Neubert, K., Chung, N.N., Lemieux, C. and Schiller, P.W., J. Med. Chem., 37(1994)1136.
6. Schiller, P.W., Nguyen, T.M.-D., Weltrowska, G., Wilkes, B.C., Marsden, B.J., Schmidt, R., Lemieux, C. and Chung, N.N., In Hodges, R.S. and Smith, J.A. (Eds.) Peptides: Chemistry, Structure and Biology (Proceedings of the 13th American Peptide Symposium), ESCOM, Leiden, 1994, pp. 483–486.
7. Temussi, P.A., Salvadori, S., Amodeo, P., Bianchi, C., Guerrini, R., Tomatis, R., Lazarus, L.H., Picone, D. and Tancredi, T., Biochem. Biophys. Res. Commun., 198(1994)933.
8. Schiller, P.W., Nguyen, T.M.-D., Berezowska, I., Weltrowska, G., Schmidt, R., Marsden, B.J., Wilkes, B.C., Lemieux, C. and Chung, N.N., In Yanaihara, N. (Ed.) Peptide Chemistry 1992 (Proceedings of the 2nd Japan Symposium on Peptide Chemistry), ESCOM, Leiden, 1993, pp. 337–340.
9. Hansen, Jr., D.W., Stapelfeld, A., Savage, M.A., Reichman, M., Hammond, D.L., Haaseth, R.C. and Mosberg, H.I., J. Med. Chem., 35(1992)684.

Topographical design of message domain pharmacophores of δ opioid agonists provide insights into the stereochemical requirements for recognizing δ opioid receptor subtypes

Xin-Hua Qian[a], Katalin E. Kövér[a], Mark D. Shenderovich[a], Edward J. Bilsky[b], Robert N. Bernstein[b], Bih-Show Lou[a], Robert Horváth[b], Teresa Zalewska[b], Peg Davis[b], Frank Porreca[b], Henry I. Yamamura[b] and Victor J. Hruby[a]

[a]Department of Chemistry and [b]Department of Pharmacology, University of Arizona, Tucson, AZ 85721, U.S.A.

Introduction

In order to obtain highly potent, selective and efficacious peptide hormone and neurotransmitter analogues, topographical design of peptide ligands is an especially important and useful tool that has proven extremely fruitful over the past decade in our and other laboratories. The approach we have taken has been to apply global and local constraints to native or well defined peptide ligands. In the past few years, side chain constraints has been of increasing importance in our laboratory, not only because they can provide potent and selective peptide analogues [1], but also because they can provide compounds with special biological properties such as receptor subtype recognition [2] and prolongation of activity [3]. To achieve the goal of applying local constraints on the side chain topology of various aromatic amino acid residues, the main strategy we have used has been the design and synthesis of a series of *chi* (ξ) space restricted aromatic amino acids. We have published extensively on the asymmetric synthesis of these unusual amino acids [4]. Some of the designer amino acids we have prepared and incorporated into peptide ligands are shown in Figure 1. The application of these designer amino acids to δ opioid agonists will be discussed here. Since there are several unique clinical advantages for δ opioid agonists, such as analgesia without respiratory depression, constipation or other adverse gastrointestinal effects, and with minimal development of physical dependence, etc., they remain important in developing novel non-addictive pharmaceuticals as pain relievers. During the past decade, we have developed a series of highly selective and potent *δ* opioid agonists using global constraint, among which cyclic [D-Pen^2, D-Pen^5]enkephalin (DPDPE) was representative [5]. Recently, the x-ray crystal structure of DPDPE has been reported [6] providing, along with the NMR structure [7], a well defined backbone conformation of this compound and its derivatives. In this paper we incorporate ξ-space-constrained aromatic amino acids to probe the bioactive topology of the side chain moieties of DPDPE and Deltorphin I that are selective agonists for the δ opioid receptors.

Results and Discussion

The message domain of δ opioid agonists contain two important aromatic amino acid residues, Tyr and Phe. The aromatic moieties in these two residues also are important pharmacophores for the function of δ opioid ligands. Any undesirable topology of these two amino acid side chains should cause significant loss of potency and/or selectivity of the agonists. The first series of constrained amino acids we incorporated into DPDPE were the four isomers of β-MePhe and *p*-NO_2-β-MePhe (Fig. 1 (c)). We systematically examined the ξ_1 space of the side chain in Phe^4 residue of DPDPE on binding and bioassays potencies [1(b)]. It was found that the best analogue contained a (2*S*,3*S*)-β-MePhe isomer that has a preferred *guache* (-) conformation about the ξ_1 torsional angle based on NMR studies. These studies provided possible bioactive conformations of the Phe^4 residue in DPDPE and other bioactive peptides [7-9]. The second series of constrained amino acids we incorporated into DPDPE were racemic 6-hydroxy-2-aminotetralin-2-carboxylic acid (Hat) (Fig. 1 (a)) which constrained both the ξ_1 and ξ_2 torsional angles of the Tyr^1 residue [1(c)]. Although the $[Hat^1]$DPDPE analogue was potent and selective [1(c)], utilization of the racemic amino acid and the unknown side chain conformations of Hat made it difficult to define the ξ space requirements of Tyr^1 in DPDPE. Studies using the four isomers of β-MeTyr to systematically constrain the ξ_1 space of the side chain in Tyr^1 did not provide a potent and selective analogue, but did indicate that (2*S*,3*R*)-β-$MeTyr^1$ is the preferred isomer [1(c)]. The advantage of using β-branched aromatic amino acids is that each isomer will provide a different predominant side chain conformation because of the constraint in ξ_1 space. Because it has been difficult to modify the Tyr^1 residue in δ opioid agonists with maintenance of potency and selectivity of the parent agonists [10], we decided to design a new class of unusual tyrosine derivatives with systematic constraints in the ξ_1 space and ξ_2 spaces at the same time. The 2',6'-dimethyl-β-methyltyrosines (TMT) were our choice, since the β-methyl group serves to restrict ξ_1 rotation, while the 2',6'-dimethyl groups and β-methyl groups restrict rotation about the ξ_2 torsional angle (Fig. 1 (e)). All four

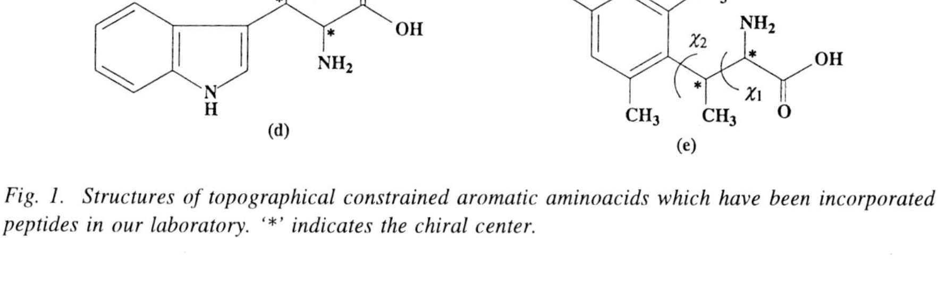

Fig. 1. Structures of topographical constrained aromatic aminoacids which have been incorporated into peptides in our laboratory. '' indicates the chiral center.*

isomers were synthesized by asymmetric synthesis methods [11] with overall yields of 30-60% and 80-98% ee/de. Modifications [4(e), 11, 12, 13] which may lead to more efficient synthetic pathways are under development in our group.

All four isomers of TMT were incorporated into two highly selective and potent δ opioid agonists: DPDPE and Deltorphin I. The initial goal was to study how the same modification of Tyr[1] would affect the activity of these two ligands, which have significantly different backbone conformations (one cyclic and one linear). After we incorporated (2*S*, 3*S*)-TMT[1] into DPDPE and DELT I, we found there was a significant decrease in the potency and selectivity of DPDPE analogue, but hardly any change of potency/selectivity of the DELT I analogue (Table 1). The NMR studies suggested that *gauche* (-) was the dominant ξ_1 rotamer of the (2*S*, 3*S*)-TMT[1] residue in both cases. Therefore if both analogues bind to the same δ opioid receptor, as was suggested by similar binding and bioassay data for DPDPE and DELT I (Table 1), we would not expect to see dramatic differences in bioactivity for these two analogues with exactly the same modification at the Tyr[1] residue (Table 1). This lends more evidence for the existence of δ opioid receptor subtypes, which were previously suggested from pharmacological studies [14].

Table 1 *Binding affinities of [TMT[1]]DPDPE and [TMT[1]]Deltorphin I analogues*

Peptide	IC_{50} (nM)		Selectivity
	[^{3}H][*p*-Cl-Phe^4]DPDPE	[^{3}H]CTOP	(*m* /δ)
DPDPE	1.6	610	380
[(2*S*, 3*S*)-TMT[1]]DPDPE	210	720	3
[(2*S*, 3*R*)-TMT[1]]DPDPE	5.0	4, 300	860
Deltorphin I (DELT I)	0.6	2, 100	3, 500
[(2*S*, 3*S*)-TMT[1]]DELT I	3.0	17, 100	5, 700
[(2*S*, 3*R*)-TMT[1]]DELT I	1.0	610	610

The binding affinities of both DPDPE and DELT I analogues containing (2*R*, 3*R*)-TMT[1] and (2*R*, 3*S*)-TMT[1] are in the process of determination and need further varification. The results of the biological assays of these δ opioid agonists follow the same profile.

It is possible that these two analogues may bind selectively to two different δ opioid receptor subtypes. Functional *in vivo* antinociception assays strongly support this suggestion. These assays use animals pretreated with antagonists selective for each δ opioid receptor subtype, followed by i.c.v. administration of the peptide analogue to identify which antagonist may significantly antagonize the analgesic effect of the analogue. It was found that DALCE, an irreversible δ_1 opioid agonist, antagonized the analgesic effect of [(2S,3S)-TMT^{1}DPDPE only, [Cys^4]Deltorphin I (a δ_2 opioid antagonist) could only antagonized the analgesic effect of [(2*S*,3*S*)-TMT^1]DELT I. In addition, since DPDPE was characterized as a δ_1 opioid agonist, and the backbone conformation in the x-ray structure of DPDPE is very similar to the backbone conformation of [(2*S*,3*S*)-TMT^1]DPDPE as indicated by a combination of NMR studies and theoretical calculations/computer modeling [2], these results suggested that the *gauche* (-) conformation may not be a proper topology of Tyr^1 in DPDPE for the recognition of δ_1 opioid receptor subtype. However, it may be a good topology for Tyr^{1}in DELT I for recognition of the δ_2 opioid receptor subtype [2]. From our prilimary results, the most potent and selective DPDPE analogue is [(2*S*,3*R*)-TMT^1]DPDPE. [(2*S*,3*R*)-TMT^1]DPDPE also was shown by the *in vivo* functional assays to be a δ_1 agonist. NMR studies indicated that the most populated ξ rotamer of (2*S*, 3*S*)-TMT^1 in DPDPE and DELT I analogues was *gauche* (-), and was *trans* for (2*S*, 3*R*)-TMT^1 in both analogues. In the series of DELT I analogues, the most potent and selective analogue is [(2*S*,3*S*)-TMT^1]DELT I, which was demonstrated to be a pure δ_2 agonist *in vivo*. Thus, the topologies in ξ_1 space for the Tyr^1 of a δ opioid ligand to selectively recognize δ opioid receptor subtypes have been defined.

We did not obtain a potent and selective ligand by only constraining the ξ_1 angle of Tyr^1 [1(c)], and it was suggested by the [Hat^1]DPDPE analogues that constraint of the ξ_2 torsional angle of Hat^1 may have a positive effect on maintaining the bioactivities of the DPDPE anlogues. In the TMT case, the ξ_2 torsional angle is about 100° from the x-ray crystallographic structures of TMT precursors. This torsinal angle also is highly constrained because of the high energy barrier (14 - 20 kcal/mole) for rotation around it [15]. It is noted in all DPDPE and DELT I analogues that the Phe$^{4(3)}$ residues always possess a most populated *gauche* (-) conformation about the ξ_1 angle, which should be the bioactive ξ_1 conformation, as suggested by previous studies [7,8]. All these results made it clear that the *gauche* (-) conformation (ξ_1 angle) of the Tyr^1 residue with a x_2 of about 100°and the *gauche* (-) conformation (ξ_1 angle) of Phe^4 residue should be the most suitable side chain topologies of the message domain of DPDPE analogues to recognize the opioid δ_1 receptor subtype. On the other hand, the *trans* conformation for the ξ_1 angle of the Tyr^1 residue with a ξ_2 of about 100°, and a *gauche* (-) conformation (ξ_1 angle) of Phe^3 residue should be the most suitable side chain topology for the message domain of DELT I analogues to recognize the opioid δ_2 receptor subtype.

In conclusion, incorporation of TMT isomers into the message domain of DPDPE and Deltorphin I in combination with conformational analysis of these analogues have for the first time provided the side chain topologies of the pharmacophores in the message domain of δ opioid agonists for specific recognition of opioid δ receptor subtypes. Conformational studies associated with modeling of receptor-ligand interactions should provide further information about the difference between these two δ opioid receptor subtypes.

Acknowledgement

This work was supported by U.S. Public Health Service Grant NS 19972 and by NIDA

Grants DA 06284 and DA 04248. Dr. Katalin E. Kövér thanks the Hungarian Academy of Science for partial support through Grant OTK 1144.

References

1. (a) Kazmierski, W.M., Yamamura, H.I. and Hruby, V.J., J. Am. Chem. Soc., 113(1991)2275. (b) Hruby, V.J., Toth, G., Gehrig, C.A., Kao, L. -F., Knapp, R., Liu, G.K., Yamamura, H.I., Kramer, T.H., Davis, P. and Burks, T.F., J. Med. Chem., 34(1991)1823. (c) Toth, G., Russell, K.C., Landis, G., Kramer, T.H., Fang, L., Knapp, R., Davis, P., Burks, T.F., Yamamura, H.I. and Hruby, V.J., J. Med. Chem., 35(1992)2384.
2. Qian, X., Kövér, K.E., Shenderovich, M.D., Lou, S. -L., Misicka, A., Zalewska, T., Horváth, R., Davis, P., Bilsky, E.J., Porreca, F., Yamamura, H.I. and Hruby, V.G., J. Med. Chem., 37(1994)1746. Haskell-Leuvano, C., Hruby, V.J., manuscript in preparation.
3. (a) Dharanipragada, R., Van Hulle, K., Bannister, A., Bear, S., Kennedy, L. and Hruby, V.J.,
4. Tetrahedron, 48(1992)4733. (b) Boteju, L.W., Wegner, K. and Hruby, V.J., Tetrahedron Lett., 33(1992)7491. (c) Nicolas, E., Russell, K.C., Knollenberg, J. and Hruby, V.J., J. Org. Chem., 58(1993)7565. (d) Li, G., Jarosinski, M.A. and Hruby, V.J., Tetrahedron Lett., 34(1993)2565. (e) Boteju, L.W., Wegner, K., Qian, X. and Hruby, V.G., Tetrahedron, 50(1994)2391.
5. Mosberg, H.I., Hurst, R., Hruby, V.J., Gee, K., Yamamura, H.I., Galligan, J.J. and Burks, T.F., Proc. Natl. Acad. Sci. U.S.A., 80(1983)5871.
6. Flippen-Anderson, J., Hruby, V.J., Collins, N., George, C., Cudney, B., J. Am. Chem. Soc. 1994, in press.
7. Hruby, V.J., Kao, L.F., Pettitt, B.M. and Karplus, M., J. Am. Chem. Soc., 110(1988)3351.
8. Nikiforovich, G.V., Hruby, V.J., Prakash, O. and Gehrig, C.A., Biopolymers, 31(1991)941.
9. Kövér, K.E., Jiao, D., Fang, S. and Hruby, V.J., J. Org. Chem., 59(1994)991.
10. (a) Summers, M.C. and Hayes, R.J., J. Biol. Chem., 256(1981)4951. (b) Kawai, M., Fukuta, N., Ito, N., Kagami, T., Butsugan, Y., Maruyama, M. and Kudo, Y., Int. J. Pept. Protein Res., 35(1990)452. (c) Miller, R.J., Chang, K. -J., Leighton, J. and Cuatrecasas, P., Life Sci., 22(1978)379. (d) Miller, R.J., Schultz, G.S. and Levy, R.S., Int. J. Pept. Protein Res., 24(1984)112.
11. Qian, X., Russell, K.C., Boteju, L.W. and Hruby, V.J., manuscript submitted for publication.
12. Li, G., Patel, D. and Hruby, V.J., Tetrahedron Lett., 35(1994)2301.
13. Li, G., Jarosinski, M.A. and Hruby, V.J., Tetrahedron Lett., 34(1993)2561.
14. For a recent review see in Opioids by Porreca, F., Bilsky, E.J., Raffa, R.B. and Lai, J., 1994, Academic Press, in press.
15. Jiao, D., Russell, K.C. and Hruby, V.J., Tetrahedron, 49(1993)3511.

Design and preparation of cyclopeptides mimicking the active sites of serine proteases

Zong-Jin Han, Tian-Dong Gao, Zhen-Kai Ding and Qi-Kai Zhang
Laboratory of Bioorganic Chemistry, Beijing Institute of Pharmacology and Toxicology, Beijing 100850, China

Introduction

The active site of serine proteases, such as trypsin or acetylcholinesterase, is known to be composed of a "catalytic triad "[1]. This triad is made up of His, Ser and Asp residues which are juxtaposed in the enzyme molecule. The high efficiency of serine proteases is mainly due to chemical reaction at this part [2]. Synthetic model compounds that mimic the active site of serine proteases are usful for the investigation of enzyme reactions. The design of the model compounds require a binding-catalytic site that is structurally stable and synthetically accessible. The cyclic nature of cyclopeptides reduces it's conformational freedom to those forms where cooperativity of functional side chains, which is necessary for catalysis, is favored. The small size of such structures make them synthetically accessible [3]. Six head-to-tail cyclopeptides and two linear peptides we designed and synthesized in this paper wear shown in Fig. 1. The sequences of these target peptides are based on the special arrangements of the crucial amino acids, Ser-His-Asp, that have been derived from CPK model designs and molecular dynamics calculations. The catalytic actions for the hydrolysis of p-nitrophenyl acetate are reported in this paper.

CP1	Gly-Asp-His-Gly-Ser-Asp
CP2	Gly-D-Asp-His-Gly-Ser-Asp
CP3	Asp-Asp-His-Gly-Ser-Asp
CP4	Asp-His-Gly-Ser-Asp
CP5	Asp-His-Ser-Ser-Ser-Ser-Ser-Ser-Asp
CP6	Asp-Ser-Pro-Ser-Ser-Gly-Pro-Ser-Asp
LP1	H_2N-Gly-Asp-His-Gly-Ser-Asp-COOH
LP2	H_2N-Gly-D-Asp-His-Gly-Ser-Asp-COOH

Fig. 1. The sequences of cyclopeptides and linear peptides.

Results and Discussion

The molecular dynamics calculations were performed using DISCOVER programs on Biosym. 4D-20 computer station, kindly provided by Prof. Run-sheng Chen. The target peptides as shown in Fig. 1 were chosen from many calculated peptides. Their low energy conformations show at lest a serine hydroxyl, a histidine imidazole and an

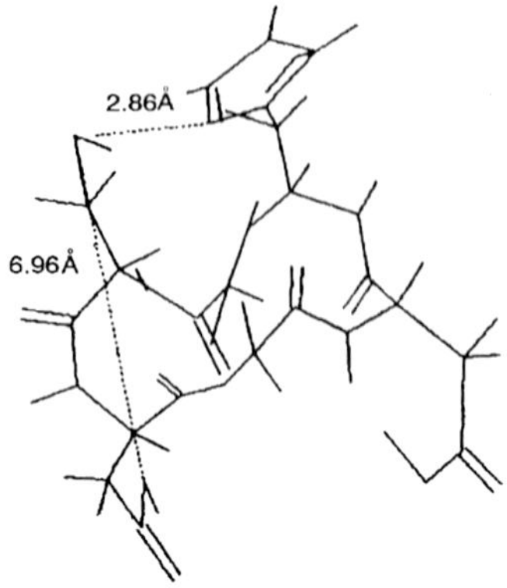

Fig. 2. The low energy conformation of CP1.

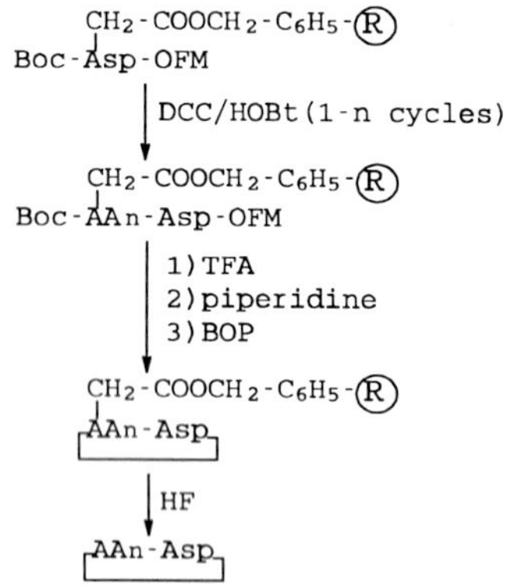

Fig. 3. Schematic representation for the synthesis of cyclopeptides.

aspartate carboxyl on the same side of the cyclic chain, and the distance of His N^{ε} to Ser O^{γ} is about within hydrogen bond length. The low energy conformation of CP1, as an example, is shown in Fig. 2.

The linear peptides were synthesized by classical N^{α}-Boc-amino acids solid phase procedure using DCC/HOBt as coupling reagent. The cyclopeptides were synthesized by side chain attachment solid phase strategy as outlined in Fig. 3. The N^{α}-Boc-Asp with the OFM as α-carboxyl protective group was linked with Merrifield resin using Cs-salt method in DMA in the presence of 1 equiv. NaI and 1 equiv. 18-Crown-6 at 40°C for 20 h (Table 1). The other N^{α}-Boc-amino acids were coupled using DCC/HOBt method. The solid phase head-to-tail cyclization were carried out using benzotriazol-1-yl-oxy-tris-(dimethylamino)-phosphonium hexafluorophosphate (BOP reagent). All the cyclization reaction except CP4 are rapid (>99% complete in 12 h, Table 2) and resulted in purer cyclic products and high yield (Table 2). Catalytic activities of the cyclopeptides and linear peptides for the hydrolysis of p-nitrophenyl acetate and acetylchline (ACh) were investigated. The results are shown in Table 3.

Table 1 *The yields(%) of Boc-Asp(resin)-OFM in different conditions*

No.	DMF	DMA	NaI	18-crown-6	yield(%)
1	+				<10
2	+		+		25
3	+		+	+	56.9
4		+			22.1
5		+	+		37.5
6		+	+	+	89.8

By picric acid determination of CP1-CP6.

Table 2 *Cyclizations and yields of CP1-CP6*

Peptides	time (h)	% Cyclization[a]	yields(%)[b]
CP1	6	99.5	41
CP2	6	99.9	24
CP3	6	99.2	13
CP4	12	22.4	--
CP5	12	99.7	7.8
CP6	12	99.9	13.5

[a] By quantitative ninhydrin determination of peptide-resin.
[b] After HPLC purification.

Table 3 *The catalytic hydrolysis activities of CP1-CP6 and LP1, LP2*

substrate	catalytic activities (%)						
	CP1	CP2	LP1	LP2	CP3	CP5	CP6
p-nitrophenyl[a] acetate	14.1	2.4	3.4	2.8	2.1	1.0	0.1
ACH[b]	*	-4.4	-5.2	-4.6	*	*	-5.7

[a] p-nitrophenyl acetate 0.1 mmol, peptides 0.1 mmol, pH 7.73, 30 min.
[b] ACH 0.875 mmol, peptides 0.625 mmol, pH7.21, 30min.
* CP1, CP3, CP5 may hydrolyze ACH but need to be confirmed.

Acknowledgements

The authors thank Prof. Run-sheng Chen in assistance of molecular dynamics calculations.

References

1. Quinin, D.M., Chem. Rev., 87(1987)955.
2. Nish, N., Int. J. Biol. Nacromol., 4(1982)281.
3. Seltzman, M.M., AD-A204 765.

The effect of synthesized oligopeptides on corpus luteum function and their mechanism of action

Ni Wang[a], Yu-Lian Zhao[a], Xiao-Ning Wang[a], Chi-Ping Cheng[a], De-Xin Wang[b] and Gui-Shen Lu[b]

[a]Harbin Medical Univercity, Harbin 150086, China

[b]Institute of Materia Medica, Chinese Academy of Medical Sciences, Beijing 100050, China

Introduction

In screening 17 tyrosine derivatives we discovered that one of them, a dipeptide Glycyl-tyrosine, prossessed a much higher *in vitro* anti-progesterone production effect on corpus luteum (CL) cells. This led us to synthesize a series of oligopeptides consisting of tyrosine and some other common amino acids. Their inhibitory effect on progesterone production by the CL cell of the rat was compared and their mechanism of action was studied. It was found that some of them decreased and others stimulated basal/hCG-induced progesterone production. The inhibitory oligopeptides were characterized by carrying a positive charge in media of pH 7.2-7.5 and have intermolecular linkages. A preliminary survey of the effect of oligopeptides on signal systems showed: 1) GY-NH_2 and GSK decreased forskolin induced progesterone production, indicating that they probably affected the PKA (protein kinase) system. 2) GY inhibited the PLC (phospholipase C) system. 3) GSK decreased and GYK increased TPK activity in hCG-treated CL cells. Although both had inhibitory effects on progesterone production, GSK also stimulated PKC (protein kinase C) and supressed PKA activity in CL cells. The anti-progesterone effect of the oligopeptides so far synthesized may influence the PKA, PKC, or TPK (tyrosine protein kinase) systems. The manner of action at the molecular level awaits further investigations.

Results and Discussion

1. The in vitro effects of oligopeptides on progesterone production by rat CL cells

The action of thirty synthetic oligopeptides, including Tyr, Ser and Thr residues, on progesterone production were studied. The inhibitory effects of the oligopeptides on basal/hCG-induced progesterone production by rat CL cells in three concentrations (0.02mM, 0.2mM, 1mM) are shown in Tables 1 and 2. Preliminary structure analysis of these oligopeptides revealed that peptides with inhibitory effects have in common one positive charge at pH 7.2-7.5 and form intermolecular linkages with hydrogen band.

2. Effects of Gly-Tyr-NH_2(GY-NH_2) and Gly-Ser-Lys (GSK) on forskolin action on CL cells of the rat

It is known that hCG stimulates progesterone secretion by activating the adenyl cyclase system of the corpus luteam cell membrane. Forskolin, the activator of adenyl cyclase, mimics the action of hCG. The results show that both GY-NH_2 and GSK can inhibit hCG-induced as well as forskolin-simulated progesterone production of CL cells (Table 2, Figs. 1 and 2), suggesting that they affect progesterone production through the cAMP messenger pathway.

Table 1 *Peptides showing more than 50% inhibitory effect on basal progesterone production by CL cells of the rat*

Synthetic Peptides	Concentration (mM)	Change in Progesterone (%)
H-Gly-Tyr-Ala-Lys-OH	0.2	-78.1
(H-Sar-Ser, H-Sar-Sar) > Lys-OH	0.2	-58.3
(H-Sar-Ser, H-Sar-Ser) > Lys-OCH_3	0.2	-57.4

Table 2 *Oligopeptides with more than 30% inhibitory effect on hCG-induced progesterone production by rat CL cells*

Synthetic Peptides	Concentration (mM)	Change in Progesterone (%)
H-Asp-His-Pro-Thr-Phe-Lys-OH	0.02	-78.1
H-Tyr-Thr-Pro-Phe-Lys-OH	0.02	-66.6
H-Tyr-Thr-Pro-Arg-Lys-OH	0.02	-43.5
H-His-Tyr-OH	0.02	-37.8
H-Thr-Pro-Tyr-Lys-NH_2	0.02	-37.3
(H-Sar-Ser, H-Sar-Ser) > Lys-OCH_3	0.02	-35.7
H-Gly-Tyr-NH_2	0.2	-48.8
H-Gly-Tyr-Ala-Lys-OH	0.2	-43.3
H-Lys-Tyr-NH_2	0.2	-35.0
H-Gly-Ser-Lys-OH	0.2	-33.2
H-Gly-Tyr-Lys-OH	0.2	-30.0

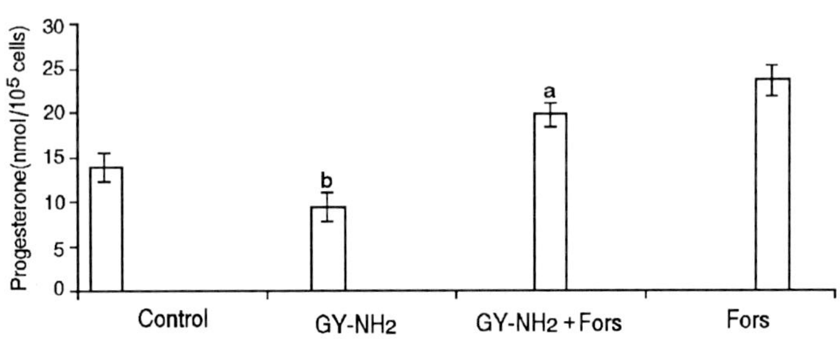

Fig. 1. Effect of GY-NH_2 on forskolin action on rat CL cells.
[a] $P<0.05$ vs Fors. [b] $P<0.01$ vs control.
GY-NH_2(0.2mM). Fors (10^{-4}M).

3. Effects of Gly-Tyr (GY) phospholipase C (PLC) and neomycin on hCG induced progesterone production in rat CL cells

PLC is one of the key enzymes in the phosphoinositide system; Neomycin is a PLC inhibitor. GY was affected by PLC in the present experiment.

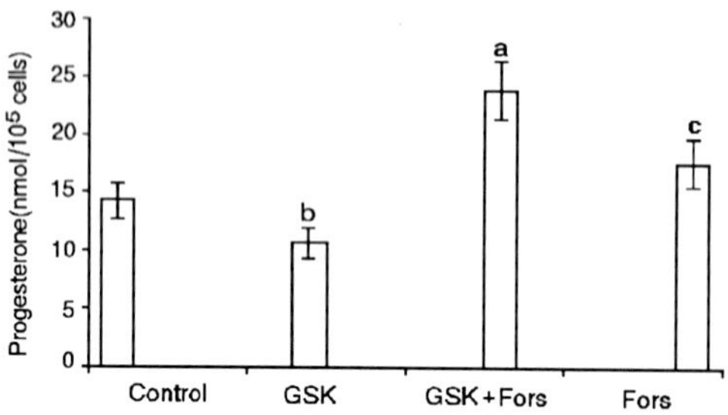

Fig. 2. Effect of GSK on forskolin action on CL cells of the rat
[a] P<0.001. [b] P<0.05 vs Control. [c] P<0.05 vs Fors.
GSK (0.2mM). Fors(0.1mM).

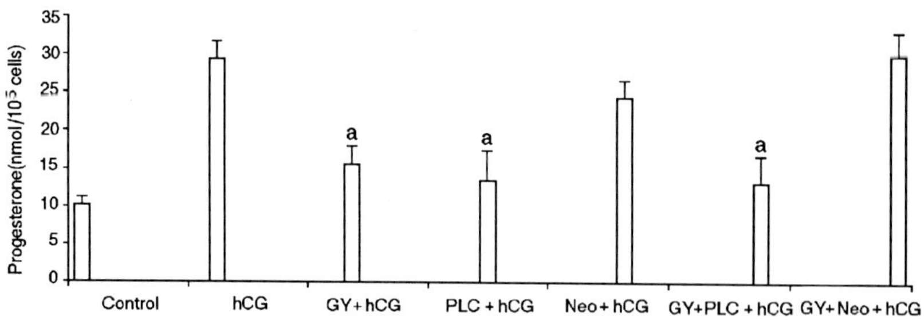

Fig. 3. Effects of GY PLC and neomycin on hCG-induced progesterone production rat CL cells.
[a] p<0.02, vs hCG. PLC (1.5IU/ml). Neomycin (5mM).
GY (0.2mM). hCG (0.01IU/ml).

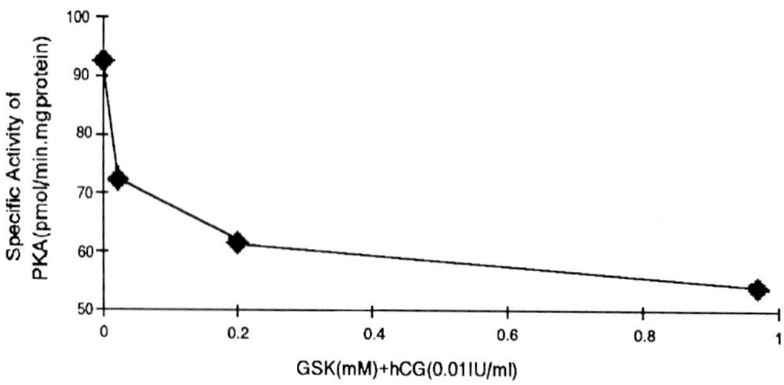

Fig. 4. Effect of GSK on the activity of PKA in hCG treated CL cells of the rat.

4. Effects of synthesized oligopeptides Gly-Ser-Lys (GSK) and Gly-Tyr-Lys (GYK) on the activity of protein kinase (PKA, PKC, TPK)

Results are shown in Figs. 4-7. Both peptides GSK and GYK decreased hCG-induced progesterone production as shown in Table 2. The activities of PKA, PKC and TPK were detected in CL cells after incubation with GSK and GYK. GSK increased the activity of PKC which belongs to the phosphoinositide system, and inhibited that of PKA, which belongs to the cAMP signalling pathway in rat CL cells. It may be one of mechanisms of GSK function. GSK reduced but GYK enhanced the activity of TPK in rat CL cells. The reason for this difference remains to be investigated.

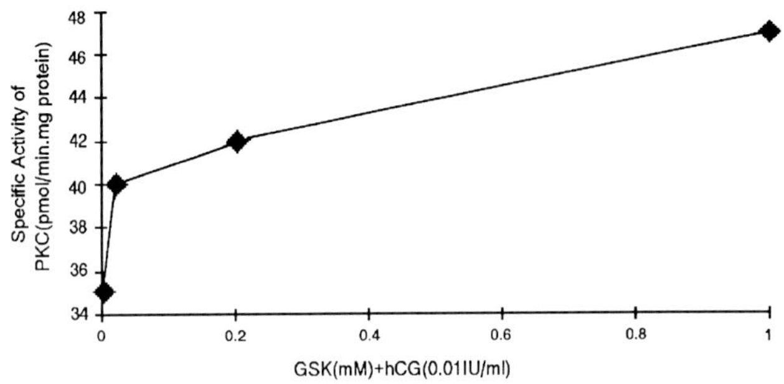

Fig. 5. Effect of GSK on the activity of PKC in hCG treated rat CL cells.

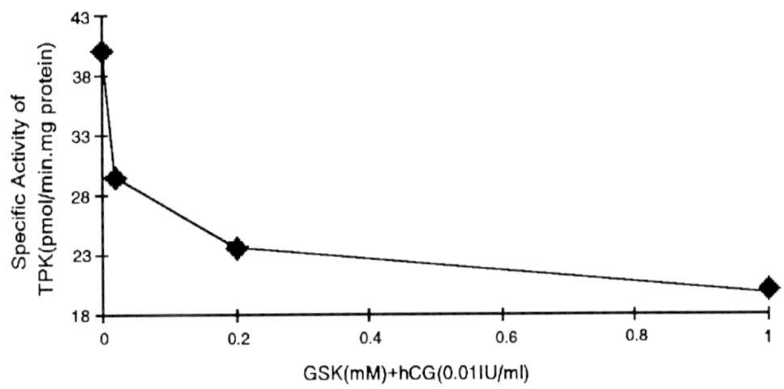

Fig. 6. Effect of GSK on the activity of TPK in hCG treated rat CL cells.

In this experiment we studied the effect of synthesized oligopeptides on progesterone production by rat corpus luteum cells *in vitro* and on the activity of protein kinase (PKA and TPK) in CL cells, as well as the relationship between some active oligopeptides and cAMP and the phosphoinositide signaling pathway. Our results showed that 1). The peptides that caused a decrease in progesterone production have common characteristics, that is, carry a positive charge in media of pH 7.2-7.5 and have intermolecular linkage by a hydrogen bond. 2) Peptides GY-NH_2 and GSK decreased forskolin-induced progesterone production, indicating that they appeared to affect the cAMP messenger system. 3) PLC enhanced the inhibition caused by GY on hCG-induced progesterone production, while neomycin, the inhibitor of PLC, attenuated this inhibition. We included that GY probably acted through inhibiting the PLC system. 4) GSK increased the activity of PKC and suppressed that of PKA in the presence of hCG in CL cells. GSK decreased the activity of TPK in hCG-treated CL cells, while GYK enhanced it. The difference between their actions remains to be investigated. These findings suggest that the mechanism of the anti-progesterone effect of our synthesized oligopeptides acted through different cellular signaling pathways. Their structure, function relationships at the molecular level are quite complicated. More work should be carried out to elucidate this problem.

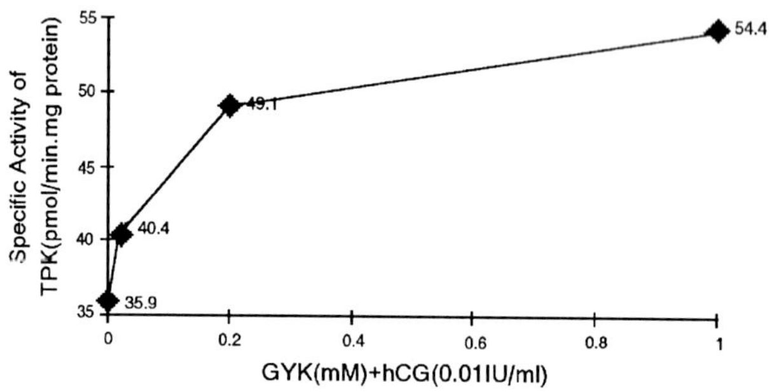

Fig. 7. Effect of GYK on the activity of TPK in hCG treated CL cells of the rat.

References

1. Wang, X.N., Zhang, Q., Cheng, C.P., Chinese Medical Jourual, 100(1987)859.
2. Merrifield, R.B., J. Am. Chem. Soc, 85(1963)2149.
3. Hoyer, P.B. and Kong, W.Y., Mol. Cell. Endocrinol., 62(1989)203.
4. Takai, Y., J. Biol. Chem, 254(1979)3692.
5. Takai, Y., Biophy. Res. Commun., 91(1979)1218.
6. Casnellie, J.E., Proc. Natl. Acad. USA., 79(1982)282.
7. Swarp, G., J. Biol. Chem, 258(1983)10341.

Stability study on LHRH antagonists(TX44 and TX53) toward heating and enzyme degradation

Li-Lin Shen, Yong-Qang Ge, Miao-Miao Liu and Shao-Bo Xiao
Tianjin Municipal Research Institute for Family Planning, Tianjin 300191, China

Introduction

In order to develop luteinizing hormone releasing hormone (LHRH) antagonist(s) for contraceptive or therapeutic applications, the two most protent LHRH antagonists invented by our group, TX44 and TX53, were selected for stability studies because of their high antifertility activity and negligible side effects such as histamine releasing activity (HRA). Their heat stability and resistance to enzyme (protease) degradation were investigated. All samples were analyzed by reverse phase high performance liquid chromatography (RP-HPLC) and compared to LHRH and LHRH-A.

Results and Discussion

The LHRH antagonists, TX44 and TX53, were purified by Sephadex G25 and their purity was confirmed by analytical HPLC. Their structures are as follows:

TX44: NAc-D2-Nal^1-D-pClPhe^2-D3-Pal^3-Ser^4-Mop^5-D3-Pal^6-Leu^7-Arg^8-Pro^9-D-Ala^{10}-NH_2
TX53: NAc-D2-Nal^1-D-Phe_2-D3-Pal^3-Ser^4-Mop^5-D3-Pal^6-Leu^7-Arg^8-Pro^9-D-Ala-NH_2
LHRH: pGlu-His-Trp-Ser-Tyr-Gly-Leu-Arg-Pro-Gly-NH_2 (Peninsula Lab. Inc.)
LHRH-A: [D-Ala^6, DesGly^{10}]-LHRH ethyl amide (Peninsula Lab. Inc.)
Protease (Sigma Chemical Company)

1. Stability toward heating
Four peptides were dissolved separately in deionized water. In the first experiment, 50µl (5µg) of aqueous solution was transferred into a test tube (5×50mm), which was then sealed by Teflon tape and parafilm. Those tubes were then heated at 100°C for 1 h, or 4 h at 126°C for 0.5 h in a special oven designed for peptide sample hydrolysis and amino acid analysis (Waters). In the second experiment, 50µl (5µg) aqueous solution was immediately applied onto the HPLC system. The remaining samples were sealed by parafilm, stored at 4°C for three months and re-analyzed.

2. Stability toward protease
Four peptides were digested with protease [1]. Digests of TX44, TX53, LHRH and LHRH-A were carried out at an appropriate molar ratio of protease (0.25µg)/peptide (3.0µg) in two buffer systems, 0.1 M phosphate buffer pH 8.0 or 0.1 M Tris buffer pH6.0 (50µl), at 37°C for 24 h, respectively.

3. Stability toward vehicles

Peptide solid powders were mixed with two vehicles widely used in manufacturing preparations, i.e., 0.9% benzyl alcohol or 5% mannitol solution [2].

4. Analysis

After heating or enzyme treatment, the samples were immediately applied onto an HPLC system [3]. The samples remaining in the test tubes were stored at 47°C for 3 months and re-analyzed.

HPLC conditions:

Column: μ-Bondapak C18 (3.9×300 mm)
Solvent system A: 0.02% TFA
B: 20% A in CH3CN
Flow rate: 1.0 ml/min
Gradient eluent: B from 20% to 100% in 20min
Detector: UV 210nm

Under these conditions, sharp and reproducible sample peaks were obtained.

The HPLC analysis of TX44 and TX53 showed no new peak after treatment in a sealed tube at 100°C for 1 h, or 4 h at 126°C for 0.5 h in the aqueous or solid state (Table 1). After being treated by protease in either phosphate or Tris buffer at 37°C for 24 h, they showed only the same single peak as the untreated sample (Table 2).

Table 1 *The retention time (RT) of peptides samples on HPLC*

sample	100°C			126°C	126°C[a]
	0 h	1 h	4 h	0.5 h	0.5 h
LHRH	11.62	11.73	11.60	11.62	11.73
LHRH-A	12.55	12.65	12.52	12.66	12.69
TX44	14.71	14.82	14.69	14.76	14.53
TX53	13.83		13.81	13.81	13.86

[a] Dry peptide.

Table 2 *The retention time (RT) of peptides treated with protease on HPLC*

sample	at 37°C for 24 h		aqueous[a] solution
	0.1 M phosphate buffer pH 8.0	0.1 M Tris buffer pH 6.0	
LHRH	3.33, 8.89, 10.81	11.46	11.33
LHRH-A	3.31, 3.59	10.02, 11.43	12.52
TX44	14.49	14.50	14.58
TX53	13.80	13.73	13.70

[a] Room temperature.

Table 3 *The retention time (RT) of remaining peptide samples on HPLC*

sample	After treatment at 100°C for 4 h stored at 4°C for 3 months	at 100°C for 4 h
LHRH	7.08, 10.07, 12.1, 13.71	11.62
TX44	14.50	14.71

In contrast, both LHRH and LHRH-A were degraded significantly, and several new peaks were observed after treatment with protease in phosphate buffer at 37°C for 24 h. LHRH-A also showed a small, new peak on HPLC spectra after being treated with protease in Tris buffer (Table 2).

Although the HPLC analysis of both heated LHRH antagonist TX44 and LHRH itself showed no new peak after storage at 4°C for 3 months, heated LHRH seemed to decompose, the original peak disappeared, and several new peaks were observed on HPLC. TX44 did not show any change under the same conditions (Table 3).

HPLC analysis demonstrated that TX44, TX53 and LHRH were stable toward benzyl alcohol, and mannitol solution, which are vehicles that are widely used in manufactured preparations (Table 4).

Table 4 *The retention time (RT) of peptide samples in vehicles*

sample	benzyl alcohol (0.9%)	manntol (5%)	aqueous solution
LHRH	11.52	11.85	11.62
TX44	14.47	14.58	14.71
TX53	13.81	13.73	13.83

The results indicated that LHRH antagonists TX44 and TX53 were extremely stable toward either heating or enzyme digestion. This finding differs from common natural peptides. The formulation of TX44 and TX53 may include benzyl alcohol and mannitol. Therefore, formulation and sterilization of TX44 and TX53 can be carried out in a standard manner, such as at 100°C for 1 h or 126°C for 15 to 30 minutes, and thus avoid complicated procedures. The results described here can be important for medical application of TX44 and TX53 or other peptides with similar stabilities.

References

1. Rivier, J., Porter, J., Hoeger, C., Theobald, P., Craig, A.G., Dykert, J., Corrigan, A., Perrin, M., Hook, W.A., Siraganian, R.P., Vale, W. and Rivier, C., J. Med. Chem., 35(1992)4270.
2. Vickery, B.H., Contraception in dogs with Luteinizing hormone releasing hormone antagonists. Eur. pat. Appl. 86105372.6.
3. Xia, Q.M., Xiao, S.B., Liu, M.M., Contraception, 46(1992)139.

The influence of GLP-1(7-36) on blood glucose and some hormones concerned

Xiao-Hong Wang, Yong-Mei Zhao and Shu-Li Sheng
Polypeptide Laboratory, Xuanwu Hospital, Capital Medical University, Beijing 100053, China

Introduction

Research on preproglucagon revealed that the precursor of glucagon is processed in the A-cell of the pancreas and in the L-cell of the intestine, yielding glucagon-like peptides (GLP) 1 and 2 in addition to glucagon or glicentin [1]. Although GLP-1 (1-37) has similarities in its amino acid sequence to glucagon, this peptide did not show any glucagon effects, such as insulinotropic or glycagenolytic action [2]. However, it has been reported that a truncated GLP-1 (7-37) elicited strong insulin stimulating action [3]. Furthermore, experimental studies revealed that GLP-1 is seceted during the ingestion of nutrients in both pig and man [4]. Therefore, at present, truncated GLP-1 (7-37) or (7-36) has been considered a possible gut hormones in the enteroinsular axis [5]. An experimental study designed by us was undertaken to study whether GLP-1 (7-36) participates in the regulation of blood glucose and some hormones concerned.

Results and Discussion

GLP-1 (7-36) was synthesized by ABI peptide synthesizer and purified by HPLC.

1. Effect of GLP-1 (7-36) infusion on glucose tolerance curve

Male Wistar rats, about 250g, were anesthetized by sodium pentobarbitol (500 mg/kg) i.p. and then trachea and vein intubations were made for infusion at 6 ml/2 hrs. The rats were divided into two groups. One group was given 400 mg glucose/kg/200 ml by bolus injection within one minute, followed by venous infusion of 3 ml NS/hr for 2 hrs. After administration of glucose, 0.1 μg GLP-1 (7-36) was added to the 6 ml NS that was infused into the other group. Determining blood glucose concentration of tail bloodat different times, we observed the effect of GLP-1 (7-36) on the glucose tolerance curve.

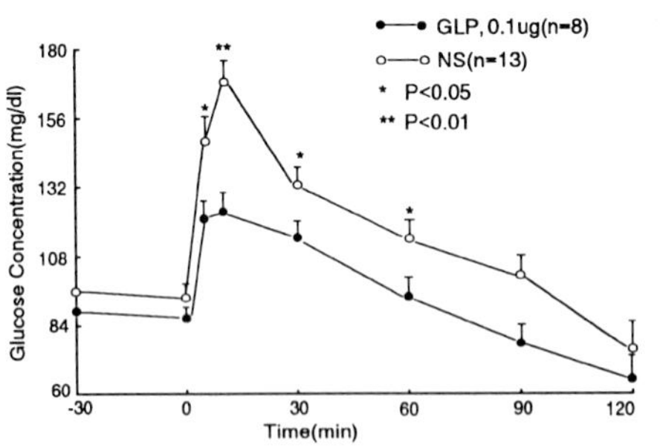

Fig. 1. The effect of insulin of GLP−1(7–36) 0.1μg on blood glucose tolerance.

Fig. 1 shows that blood glucose concentrations doubled 5 min after the intravenous bolus injection of glucose compared with the basic concentration. The level of blood glucose started to decline at 30 min. Observations from the glucose tolerance test of GLP-1(7-36) (i.v. 0.1 μg) revealed that glucose concentrations declined faster and reached a significantly lower concentration ($P<0.05$) at 30 min and 60 min than that of the control group. According to these results, GLP-1 (7-36) is able to influence the glucose tolerance curve.

2. Effect of GLP-1 (7-36) given by i.v. on the concentration of blood glucose regulating hormones

The rats were divided into two groups. One group was given NS 6 ml/2 hr, the other group was give GLP-1 (7-35) 0.1 μg/6 ml NS/2 hr. The concentrations of insulin, C-peptide, and glucagon in plasma was determined by RIA at different times.

Figures 2–4 showed that i.v.GLP-1 (7-36) could make insulin and C-peptide increase significantly, but it had no effect on glucagon secretion.

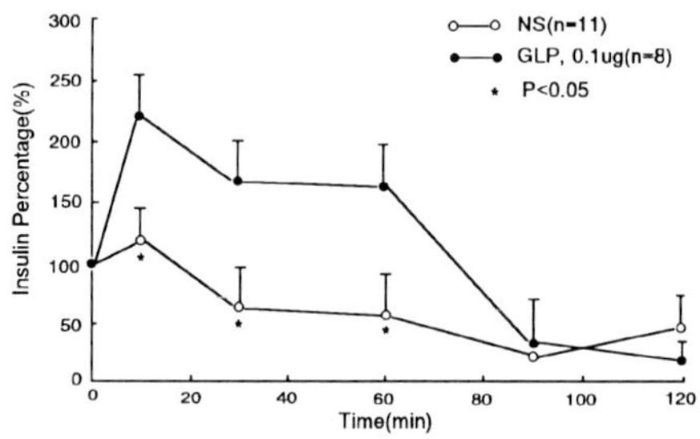

Fig. 2. The effect of intravenous infusion of GLP–1(7–36) 0.1μg on blood insulin concentration.

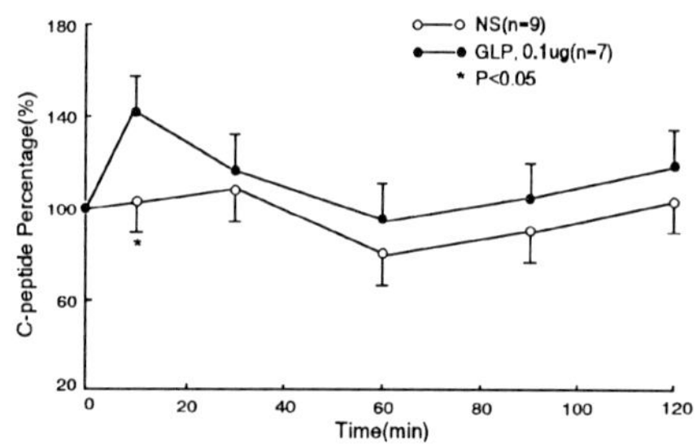

Fig. 3. The effect of intravenous infusion of GLP–1(7–36) 0.1μg on C–peptide concentration.

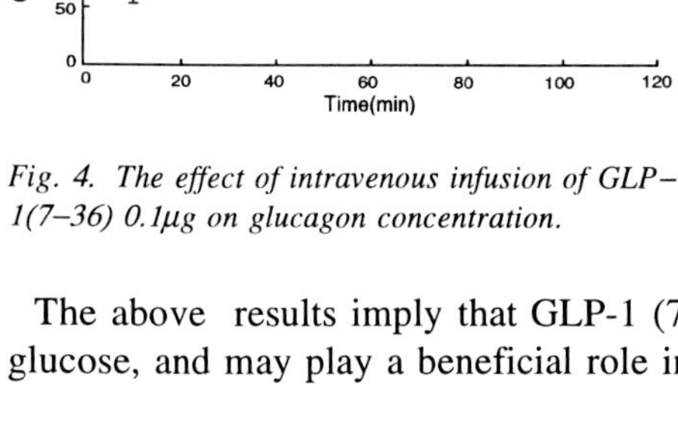

Fig. 4. The effect of intravenous infusion of GLP–1(7–36) 0.1μg on glucagon concentration.

Our results showed that (1) GLP-1 (7-36) could make plasma insulin rase significantly. Because GLP-1 receptor exists only in islet B cells, the action of GLP-1 (7-36) we have described must be associated with direct stimulation of pancreatic B cell secretion. (2) GLP-1 (7-36) infusion has an effect on the glucose tolerance curve. This phenomenon can be probably explained by the combined effect of GLP-1 (7-36) and glucose on pancreatic B-cell secretion.

The above results imply that GLP-1 (7-36) participates in the regulation of blood-glucose, and may play a beneficial role in the pathogenesis of diabetes mellitus.

References

1. Mojsov, S., Heinrich, G., Wilson, I.B., Ravazola, M., Orci, L. and Šhabener, J.F., J. Biol. Chem., 261(1986)11880.
2. Ghiglione, M., Uttenthal, L.O., George, S.K. and Bloom, S.R., Diabetologia, 27(1984)599.
3. Holst, J.J., Orskov, C., Nielsen, V. and Schwartz, T.W., FEBS Lett., 211(1987)169.
4. Orskov, C., Holst, J.J., Knuhtsen, S., Baldissera, F.G.A., Endocrinology, 119(1986)1467.
5. Mojsov, S., Weir, G.C. and Habener, J.F., J. Clin. Invest, 79(1987)616.

Synthesis of N^{α}-Ac-Arg-Gly-Asp-$NHCH_2CH_2C_6H_5$ as an antiaggregatory agent

Shou-Ming Wen, Jian-Chang Wang, Ying Li, Yu-Ling Li, Tao Jin and Xiao-Li Ji

Department of Clinical Pharmacology, General Hospital of Air Force, PLA, Beijing 100036, China

Introduction

The tetrapeptide Arg-Gly-Asp-Ser (RGDS) is one of the natural sequences of the plasma adhesive protein receptor (the glycoprotein complex GPIIb/IIIa) and is capable of inhibiting fibrinogen binding to activated platelet aggregates in vitro. RGDS is relatively unstable in plasma and has low potency as an antithrombotic agent [1,2]. Thus, we synthesized N^{α}-Ac-Arg-Gly-Asp-$NHCH_2CH_2C_6H_5$ (Ac-RGD-PEA), a RGDS analogue in which serine is replaced by enethylamine, and exmined its anti-aggregatory effect in human platelet-rich plasma compared with that of RGDS.

Results and Discussion

Peptide Synthesis in Solution

All L-amino acids were protected as t-butyloxy-carbonyl derivatives on the N^{α}-amino group and were coupled using N, N'-dicyclo-hexyl-carbodii-mide/1-hydroxybenzo-triazole(DCC/HOBT). Boc-Asp (Bzl) was condensed with phenylethyl-amine, then N-deblocked in 4M HCl/dioxane at room temperature for 30 min and coupled sequentially with Boc-Gly and Boc-Arg (Tos). The resulting tripeptide derivative (0.35mmol) was acetylated with acetic anhydride (0.5ml) and diisopropyl-ethylamine (DIPE,0.1ml) and then treated with anhydrous HF (15ml) / anisole (2.0ml) at 0°C for 90 min. After evaporation of HF in vacuo to dryness, the residue was dissolved in 1M acetic acid (20ml), washed with Et_2O, and lyophilized to an amorphous solid. The crude peptide was purified by Sephadex-LH20 gel fitration using 0.2 M acetic acid and then by HPLC on a C-18 silica gel column with an appropriate mixture of acetonitrile in 0.1% aqueous TFA as eluent. Fractions containing product were pooled and reduced in volume by rotary evaporation and lyophilized to give the final product (130mg).

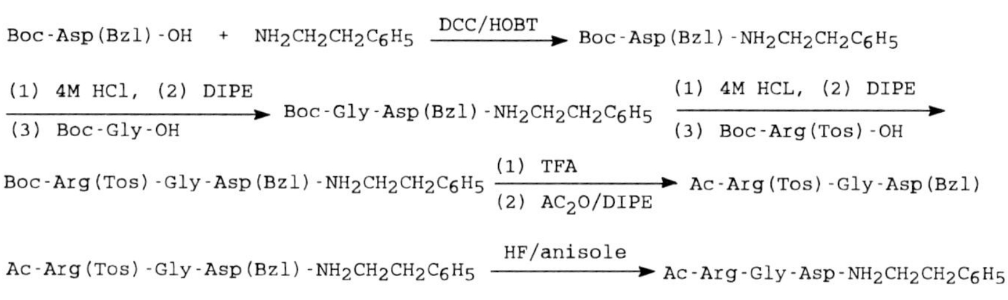

The structure and purity of the peptide was confirmed by TLC, HPLC, fast-atom-bombardment mass spectrometry (FABMS), and amino acid analysis. The peptide was more than 90% pure, as judged by reversed-phase HPLC (WATERS600E employing a 1×25 cm Vydac C-18 column with a 0~60% acetonitrile gradient in 0.1% trifluoroacetic acid, at a flow rate of 1.0ml/min, analyzing by UV at 220nm). The amino acid composition of the peptide was determined with an amino acid analyzer (model 835-50, Hitachi) after 24-hr hydrolysis in 6M HCl at 110°C (Arg1.00, Gly1.00, Asp0.88) and the sequences were consistent with those resolved by FAB-MS (m/z: 492 [M+H], 448 [RGDPEA], 370 [AcRGD-H], 292 [GDPEA], 255 [AcRG], 199 [AcR]).

Table 1 *Yields and physicochemical properties of the peptide and its protected intermediates*

Compound	Yield (%)	mp °C	TLC[a] Rf	Formula	FAB-MS (m/z)	HNMR[b] δ
Boc-Asp(OBzl)-$NHCH_2CH_2C_6H_5$	80.2	131-2	0.76	$C_{24}H_{30}N_2O_5$	427[M+H]	7.4~7.3(1OH) 1.40(s, 9H)
Boc-Gly-Asp(OBzl)-PEA	88.5	142-3	0.54	$C_{26}H_{33}N_3O_6$	484[M+H]	7.4~7.2(m, 1OH) 1.4(s, 9H)
Boc-Arg(Tos)-Gly-Asp(OBzl)-PEA	50.4	-	0.52	$C_{39}H_{52}N_7O_8S$	795[M+H]	7.4~7.2(14H) 1.8~1.4(14H)
Ac-Arg(Tos)-Gly-Asp(OBzl)-PEA	90.3	-	0.34	$C_{37}H_{46}N_7O_7S$	-	7.4~7.2(14H) 1.85(3H, Ac)
Ac-Arg-Gly-Asp-PEA	85.5	-	0.58	$C_{22}H_{33}N_7O_6$	492[M+H]	7.4~7.2(5H) 1.85(3H, Ac)

[a] TLC systems: $CHCl_3$:MeOH:AcOH (95:4:1) for protected intermediates; nBuOH:AcOH:H_2O:EtOAc(1:1:1:1) for the final peptide.
[b] NMR solvents: $CDCl_3$ for the Boc-protected peptides; DMSO for the acetylated peptides.

In Vitro Inhibition of Aggregation of Human Platelet-rich Plasma

Blood from normal human volunteers who had not taken any antiplatelet drugs for at least 14 days was collected into trisodium citrate. Platelet-rich plasma (PRP) was prepared by low speed centrifugation (600 rpm) for 3 min. The PRP was removed and the remaining blood was centrifuged at 2000 rpm for 10 min to obtain platelet-poor plasma (PPP). The PRP was adjusted with PPP to give a final platelet count of 250,000/μl, and preincubated in siliconized glass cuvettes with various concentrations of RGDS (Sigma), Ac-RGD-PEA or saline control for 2 min at 37°C, with stirring at 900rpm. Aggregation was recorded for 10 min after the addition of ADP (5μM, Sigma). Aggregation was quantitated by monitoring increase in light transmittance through the stirred platelet suspension on a aggregometer (model PA3210, Japan). Inhibitor potency was expressed as the concentration of peptide that resulted in 50% inhibition of aggregation (IC_{50}). The peptide (300μM) was incubated with human PRP for 3 hours and then ADP (5μM) was added to elicit platelet aggregation and biological stability of the peptides in plasma. At 3 hr, 26% aggregation was restored with RGDS, whereas only 5% aggregation was restored with Ac-RGD-PEA. The results showed that Ac-RGD-PEA is more potent in preventing ADP-induced aggregation of human platelet-rich plasma and is more stable in plasma than RGDS. However the inhibitory potency of both peptides were different from those reported previously by the others [1,2].

Table 2 *Inhibition of platelet aggregation in human PRP induced by ADP*

Peptide	No. of assays	IC_{50}(μM)	% Inhibition at 3 hr
RGDS	6	156.47±32.94	74.30
Ac-RGD-PEA	6	122.16±41.69	94.87

References

1. Nicholson, N.S., Panzer-nodle, S.G., Salyers, A.K., Taite, B.B., King, L.W., Miyano, M., Gorczynski, R.J., Williams, M.H., Zupec, M.E., Tjoeng, S.P., Feigen, L.P., J. Pharm. Exp. Ther, 256(1991)876.
2. Samanen, J., Ali, F., Romoff, T., Calvo, R., Sorenson, E., Vasko, J., Storor, B., Berry, D., Bennett, D., trohsacker, M., Powers, D., Stadel, J., Nichols, A., J. Med. Chem., 34(1991)3114.

Studies on intra-A chain disulfide bond (A6-11) of insulin

Yong Dai and Jian-Guo Tang
Department of Biochemistry and Molecular Biology, College of Life Sciences, Peking University, Beijing 100871, China

Introduction

A statistical method for quantifying the relation between the extent of functional group modification and the inactivation of the protein concerned was established thirty years ago [1] and is still widely used. It was also employed to investigate the effect on the activity of insulin of splitting its disulfide bonds and showed that only two of the three disulfide bonds are probably essential for insulin activity [2]. So far, however, this finding has not been verified. Although there are only three disulfide bonds in the insulin molecule, their biological importance is still a question. Only a few works have been done on these disulfide bonds. By analyzing the available results in insulin research, we thought that the A6-A11 intra-chain disulfide bond might be non-essential and this was selected as our first target.

The mutant proinsulin gene was constructed with the codens for A6 and A11 Cys changed into Ser to delete the intra-A chain disulfide bond. After expression and purification, the mutant proinsulin and mutant insulin were confirmed by correct amino acid composition. The electrophoretic behavious of mutant proinsulin is quite similar to that of human proinsulin, as is products of tryptic digestion. The mutant Met-proinsulin, which retains nearly full radioimmunoactivity, has only 5.4% of receptor binding activity of the human Met-proinsulin. The mutant Met-insulin exhibited receptor binding activity of 1.3% and radioimmunoactivity of 113% of Met-insulin. It is suggested that although the intra-A chain disulfide bond disappears,the other two inter-chain disulfide bonds are still correctly paired, and its three dimensional structure is not significantly changed. It seems that this intra- chain disulfide bond is still essential for insulin activity.

Results and Discussion

The synthetic gene encoding human proinsulin was cloned into M13mp18 RF DNA. The mutant proinsulin gene was then obtained by Kunkel's method for site-directed mutagenesis and confirmed by DNA sequencing (data not shown) with codens for A6 and A11 Cys changed into Ser to delete this intra-A chain disulfide bond. The mutant gene was cloned into polycloning site of pBV220 and successfully expressed in the cytoplasm of E.coli the same as in our previous report [3] for production of human proinsulin and essentially for processing expressed mutant proinsulin. Fig. 1 shows the gel filtration of the mutant human proinsulin on Sephadex G50 column. The purified mutant proinsulin could be changed into mutant insulin by trypsin digestion. The above obtained mutant insulin, in the form of Met-insulin, was separated by DEAE Sephadex A-25 column. Fig. 2 shows the purification of the mutant insulin on DEAE Sephadex A25 column.

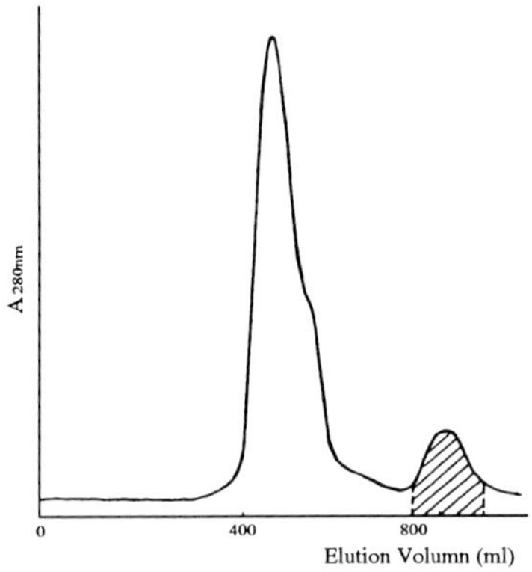

Fig. 1. Gel filtration of the mutant human proinsulin. The sample was loaded onto a Sephacex G50 column (4×100cm) eluting with 0.05M Gly-NaOH buffer, pH 10.8. The second peak was pooled, ultrafiltrated, desalted and lyophilized.

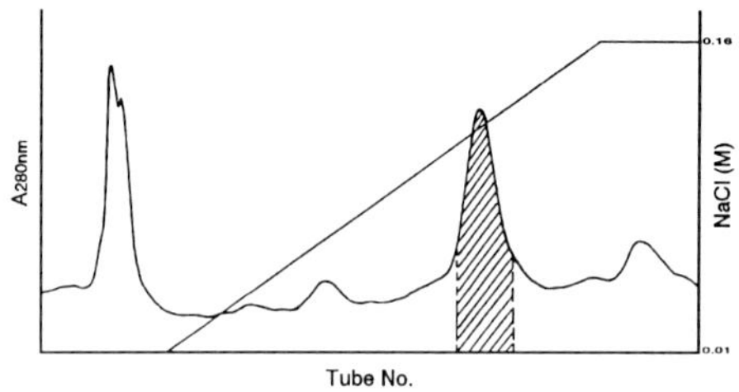

Fig. 2. Purification of mutant human insulin on DEAE Sephadex A25 column (2.2×18cm). Starting buffer: 0.01M NaCl in 0.05M Tris buffer with 40% isopropanol, pH7.5; Eluent: Starting buffer with a linear gradient from 0.01 to 0.16M NaCl. The main peak is mutant insulin.

The amino acid composition analysis of obtained mutant proinsulin and mutant insulin are shown in Table 1. The data are in a good agreement with expected values except for the existence of one methionine. The methionine residue at the N-terminus of mutant proinsulin has no influence on our data on the A6-A11 intra-A chain disulfide bond with Met-human proinsulin and Met-insulin as the standard.

The mutant proinsulin was analyzed by polyacrylamide gel electrophoresis as demonstrated in Fig. 3. It shows a almost mobility rate identical to that of the molecule with intra-A chain disulfide bond. It should be noted that there are two major bands with an undifferentiated contents in each purification for the mutant molecule probably because of the oligomerization. Fig. 3 also shows the results of the trypsin digested mutant molecule as compared with recombinant Met-human proinsulin. It can be seen that the digested products of mutant Met-proinsulin are quite similar to that of met-human proinsulin. Both of them have two dominant products, Met-des-B30 insulin and Met-des-octapeptide insulin with almost the same mobility rate as that of porcine insulin and des-octapeptide insulin. There is another product indistinguishable from both of them with more a positive charge close to proinsulin, as shown in Fig.3. It is suggested that although the A6-A11 disulfide bond is deleted, the other two disulfide bonds are still correctly paired and there is not a significant change in the mutant insulin compared with insulin. The mobility rate of purified mutant insulin on PAGE (Fig. 4) is lower than that of human insulin, probably because the deletion of the intra-A chain disulfide bond produces a slightly looser structure or a more positive charge of the mutant molecule.

Table 1 *The amino acid composition analysis of mutant proinsulin with deleted intra-A chain disulfide bond*

	Met-human proinsulin	Met-mutant proinsulin	Met-mutant insulin
CysH	(6)	4.0(4)	4.1(4)
Asx	3.7(4)	3.8(4)	3.0(3)
Thr	3.0(3)	3.1(3)	2.8(3)
Ser	4.2(5)	6.0(7)	5.2(5)
Glx	15.3(15)	15.6(15)	6.9(7)
Pro	2.9(3)	3.8(3)	1.4(1)
Gly	12.4(11)	10.1(11)	4.2(4)
Ala	4.0(4)	4.6(4)	1.3(1)
Val	5.7(6)	5.9(6)	3.8(4)
Met	1.2(1)	1.2(1)	1.0(1)
Ile	1.8(2)	1.9(2)	2.3(2)
Leu	10.8(12)	10.6(12)	6.2(6)
Tyr	3.5(4)	3.2(4)	4.3(4)
Phe	2.9(3)	2.9(3)	3.0(3)
Lys	2.0(2)	2.5(2)	0.8(1)
His	1.9(2)	2.0(2)	2.1(2)
Arg	3.9(4)	3.9(4)	1.2(1)

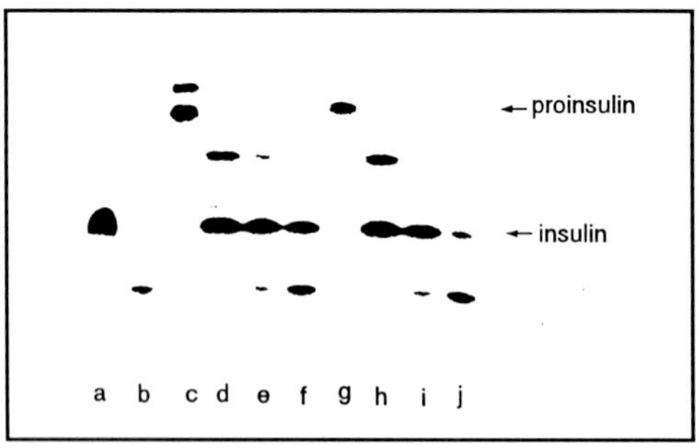

Fig. 3. Electrophoretic analysis of mutant proinsulin with deleted intra-A chain disulfide bond and its trypsin digested products on 15% polyacrylamide gel as compared with that of wild molecule andinsulin. (a) crystalline porcine insulin; (b)trypsin digested porcine insulin with enzyme and substrate weight ratio of 1:100; (c) met-mutant proinsulin; (d)(e)(f) trypsin digested met-mutant proinsulin with enzyme and substrate weight ratios of 1:200, 1:100 and 1:50 respectively; (g) met-human proinsulin; (h)(i)(j) trypsin digested met-human proinsulin with the ratios of 1:100, 1:50 and 1:25 respectively.

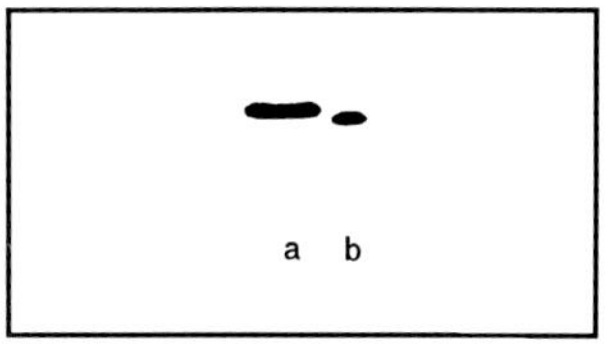

Fig. 4. Electrophoretic analysis of purified mutant insulin on 15% polyacrylamide gel. (a) human insulin; (b) mutant Met-insulin.

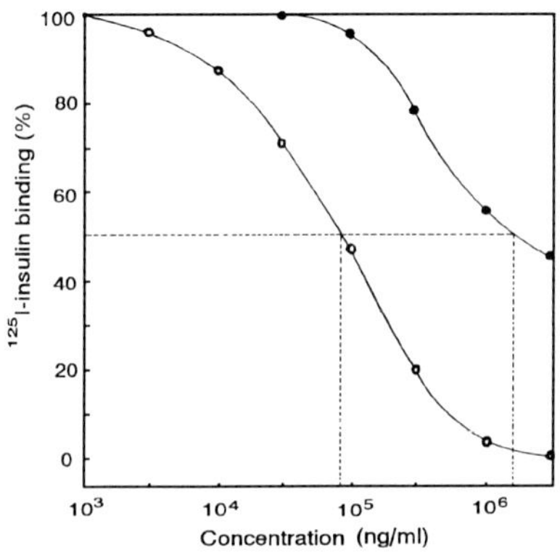

Fig. 5. Displacement curve of mutant proinsulin in the binding of ^{125}I-insulin to human placental membrane. Left-recombinant human proinsulin (Met-BCA), Right-mutant proinsulin (Met-BCA').

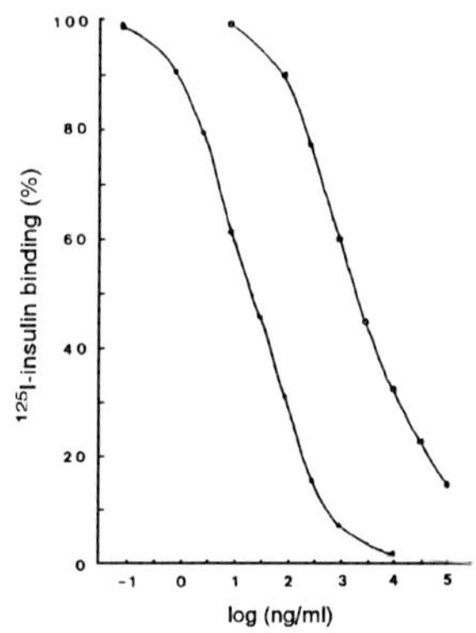

Fig. 6. Binding of mutant insulin with insulin receptor on human placental membrane. Left-Met-insulin, Right-mutant Met-insulin.

Table 2 *The biological activity analysis of mutant proinsulin*

	RBA	RIA
Met-human proinsulin	100%	100%
Met-mutant proinsulin	5.4%	110%

Table 3 *The biological activity analysis of mutant insulin*

	RBA	RIA
Met-human insulin	100%	100%
Met-mutant insulin	1.3%	113%

The biological activities of mutant products were also determined by receptor binding and radioimmunoassays as shown in Tables 2 and 3. The mutant products show radioimmunoactivity nearly identical to that of Met-proinsulin and Met-insulin which agrees with the results of the electrophoretic analysis. However, the mutant molecules demonstrate only 5.4% and 1.3% receptor binding activities, respectively (Fig. 5 and Fig. 6). These findings strongly indicate that although deletion of the intra-A chain disulfide bond makes almost no conformational change in the insulin molecule, it has great influence on the receptor binding activity and seems essential for insulin biological activity. Deletion studies on other two disulfide bonds are needed to determine which one is not essential.

Acknowledgements

We thank Prof. Yunde Hou for the expression vector pBV220, and Prof. Tongjian Shen for the synthetic gene encoding human proinsulin.

This work was supported in part by Grant No.881266 from the National Laboratory for Biomacromolecules.

References

1. Tsou, C.L. Sci. Sin. (Engl. ed.), 11(1962)1535.
2. Jiang, R.Q., Du, Y.C., and Tsou, C.L. Acta Biochim. Biophys. Sin., 3(1963)176.
3. Tang, J.G., and Hu, M.H. Biotechnology Letters, 15(1993)661.

The study on S(T)PKK motif peptides

Yan Hao, Ruo-Heng Zhang and Chong-Xi Li
Department of Chemistry, Peking University, Beijing 100871, China

Introduction

Many DNA binding drugs have great interest for investigations in biology, medicinal chemistry and molecular pharmacology. Netropsin is a type of DNA-binding antibiotic. This pyrrole-amidine antibiotic shows antibacterial and antitumor activity. The precise data on its geometry was obtained by single-crystal X-ray diffraction. Interaction of netropsin with dA-dT pairs has been directly monitored by NMR and X-ray crystal analysis. In DNA-binding proteins, S (T) PKK motif is very interesting because of its small number of amino acid residues. To elucidate the structure-function relationship of that motif, three undeca-peptides were synthesized by SPPS method.

```
PI :  T P K R P R G R P K
PII:  S P R K R G R P R K
PIII: A P K R P R G R P K
```

The models of these peptides are similiar to that of netropsin, and these peptides also have antibacterial and antitumor activities.

Results and Discussion

The peptides were synthesized by Merrifield solid-phase method with tert-butyloxy-carbonyl (Boc) group for temporary amino-terminal protection and other acid-stable groups for the protection of side chains. The following side-chain protection groups

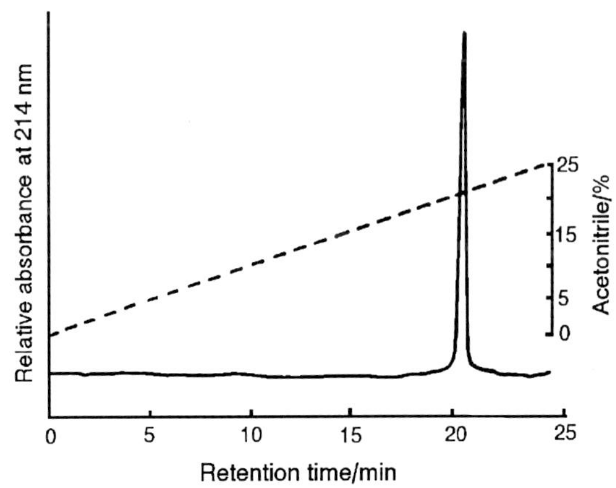

Fig. 1. HPLC profile of the peptides Instrument: Waters 510; Column: cosmosil 10x250mm; Mobile phase: A:0.1%TFA in water; B: 0.1%TFA in acetonitrile; Gradient: 0-25% B in 25min; Detection:214nm.

were used: O-benzyl for Thr, Ser; tosyl for Arg; N^{ε}-benzyloxycarbonyl for Lys. Peptides resins were cleaved by hydrogen fluoride and extracted by 10% acetic acid in water. Peptides were purified by RP- HPLC using a linear gradient from 0 to 25% of solution B (0.1% trifluoroacetic acid in acetonitrile) into solution A (0.1%trifloroacetic acid in water) in 25min. Composition of the peptides was confirmed by amino acid analysis. The molecular weights of the peptides were confirmed by FAB-MS. The analytical HPLC of the peptides is represented in Fig. 1.

We selected the dihedral angles of PII as follows: (11^0, 141^0), (-75^0, -30^0), (-90^0, 0^0), (60^0, 150^0), (-139^0, 84^0), (172^0, 68^0), (-169^0, 141^0), (-75^0, -30^0), (-90^0, 0^0), (60^0, 160^0) to predict its conformation. We compared the model of PII with that of netropsin (which was taken from PDB), and built up the model of PII binding with DNA.

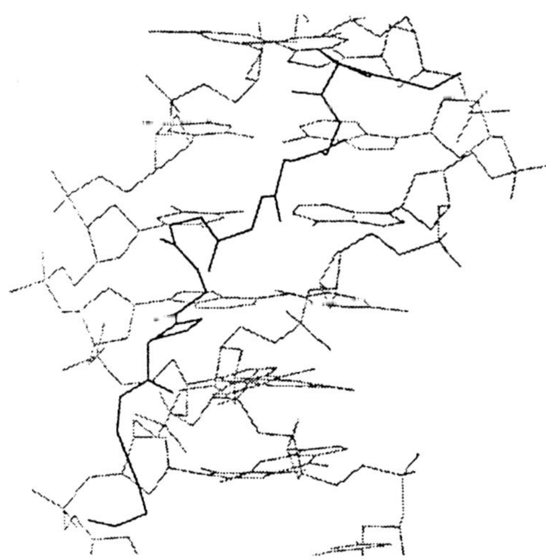

Fig. 2. The model of PII binding with DNA.

Table 1 *Antibacterial activities*

	Bacillus subtilis	*Staphylococcus aureus*	*E. coli*
PI	7	22	6
PII	13	20	8
PIII	8	15	7

We selected *E. coli*, *Bacillus subtilis* and *Staphylococcus aureus* to test the antibacterial activities of the peptides. The concentration of the peptides was 50mg/ml. All peptides showed antibacterial activities (Bacteria: grampositive and gramnegative). The results are presented in Table 1.

In order to test the antitumor activity of the peptides, we selected KB cancer and MGC cancer. The concentrations are 10^{-7} M, 10^{-6} M, and 10^{-5} M. PI and PII show anti-KB cancer activity at 10^{-5} M (shown in Table 2).

Table 2 *Activity on anti KB cancer*

Sample	Concentration (M)	OD	Percent	Results
PI	10^{-7}	1.624±0.068	-1.78%	+
	10^{-6}	0.799±0.265	49.97%	
	10^{-5}	0.615±0.024	61.45%	
PII	10^{-7}	1.689±0.103	-5.81%	+
	10^{-6}	1.230±0.207	22.93%	
	10^{-5}	0.651±0.013	59.21%	
PIII	10^{-7}	1.581±0.007	0.94%	-
	10^{-6}	1.376±0.204	13.76%	
	10^{-5}	0.833±0.028	47.83%	

Acknowledgement

This work was supported by the National Natural Science Foundation of China.

References

1. Finlay, A.C., J. Am. Chem. Soc., 73(1951)341.
2. Suzuki, M., Protein Engineering, 6(1993)565.
3. Suzuki, M., Churchill, M., EMBO. J., 8(1989)4189.
4. Suzuki, M., EMBO. J., 8(1989)797.

Homology modeling of the human brain and human serum cholinesterases

Song Li and Ke-Fang Jiao
Institute of Pharmacology and Toxicology, Academy of Military Medical Sciences, Beijing 100850, China

Introduction

Human brain and human serum cholinesterases (H.AChE and H.BuChE) belong to the serine protease family [1-2], and have the specific function of hydrolyzing acetylcholine (ACh) and butyrylcholine (BuCh). Because there is a close relationship between acetylcholine and human nervous system transmitions, systematic study of the relationship between chemical substances and enzymes may facilitate elucidation of the composition of the active site and three-dimensional structure of human serum cholinesterases. It is very important not only in the study of protective medicine against chemical weapons, but also in the study of the prevention of senescence, and in providing a target for the understanding and designing of effective drugs. Up to now, however the three-dimensional structures of H.AChE and H.BuChE have not been determined. The large H.AChE molecule (with 583 residues) and H.BuChE molucule (with 574 residues) are difficult to purify and crystallize. It is therefore difficult to determine their three-dimensional structures by x-ray and NMR methods.

In 1991, the Israeli biologist J.L.Sussman determined the three-dimensional structure of Torpedo california cholinesterase (T.AChE) [3], which is highly homologous to H.AChE and H.BuChE. Computer modeling based on this homologous protein enabled us to predicate the composition of the active site and the three-dimensional structure of H.AChE and H.BuChE.

Results and Discussion

The Protein Sequence Data Bank (PIR) [4] was systematically searched for the homologous cholinesterase family (ChE). The homologous ChE's and the percentage of identical residues are listed in Table 1.

Among these seven homologous proteins, the three-dimensional structure of T.AChE had already been determined by Sussman by the x-ray method. By comparing of the spatial structures of similar proteins, the three-dimensional structure of protein is more conservation than its primary structure. For a protein with only 50% identical residues, about 90% of the atomic deviation may not exceed 3Å, and the mean square root deviation (RMSD) may be about 1Å. As amino acid residue replacement occurs usually at the loop region of the protein surface, the protein main chain structure, especially the structure of the hydrophobic center, is very little affected by sequence variation [5]. Therefore we used the three-dimensional structure of T.AChE as a template protein to predict the three-dimensional structures of H.AChE (58.5%) and H.BuChE (53.1%).

Table 1 *The homology of cholinesterases*

ChE[a]	H.AChE(%)	H.BuChE(%)
Bov.TG (2183-2750)	31.7	29.0
D.AChE	36.4	36.8
T.AChE	58.5	53.1
M.AChE	89.2	52.1
TM.AChE	58.3	53.2
H.AChE	--	52.3
H.BuChE	52.3	--

[a] Bov.TG, bovine thyroglobulin. D.AChE, drosophila AChE. T.AChE, torpedo california AChE. M.AChE, mouse AChE. TM.AChE, T. marmorate AChE. H.AChE, human brain AChE. H.BuChE, human serum BuChE.

When multiple sequence alignment was carried out with these seven proteins and the statistical score of the alignment sequence block was calculated, the common structural conserved region of this protein family was determined [6]. based on the three-dimensional structure coordinates in T.AChE conserved region, the coordinates of the conserved region residue of H.AChE and H.BuChE were used to establish the three-dimensional structure for the predicted of protein conserved region. Amino acid residues in conserved regions may not necessarily be identical, while in protein conserved regions, residues with similar physical and chemical characters are usually interchangeable, as F and K and so on.

For the loop region (variable regions) between two conserved regions [7], the Protein Data Bank of all known three-dimensional structures is searched according to the α-carbon distance matrix of the conserved regions before and after the loop region. The three-dimensional structure segments retrieved from the conserved regions on the two ends were evaluated through comparing the mean square root deviations of their α-carbon matrix distances, and the three-dimensional structure segment with the best superimposition was determined to be the given coordinates of the loop region of the predicted protein.

Because connecting peptide chain between the N-terminal and C-terminal and the structurally conserved regions and the loop region of the predicted protein might be too long or too short, and its dihedral angle might also be twisted, we optimized the peptide chain at the connecting point with the AMBER forcefield of molecular mechanics by making make the length of the peptide chain and the dihedral angle more reasonable. Then loose calculation structurally conserved region and loop region was carried out with molecular mechanics method, so that the three-dimensional structure of the unknown protein might be preliminarily predicted.

Statistical study showed that the direction of the side chain of the residue in the protein was not irregular and in fact the side chain of an amino acid residue might have only several possible conformations [8]. In the process of prediction of H.AChE and H.BuChE, we systemically searched for all possible side chain conformation combinations. The conformation combinations of optimum side chain was found by calculating the nonbond interaction energy of each conformation combination. Finally molecular mechanics and molecular dynamics modeling of the whole H.AChE and H.BuChE was carried out to get the three-dimensional structure of these two proteins corresponding to their optimum conformations. each of these two enzyme molecules contains about ten thousands atoms, therefore it is very difficult to make the molecular mechanics and molecular dynamics modeling of such large and complex systems, and large amount of CPU time is necessary. The prediction utilized the *Homology* and *Discover* Program package of *Biosym,* and was completed under the *InsightII* interface[9], the calculation used was *SGI INDIGO* graphic workstation.

Determination of the homologous ChE multiple sequence alignment and structure conserved region

Protein multiple sequence alignment is an important method for the study of protein homology, however multiple sequence alignment is difficult even with the aid of a computer. Multiple sequence alignment had been carried out on a part of homologous ChE. We carried out multiple sequence alignment of the above mentioned seven homologous ChE with the rapid multiple sequence alignment method recently reported by Schuler et al [6]. The advantages of this method are not only the possibility of differentiating inter-related sequence segments, finding out the common sequence block in the homologous protein family and evaluating statistically the confidence limit of the selected block. In those blocks with a high correlation score, the structural conserved region of the protein may be included.

The results of the sequence alignment may find out that in seven enzymes, except the bovine thyroglobulin, the composition of the catalytic triad in their active site is entirely identical, they are all composed of serine, glutamic acid and histidine. sequence alignment showed that all the seven proteins had three disulfide bonds and 12 identical sequence blocks. Except the bovine thyroglobulin, all the other six proteins have two other identical sequence blocks, the three active site residues are included in these sequence blocks.

Comparison of the three-dimensional structures of T.AChE, H.AChE and H.BuChE

The predicted three-dimensional structures show that H.AChE and H.BuChE belong to α/β type protein, they all have the spatial structures of 12 strands β-sheet and 14 strands α-helix. The superimposition of α-helix and β-sheets of H.AChE and T.AChE shows that the α-helix and β-sheets of H.AChE and T.AChE are completely similar. H.AChE and H.BuChE may form the same three disulfide bonds as T.AChE, all the three disulfide bonds make important contributions to the stability of the spatial structure and conformation of this protein family. Simulating prediction shows that the residues between the three disulfide bonds are all at the nonconserved region. Between H.AChE and T.AChE, great difference is found in the structure area of the second disulfide bond, while the other two structure areas are very similar. Sequence alignment shows that the structure areas between the three disulfide bonds of H.BuChE and T.AChE are very similar (Table 2), while their three-dimensional structures show some difference between them.

Table 2 *The characters of the spatial structures of T.AChE, H.AChE and H.BuChE*

	T.AChE	H.AChE	H.BuChE
Active site catalytic triad	Ser200 Glu327 His440	Ser203 Glu334 His447	Ser198 Glu325 His438
Number of residues between disulfide bonds	C67-C94(27) C254-C265(11) C402-C521(119)	C69-C96(27) C257-C272(15) C409-C529(120)	C65-C92(27) C252-C263(11) C400-C519(119)
Residues around the active site	Y70 W84 W114 Y121 Y130 W233 W279 F288 F290 F330 F331 Y334 W432 Y442	Y72 W86 W117 Y124 Y133 W236 W286 F295 F297 Y337 F338 Y341 W439 Y449	N68 W82 W112 Q119 Y128 W231 A276 L285 V288 A328 F329 Y332 W430 Y440

[a] The code number of all residues is counted from the starting point of the mature protein.

The conditions of the active sites of the three enzymes are listed in Table 2. Analytically predicted three-dimensional structures of H.AChE and H.BuChE show that the active sites of these enzymes are all distributed between α-helix and β-sheets and inside the enzymes. The active sites, catalytic triads and catalytic pockets of H.AChE and H.BuChE are similar to T.AChE in having a narrow gorge going to the active site. For H.AChE and T.AChE, 14 aromatic residues are distributed around this gorge, the only difference is that corresponding to the F330 residue of T.AChE is the Y337 residue of H.AChE, while the 14 residues around the gorge in H.BuChE differ much from those in H.AChE and T.AChE. among the 14 residues in H.BuChE, only eight (W82, W112, Y128, W231, F329, Y332, W430, Y440) are aromatic residues with conjugated π electrons. According to the classification of T.AChE and H.AChE belong to true enzyme, while H.BuChE belong to false enzyme. The difference in the aromatic residues may be the essential difference between the true H.AChE and the false H.BuChE.

Experimental results show that T.AChE is very unstable. In its three-dimensional structure, there is a free -SH group (Cys231) at the bottom of the active site. Therefore, if another -SH group were introduced in its proximity to form a disulfide bond, the stability of the whole enzyme would be improved while not destroying the activity of the enzyme. Its spatial structure shows that residues suitable for mutation are Leu404, Pro403, Val400 and Asn399, their distances from the free sulfhydryl group are 5.27, 4.56, 4.57 and 5.55 Å respectively. After the mutation of these residues into Cys, the spatial distance is favorable for the formation of disulfide bond with the free sulfhydryl group.

After the docking of ACh into H.AChE. It shows that beside the catalytic triad, the tryptophan (Trp86) with a large conjugated π electron system will also play an important role, while the glycine (Gly) and alanine (Ala) near the active site may interact with the carbonyl group to play an auxiliary role in the catalytic hydrolysis of ACh.

In this paper we report our results of predicting three-dimensional structures of H.AChE, and H.BuChE with the protein structure prediction method, and point out the difference in the type and number of aromatic residues around the gorge of the active sites of the true H.AChE and the false H.BuChE and H.AchE during catalytic hydrolysis. The determination of the three-dimensional structures of H.AChE and H.BuChE may provide targets for the further study of the poison mechanism and new rational drug design.

Acknowledgements

The project was supported by the National Science Foundation of China.

References

1. Quinn, D. M., Chem. Rev., 87(1987)955.
2. Soreq, H., Proc. Natl. Acad. Sci. USA, 87(1990)9688.
3. Sussman, J.L., Harel, M., Frolow, F., Oefner, C., Goldman, A., Toker, L and Silman, I., Science, 253(1991)872.
4. PIR/NBRF database, National biomedical research foundation, Georgetown University medical center, 3900 Reservoir Rd., NW, Washington, D.C. 2007.
5. Blundell, T.L., Sibanda, B.L., Sternberg, M.J.E. and Thornton, J.M., Nature, 326(1987)347.
6. Schuler, G.D., Proteins Struct. Func. Gen., 9(1991)180.
7. Shenkin, P.S., Biopolymers, 26(1987),2053.
8. Summers, N.L., Carlson, W.D. and Karplus, M., J. Mol. Biol., 196(1987)175.
9. Homology User Guide, Version 2.2.0. San Diego: Biosym Technologies, 1993.
10. Luthy, R and Bowie, J.U., Nature, 356(1992)83.

Research of synthesis and biological activity of analogs of two kinins

Shang-Yi Liu, Ming-Nai Zhong, Yu-Xuan Dong, Yu-Lin Li and Xiao-Mei Yuan
Research Institute of Pharmaceutical Chemistry, Beijing 102205, China

Introduction

Bradykinin (BK) can be included in the kinin system of the tissue-kinins. It is not secreted a specialized gland cell, but is an active peptide which originates from the α-globulin of blood when specialized proteases are activated. It is released in large quantities in the pathological or special physiological conditions, although under normal conditions its concentration is very low. Bradykinin is a potent pain-producing substance that mainly relates to the cardiovascular system and exerts powerful physiological effects. In the inflammatory process, it relates to rubor, heat edema (the increase of capillary permeability) and pain. Substance P (SP) is a neuropeptide participating in nervous action and also has kinin effects. It is mainly present in the grey matter of the brain and may be a non-cholinergic nerve transmitter. Acting on the peripheral sensory nerve ending, it produces pain similar to BK but it has potent analgesic effects when acting on the central nervous systerm. These two kinins have been studied to understand the mechanism of pain.

We have synthesized a series of analogs of BK and SP (Table 1) in order to study their structure-activity relationships and to approach the special physiology mechanism.

Table 1 *Amino acid sequences of BK and SP analogs*

SP:	Arg-Pro-Lys-Pro-Gln-Gln-Phe-Phe-Gly-Leu-MetNH$_2$
SP-1: [AcGln^5, Pro^9]SP$_{5\text{-}11}$	AcGln-Gln-Phe-Phe-Pro-Leu-MetNH$_2$
SP-2: SP$_{4\text{-}11}$	Pro-Gln-Gln-Phe-Phe-Gly-Leu-MetNH$_2$
SP-3: [Cys^{5,9}]SP$_{4\text{-}11}$	Pro-Cys-Gln-Phe-Phe-Cys-Leu-MetNH$_2$
BK:	Arg-Pro-Pro-Gly-Phe-Ser-Pro-Phe-Arg
BK-1: Lys-BK$_{1\text{-}8}$	Lys-Arg-Pro-Pro-Gly-Phe-Ser-Pro-Phe
BK-2: [D-Phe^9]LysBK$_{1\text{-}8}$	Lys-Arg-Pro-Pro-Gly-Phe-Ser-Pro-D-Phe
BK-3: [Ala^{3}Trp(CHO)^8]BK	Arg-Pro-Ala-Gly-Phe-Ser-Pro-Trp(CHO)-Arg
BK-4: [des-Pro^2, Arg^8]BK-ProNH$_2$	Arg-Pro-Ala-Gly-Phe-Ser-Pro-Arg-Arg-ProNH$_2$
BK-5: BK-NH$_2$	Arg-Pro-Pro-gly-Phe-Ser-Pro-Phe-ArgNH$_2$

Results and Discussion

Ten active peptides, BK and SP analogs, were synthesized by solid-phase peptide. Synthesis method [1]. MBHA resin (the equivalent of replacement: 0.50mmol/g) and Merrifield resin (the equivalent of replacement: 1.23 mmol/g) were used, respectively, for the six C-terminal amide peptides and the four C-terminal free-acid peptides. Couplings were performed by the active ester and the symmetrical anhydride procedure, and were monitored quantitatively by the ninhydrin method [2]. Nα-Boc deprotection was performed by 50% TFA/CDM-indole. The peptidyl resin products were cleaved with 90% anhydrous HF (containing 7% anisole and 3% thio-anisole). Then the products were purified on C_{18} RP-HPLC, using 0.055 TFA/H_2O and CH_3OH gradient systems for elution. In the separation process, we achieved satisfactory results using CH_3OH instead of CH_3CN. The group of -SH of Cys in SP-3 was protected with Acm, then ACM was deprotected with iodine oxidation to form a bisulfide bridge. All ten peptides were analyzed with amino acid analysis; BK-2, BK-3, BK-4, BK-5, SP-1 and SP-3 were also analized with FAB-MS. The two analytic results agreed.

The results from the is+wated guinea-pig ileum (GPI) tests and the writhing of mice tests on ten peptides are shown in Tables 2, 3, and 4.

Table 2 *Effects of BK analogs on GPI and the writhing of mice (i.p.)*

	GPI		The writhing of mice(i.p)	
	IC_{50}(nM)	R.A.[a]	ED_{50}(nmol/kg)	R.A.
BK	8.13	100	25	100
BK-1	301	2.70	26.03	96.7
BK-2	1.22×10^3	0.67	98.35	26.4
BK-3	1.40	581	0.205	1.26×10^4
BK-4	1.68×10^3	0.48	69.97	37.1
BK-5	2.29	355	2.98	870

[a] Relative activity.

The above mentioned results indicate that the potency of BK analogs on GPI was as follows: BK-3>BK-5>BK>BK-1>BK-2>BK-4; and on the writhing of mice: BK-3 >BK-5>BK>BK-1>BK-4>BK-2. BK-3 is the most potent among these BK analogs, it has about 5.8-fold and 126-fold of the potency of BK, respectively. All of these BK analogs have the effective characteristics of bradykinin. On GPI test, they have shown a relatively long latent period (about 10s) and slowly reached a maximum contraction. The results from the writhing mice tests show that all BK analogs have obvious analgesic effects. The results from biological tests indicate that the displacement of Pro^3, Phe^8 in BK influences its activity and these two sites may be its effective region; BK-5 is potent as the amidation of the BK C-terminal improves its potency; the structural changes in the less effective peptides, such as BK-1, BK-2 and BK-4, are concentrated in the BK C-terminal structure. Thus it can be seen that the change of C-terminal sequence in BK greatly influences its bioactivity. The results of the two tests differ from each other: BK-2 is more potent than BK-4 on GPI test, but less on the writhing of mice test. It shows that these two effects may be mediated by different mechanisms.

Table 3 *Effect of SP analogs on GPI*

	IC_{50}(nM)	R.A.[a]
SP	1.57	100
SP-1	1.12	140
SP-2	2.13	74
SP-3	13.23	12

[a] Relative activity.

Table 4 *Effect of SP analogs on the writhing of mice(i.p)*

	The lowest effective dose(nmol/kg)	The highest dose (nmol/kg)	ED_{50} (nmol/kg)
SPa	0.93	2.32	/
SP-1[a]	132	263	/
SP-2[a]	13.0	25.9	/
SP-3[b]	0.51	2.54	0.65

[a] Only 1/4 animals appeared writhing between two doses.
[b] 1/4~4/4 animals appeared writhing between two doses.

The potencies of the SP analogs in these two tests are: SP-1>SP>SP-2>SP-3 on the GPI test, and SP-3>SP>SP-2>SP-1 on the writhing of mice test.

Among these four peptides, SP-1 is the most potent on the GPI test, but it is the weakest on the writhing of mice. It is possible that N-terminal acetylation increases antienzymolysis, and/or that Pro replacement of Gly^9 increases its selectivity to NK-1 receptor [3]. On the other hand, SP-3 is the weakest on GPI but the most potent on the writhing test, indicating that the formation of bisulfide bridge cyclization on positions 5, 9 may restrict the preferential conformation of SP and increase its selectivity to the binding site. The two tests give such different results that SP may be controlled by different mechanisms in different circumstance.

On the GPI test, the latent period of SP analogs is shorter than that of BK analogs, which is characteristic of tachykinin. SP analogs bind their receptors stably, and are difficult to uncouple. These two kinds of kinin analogs also produce the writhing efect in various degrees when administered intraperitoneally.

References

1. Stewart, J.M., Young, J.D., in Solid Phase Peptide Synthesis, Pierce Chemical Company, 1984 (2nd edition), Pierce Chemical Company Rockford Illinois.
2. Sarin, V.K., Anal. Biochem., 117(1981)147.
3. Haro, I., Int. J. Physiol. Pharmacol., 59(1981)621.

Actions of GWamides and WGTamides on identified central neurones of *Helix aspersa*

M-L. Chen[a], R.P. Sharma[b] and R.J. Walker[a]
[a]*Department of Physiology and Pharmacology and* [b]*Department of Biochemistry, University of Southampton, Southampton SO16 7PX, U.K.*

Introduction

A tetrapeptide, APGWamide, was first isolated from ganglia of the mollusc, *Fusinus* [1] and subsequently in other molluscs including *Helix* [2] where it plays a role in penis eversion and dart release. This peptide has a C-terminal portion similar to the Red Pigment Concentrating Hormone (RPCH: pQLNFSPGWamide) present in crustacean nerve tissue [3] and exhibits some similarity to the Adipokinetic Hormone (AKH-1: pQLNFTPNWGTamide) of insects [4]. Recently a further family of peptides has been isolated in the mollusc, *Achatina*, which has Trp at both the N- and C-terminal, eg, WLEMSVWamide [5]. In the present study the actions of these peptides on identified central neurones of *Helix aspersa* have been investigated. Intracellular recordings were made using current and voltage clamp techniques (n=3-5; ± S.E.M.).

Results and Discussion

Both APGWamide and the synthetic dipeptide, GWamide, possess both presynaptic and postsynaptic actions on *Helix* neurones. Stimulation of the right pallial nerve evokes a dopaminergic inhibitory postsynaptic potential (IPSP) in neurone F-1. While neither peptide has a direct effect on F-1, both reduce the IPSP, eg, 1 μM APGWamide reduces the IPSP amplitude from a control value of 100% to 38.3 ± 11.2%. Neither peptide has any effect on the postsynaptic effect of dopamine on F-1. In contrast, both APGWamide and GWamide, have direct inhibitory actions on cell F-2. The peptides abolish spike activity and hyperpolarize the cell membrane potential, eg, 5 μM APGWamide hyperpolarizes this cell by up to 15 mV. Both peptides have a similar potency with an EC-50 value of around 0.5 μM. The reversal potential for this inhibitory event is -80-85 mV, close to the equilibrium potential for potassium (K) in this cell. In normal saline, 5 μM APGWamide, hyperpolarizes F-2 by 16.7 ± 1.0 mV while in K-free saline this increased to 23.3 ± 1.0 mV. In 16 mM K saline, the hyperpolarization is elimated. The ionic mechanism was further investigated using tetraethylammonium (TEA) and 4-aminopyridine (4-AP), both K channel blockers. Both compounds greatly reduce the APGWamide response and when added together, 5 mM TEA and 250 μM 4-AP, elimate the response. 10 mM Co^{++}, 1 mM La^{++} or 12 mM Cl^{-} saline had no effect on the peptide response. Presynaptic effects on F-2 were examined since stimulation of the right pallial nerve evokes a cholinergic excitatory postsynaptic potential (EPSP) in F-2. In the presence of 10 μM GWamide, this EPSP is greatly attenuated. Acetylcholine excites F-2 and this excitation is unaltered by either peptide.

RPCH, 10 μM, mimics the action of APGWamide on F-2 but is less potent. RPCH also has a very slight direct effect on F-1. In contrast AKH-1, 10 μM, fails to hyperpolarize F-2 but does slightly reduce the firing rate of F-1. RPCH, 10 μM, reversibly reduces an evoked IPSP in F-1 and an EPSP in F-2 in a similar manner to that observed for APGWamide and GWamide. In contrast AKH-1, 10 μM, has no effect on either synaptic event. The synthetic peptides APWGamide, WGamide and Wamide have no direct effect on either cell or on their associated IPSP and EPSP.

These results demonstrate that GWamide peptides have both direct actions to modify neuronal activity and modulatory actions to modify the release of transmitters which in turn influence neuronal activity and neuronal circuitry. Their major effect is one of inhibition as is the case for the recently identified WWamides [5] which inhibit *Achatina* central neurones. The literature identifies that GWamides can modulate the activity of many molluscan muscles including radula protractor, crop, pharyngeal retractor and anterior byssus retractor (ABR) [1]. Certain of these actions are presynaptic to modify transmitter release, eg, serotonin in the case of the ABRM. APGWamide has an inhibitory effect on *Achatina* neurones and this effect is mediated through an increase in K permeability [6]. APGWamide immunoreactivity has been identified in 120-150 neurones in the cerebral ganglia of *Lymnaea*, and has a role in reproduction [7]. It is likely that this peptide is a transmitter both in the peripheral and central nervous systems of molluscs. Structure-activity studies indicate a similar potency sequence for molluscan muscle [8] and central neurones as shown in the present study.

Conclusions

Current evidence strongly suggests that APGWamide is an important transmitter and modulator in molluscan nervous systems. The minimum sequence required for activation of the receptor for this peptide is a dipeptide, GWamide. These two amino acids plus the amide are essential for activation of the receptor. This contrasts with the situation for other molluscan peptide receptors which require a longer minimum C-terminal sequence for activity.

References

1. Kuroki, Y., Kanda, T., Kubota, I., Fujisawa, Y., Ikeda, T., Miura, A., Minamitake, Y. and Muneoka, Y., Biochem. Biophys. Res. Comm., 167(1990)273.
2. Li, G. and Chase, R., Soc. Neurosci. Absts., 18, 531(1992)224.7.
3. Fernlund, P. and Josefsson, L., Science, 177(1972)173.
4. Stone, J.S., Mordue, W., Batley, K. and Morris, H.R., Nature, 263(1976)207.
5. Minakata, H., Ikeda, T., Muneoka, Y., Kobayashi, M. and Nomoto, K. FEBS., 323(1993)104.
6. Liu, G.J., Santos, D.E., Takeuchi, H., Kamatani, Y., Minakata, H., Nomoto, K., Kubota, I., Ikeda, T. and Muneoka, Y., Biochem. Biophys. Res. Comm., 177(1991)27.
7. Croll, R.P., Van Minnen, J., Kits, K.S. and Smit, A.B., In Editor, Kits, K.S., Boer, H.H. and Joosse, J., Molluscan Neurobiology, North Holland Publ. Co., Amsterdam (1991), p248.
8. Minakata, H., Kuroki, Y., Ikeda, T., Fujisawa, K., Nomoto, K., Kubota, I. and Muneoka, Y., Comp. Biochem. Physiol., 100C(1991)565.

Protective effect of KA8 peptide on ethanol-induced gastric mucosal erosion in rat

Yong-Mei Zhao, Xiao-Hong Wang, Shi-Qin Dong and Shu-Li Sheng
Peptide Laboratory, Xuanwu Hospital, Capital Medical University, Beijing 100053, China

Introduction

KA8 peptide is the C-terminal octapeptide of oxyntomodulin (1-37). Oxyntomodulin (1-29) is glucagon. We previously reported that KA8 could inhibit gastric acid secretion induced by pentagastrin, cold-restraint, and TRH. But it is not known whether it can protect the gastric mucosa from chemical damage. To determine the protective effect of KA8 peptide on gastric mucosal erosion, we synthesized KA8 peptide by the solid phase method. Rats were given different doses by hypodermic injection prior to administration of 1 ml absolute ethanol by gastric catheter.

Results and Discussion

Absolute ethanol produced severe damage to the gastric mucosa. Lesions were displayed as elongated hemorrhagic striae and patches, usually parallel to the long axis of the stomach. Most lesions were located in the gastric corpus and the antrum was less affected. No gross lession was found in the forestomach (the non-secreting part covered by a squamous epithelium). This may be due to a thicker epithelial layer. The total area of the lesions was 337.2 ± 26.2 mm^2 (Table 1). Histologically, the lesions included epithelial degeneration of the gastric mucosal surface, significant dilation of capillaries and stroma swelling at the vessel periphery with red cell and plasma infiltration. The depth of the lesion was about one-third to one-half of the mucosa (involving the surface epithelium, the region of mucus nesk cells and of parietal cells).

Table 1 *Protective effect of KA8 peptide on ethanol-induced gastric mucosal erosion in rat*

	NS	KA8 (30μg/kg)	KA8 (60μg/kg)
the area of gastric lesion (mm^2)	337.2 ± 26.2 n=8	98.5 ± 18.1[a] n=8	169.3 ± 15.8[b] n=8

[a] Compare with NS group, P<0.01.
[b] Compare with KA8 (30μg/kg) group, P<0.01.

In the experimental group, when KA8 was administered at 30 μg/kg body weight, the gastric lesion was not as severe as in the control group, with an area of 98.5 ± 18.1 mm^2. When KA8 was given at 60 μg/kg, the lesion aera was 169.3 ± 15.8 mm^2. The results showed that both doses of KA8 could decrease ethanol-induced gastric erosion significantly ($p<0.01$), but protection was not dose-dependent. Histologically, the gastric mucosal epithelial cells maintained their integrity. The dilation of capillaries and stroma swelling was mild compared with the control. The depth of the lesion was smaller and shallower. KA8 cytoprotection has no relationship to the inhibition of gastric acid secretion. Although the mechanism of gastric cytoprotection is unknown, our experiments indicate that KA8 can maintain the cellular integrity of the gastric mucosa, and may be beneficial in the treatment of a variety of diseases in which gastric mucosal injury is present.

References

1. Zhang, J.J., J. of Capital Institute of Medicine, 11(1990)299.
2. Zhang, S.D., J. of Capital Institute of Medicine, 11(1990)302.
3. Shibasaki, T., Life Sci., 47(1990)925.
4. Robert, A., Gastroenterology, 77(1979)433.
5. Laszlo, F., Life Sci., 41(1987)1123.
6. Robert, A., Gastroenterology, 77(1979)761.

Memory-enhancing peptide NLPR increases NGF gene expression in memory-impaired rat brain

An-wu Zhou, Jun Guo and Yu-cang Du
Shanghai Institute of Biochemistry, Academia Sinica, Shanghai 200031, China

Introduction

The tetrapeptide NLPR is an analog of pentapeptide ZNC(C)PR, which is an enzymatic product of arginine-vasopressin. Recently, NLPR was found to be potent in enhancing learning and memory abilities in both normal and memory-impaired rats (to be published). Studies using an animal model for observing the behavioral responses of memory-impaired rats showed that the rats had significantly lower abilities in learning behavioral responses [1]. The nerve growth factor (NGF) is an essential peptide for the survival and development of sympathetic and sensory neurons. It can also induce regrowth of the damaged fibers to give an appropriate substratum for growth [2].

In this study, NGF gene expression was analyzed for levels of mRNA transcription and protein expression after memory-impaired rats were pretreated with peptide NLPR.

Results and Discussion

Northern blot analysis of NGF mRNA was performed with a ^{32}P-labeled NGF antisense RNA probe prepared from a mouse βNGF cDNA clone. Blots generated from an equivalent amount of RNA from control or experimental rats were visualized by autoradiograph and 1.35 kb hybrids representing entire βNGF mRNA were present in larger amounts in the experimental group in both the hippocampus and cerebral cortex (Fig. 1).

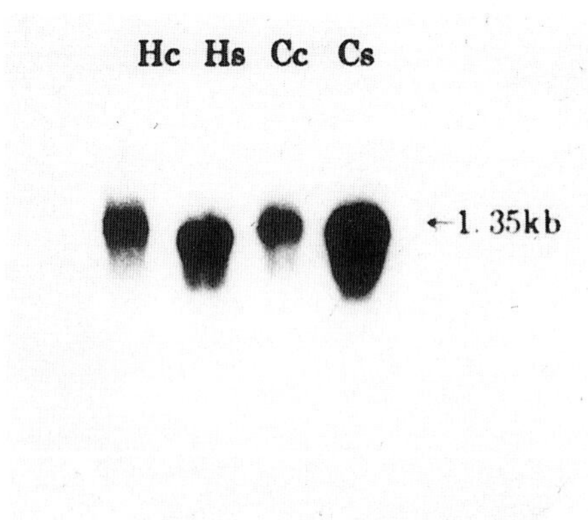

Fig. 1. Effects of NLPR on NGF mRNA level in rat hippocampus(H) and cerebral cortex(C) analyzed by Northern blot. The total RNA samples were 20μg. c:control; s:sample(treated with peptide).

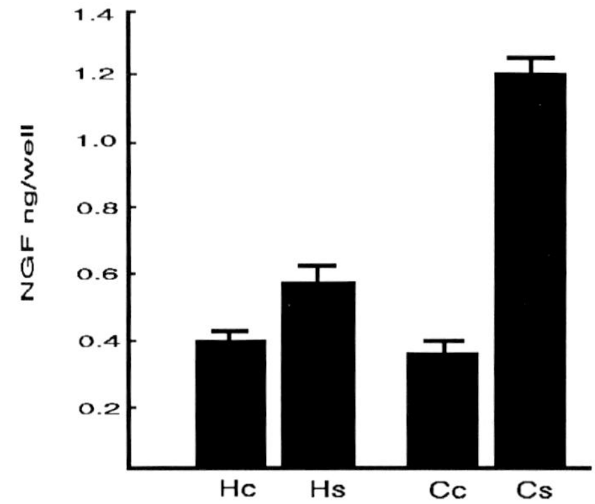

Fig. 2. Effects of NLPR on NGF level in rat hippocampus(H) and cerebral cortex(C) detected by ELISA. The samples contained 100μg protein for each well. Data are based on three parallel tests. c:control; s:sample(treated with peptide).

Furthermore, the product of NGF expression was quantitated by ELISA with rabbit anti-mouse βNGF-2.5S antiserum. Results showed that NGF expression increased significantly in memory-impaired rat brain, especially in their cerebral cortex (as high as 3 times that of the controls) (Fig. 2). These results strongly support the positive effect of NLPR in rat behavioral responses.

This is animal model for memory-impaired rats was based on findings that perinatal hypoxia can cause neurological damage in the central nervous system (CNS), especially in memory-related functional areas, including the cerebral cortex and hippocampus [3]. Considering the function importance of NGF in the CNS, we propose that NLPR can improve memory disability by promoting NGF gene expression and may therefor be a potential therapeutic apent for memory impairment.

Acknowledgements

The authors thank Dr. William J. Rutter for the mouse NGF cDNA clone and Mrs. Wen-yu Wu for kindly providing HPLC purified NLPR peptide.

References

1. Yang, F.L. and Wu, X.H., Shanghai Laboratory Animal Science, 14(1994)28.
2. Ayer-LeLievre, C., Olson, L., Ebendal, T., Seiger, A. and Persson, H., Science, 240(1988)1339.
3. Fenichel, G.M., Neonatal Neurology, New York: Churchill livingstone, 1988.

The chemical semisynthesis of $[B_{1,2}$-Ala, Ala, B_3-Lys]-insulin

Ying Ye, Xin-Tang Zhang and Shang-Quan Zhu
Shanghai Institute of Biochemistry Chinese Academy of Sciences, Shanghai 200031, China

Introduction

Based on X-ray structural analysis of porcine crystal insulin the B-chain N-terminal is located on the molecular surface and appears in the extended status. The amino acids within this region were considered to be unnecessary for interaction of insulin with its receptors. However, different results have been obtained in our laboratory. It was found that $[B_{1,2}$-Ala, Ala]-insulin and $[B_3$-Lys]-insulin possessed higher receptor binding capacity than that of native porcine insulin[1,2]. However, [B2-Lys]-insulin receptor binding activity was only 80% of native porcine insulin[3]. In order to further study the effect of the N-terminal of the insulin B-chain on biological activity, we have prepared $[B_{1,2}$-Ala, Ala, B_3-Lys]-insulin by chemical semisynthesis. The biological activity *in vivo* and receptor binding capacity of this insulin analogue have been determined.

The following scheme shows the procedure for preparation of $[B_{1,2}$-Ala, Ala, B_3-Lys]-insulin:

Porcine crystal insulin (INS)
↓ MSC•ONSu
[MSC]xINS
↓ SP-Sephadex C-25
$[MSC]_2^{A1, B29}$-INS
↓ Edman degradation for 3 cycles
$[MSC]_2^{A1, B29}$des (B1-3)-INS
↓ (1) Coupling with BOC•Ala•Ala•Lys(Boc)•OSu
↓ (2) TFA
$[MSC]_2^{A1,B29}[B_{1,2}$-Ala, Ala, B_3-Lys]-INS
↓ OH^-
crude $[B_{1,2}$-Ala, Ala, B_3-Lys]-INS
↓ (1) SP-Sephadex C-25 (Fig. 1)
↓ (2) DEAE-Sepharose Cl-6B (Fig. 2)
↓ (3) HPLC (Fig. 3)
$[B_{1,2}$-Ala, Ala, B_3-Lys]-INS

Results and Discussion

The purified $[B_{1,2}$-Ala, Ala, B_3-Lys]-insulin was homogenous in polyacrylamide gel electrophoresis (pH 8.3) and cellulose acetate electrophoresis (pH 3.0) (Fig. 4). The

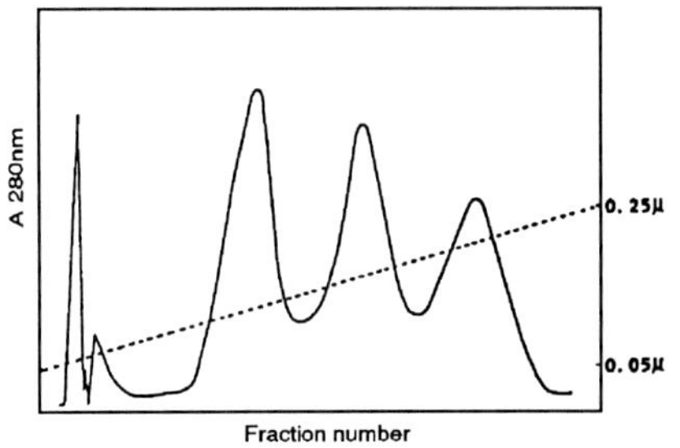

Fig. 1. Elution profile of crude [$B_{1,2}$-Ala, Ala, B_3-Lys]-insulin on SP-Sephadex C-25 (1.3x17cm). Eluent: 9% HOAc, pH 3.0, containing 40% isopropanol with salt gradient (0.05-0.25M). The last peak was collected.

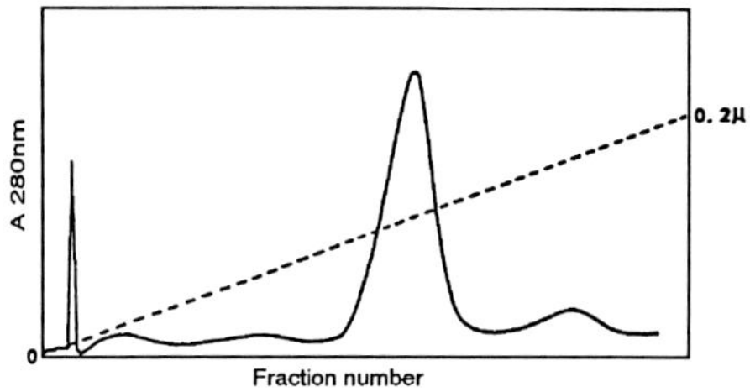

Fig. 2. Elution profile of SP-Sephadex C-25 purified [$B_{1,2}$-Ala, Ala, B_3-Lys]-insulin on DEAE-Sepharose Cl-6B (1.3x17cm). Eluent: 0.05M Tris-HCl, pH 8.3, containing 40% isopropanol with salt gradient (0-0.2M). The main peak is [$B_{1,2}$-Ala, Ala, B_3-Lys]-insulin.

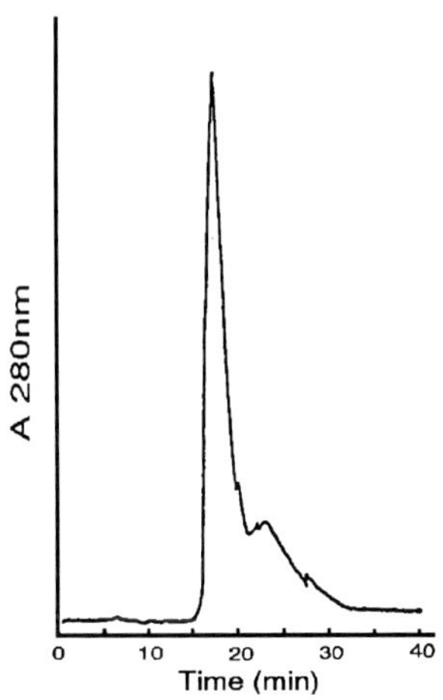

Fig. 3. Purification of [$B_{1,2}$-Ala, Ala, B_3- Lys]-insulin on HPFLC RP-300 (0.7x25cm).
Solvent system: A: 0.125% TFA; B: 60% CH_3CN, 0.15% TFA
Flow rate: 2.5 ml/min.

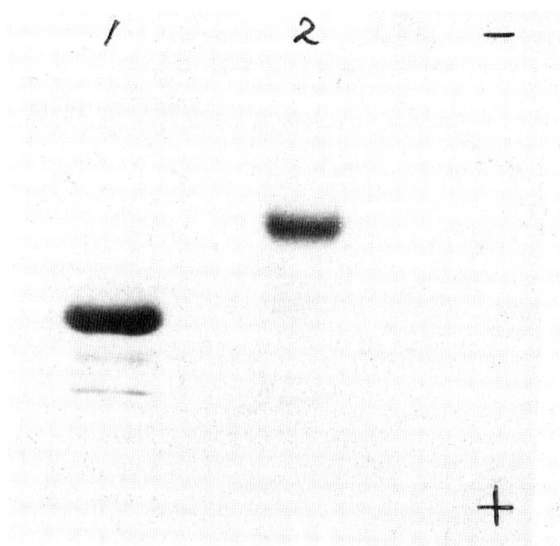

Fig. 4. Polyacrylamide gel electrophoresis of [$B_{1,2}$-Ala, Ala, B_3-Lys]-insulin.
1. MC-insulin
2. [$B_{1,2}$-Ala, Ala, B_3-Lys]-insulin.

amino acid compositions were in accordance with the theoretical values (Table 1). The biological activity *in vivo* was 50% compared with native porcine insulin. According to the IC (50) values of sample and native porcine insulin, the receptor binding capacity of [$B_{1,2}$-Ala, Ala, B_3-Lys]-insulin on human placental membrane was 72% of native porcine insulin (Fig. 5).

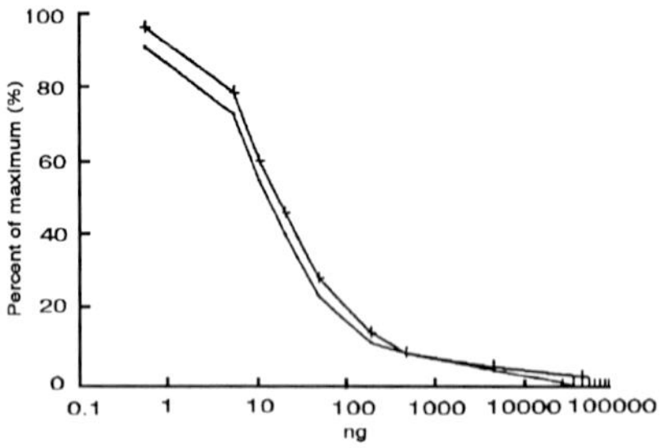

Fig. 5. Displacement curve of [$B_{1,2}$-Ala, Ala, B_3-Lys]-insulin binding in human placental membrane.
•--• *MC-insulin*
+--+ *[$B_{1,2}$-Ala, Ala, B_3-Lys]-insulin*

As mentioned above, it was found that [B_3-Lys]-insulin and [$B_{1,2}$-Ala, Ala]-insulin both have higher receptor binding capacity than that of native porcine insulin. Our purpose in preparing [$B_{1,2}$-Ala, Ala, B_3-Lys]-insulin was to obtain an insulin analogue with higher receptor binding capacity than that of [B_3-Lys]-insulin and [$B_{1,2}$-Ala, Ala]- insulin. However, from the data presented in this paper it has been shown that the N-terminal amino acids of the insulin B-chain probably have a dependent relationship in maintaining insulin activity and take part in the interaction between insulin and its receptors.

Table 1 *Amino acid compositions of porcine insulin and [$B_{1,2}$-Ala,Ala, B_3-Lys]-insulin*

AA	porcine insulin	[$B_{1,2}$-Ala-Ala, B_3-Lys]-insulin
Asp	2.91 (3)	2.08 (2)
Thr	1.92 (2)	1.87 (2)
Ser	2.88 (3)	2.75 (3)
Glu	6.56 (7)	6.76 (7)
Pro	1.00 (1)	1.04 (1)
Gly	4.00 (4)	4.00 (4)
Ala	1.98 (2)	3.56 (4)
Val	4.00 (4)	3.04 (3)
Ile	2.00 (2)	1.79 (2)
Leu	5.98 (6)	5.88 (6)
Tyr	4.00 (4)	3.83 (4)
Phe	2.73 (3)	2.43 (2)
Lys	0.94 (1)	1.94 (2)
His	1.94 (2)	1.76 (2)
Arg	1.02 (1)	1.12 (1)

Note: ( ) Theoretical value; Cys not determined.

References

1. Zhu, S.Q., Ye, Y., Xu, M.H., Zhang, X.T. and Xu, Y.G., Acta. Biochim. Biophys. Sin., in press.
2. Xu, M.H., Ye,Y. and Zhu, S.Q., Acta. Biochim. Biophys. Sin., 24(1992)213.
3. Xu, M.H., Ye, Y., Zhang, X.T. and Zhu, S.Q., Chinese. Biochemical Journal, 8(1992)410.

The importance of C terminal peptide (KA8) of oxyntomodulin (OM) on OM bioactivities

Shu-Li Sheng, Xiao-Hong Wang and Yong-Mei Zhao
Peptide Laboratory, Xuanwu Hospital, Capital Medical University, Beijing 100053, China

Introduction

Oxyntomodulin (OM) consists of 37 amino acid residues. OM 1-29 is glucagon (Gg), and C-terminal octapeptide is known as KA8 [1]. OM is synthesized in intestinal L cells and in the hypothalamus. The major function of OM is to inhibit gastric secretion [2]. To investigate the importance of KA8 in OM structure, we synthesized KA8 by synthesizer.

Results and Discussion

A. Inhibition on gastric secretion. All the rats were fasted for 24 hrs. The pylorus and cardia were ligated under light ether anasthesia. As soon as the rats recovered consciousness, KA8 (60 μg/kg) was given by iv to the rats which had pentagastrin (120μg/kg) iv administered or had been under cold-restraint or TRH (10μg/μl) administered 4 hrs previously. Gastric juice analysis indicated that KA8 significantly reduce secretion total acidity and free acid in the above three conditions (Tables 1 and 2).

Table 1 *Effect of KA8 peptide on gastric juice secretion induced by TRH administration in to the intracerebraventricle(icv)*

	NS group (n=6)	KA8 group (n=8)
Volume (X+Sml)	1.34±0.59	0.646±0.260[a]
Free acid ($x \pm S \times 10^{-3}$mmol/L)	3.31±2.96	0.718±1.360[b]
Total acid ($x \pm S \times 10^{-3}$mmol/L)	19.18±14.08	7.230±5.250[b]

[a] N<0.02. [b] P<0.05.

Table 2 *Effect of KA8 peptide on gastric juice secretion induced by cold-restraint*

	NS group (n=11)	KA8 group (n=12)
Volume (X+Sml)	1.05±0.669	0.455±0.200[a]
Free acid ($x \pm S \times 10^{-3}$mmol/L)	6.11±5.98	0.725±1.150[a]
Total acid ($x \pm S \times 10^{-3}$mmol/L)	27.36±26.43	7.620±4.040[b]

[a] P<0.01. [b] P<0.02.

When both KA8 (24μg/kg) and OM (4μg/kg) were injected in icv, the analysis of GJ was normal.

B. Effect on blood sugar (BS) and hormone levels. KA8 (130μg/kg) or OM (4μg/kg) was infused for 1 hr, with rat plasma taken before and at the end of the infusion. Neither BS nor insulin (IS) concentrations changed. The administration of KA8 by icv gave the same result. However, OM icv injection caused IS to decrease and consequently, BS to increase, although the Gg concentration was stable (Figs. 1-3).

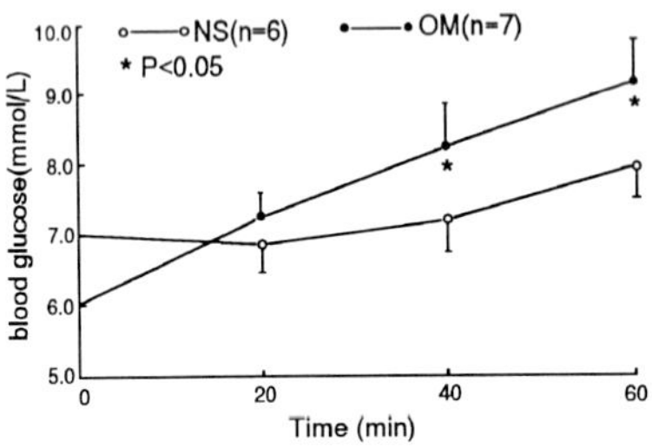

Fig. 1. Effect of OM (1ug) icv injection on blood sugar level.

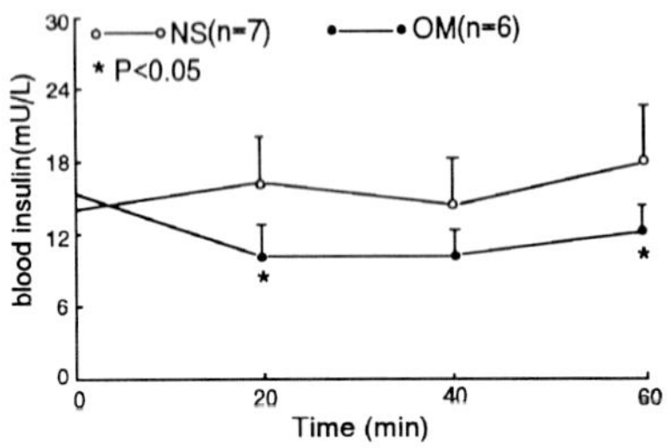

Fig. 2. Effect of OM (1ug) icv injection on insulin secretion.

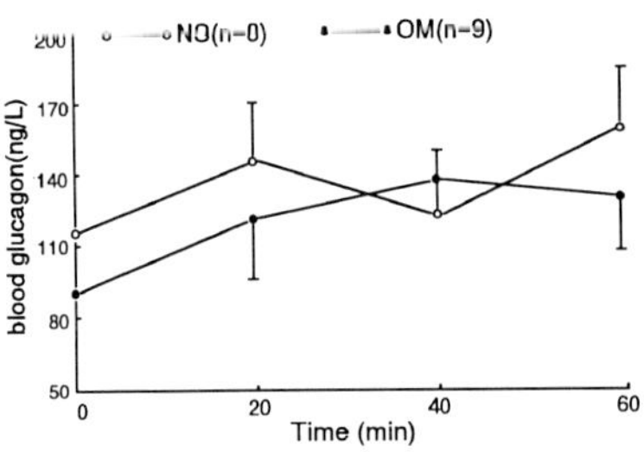

Fig. 3. Effect of OM(1ug) icv injection on glucagon secretion.

Our experiments indicate that: 1. KA8 can protect gastric secretion from three kinds of stimulating agents. Our results imply that the C-terminal peptide is the molecular basis for OM biological activity. It may involve specific receptor mediated signal transduction of stimulating agents. 2. KA8 and OM given by iv has no effect on BS and IS. In fact, C terminal peptide blocks the bioactivity of the N terminus(Gg) of OM. 3. According to the result of icv administration, it is possible that there is no receptor for the C-terminus of OM in brain tissus. Probably OM combines with Gg receptor by its N terminus to participe in control of BS concerntrations.

Our conclusion is that the C-terminus of OM is the principal molecular structure for inhibition of gastric secretion, but it does not play a role in sugar metabolism.

References

1. Shimizu, I., Endocrinology, 121(1987)1076.
2. Dubrasquet, M., Biosci Rep, 2(1982)391.

Session IV
Endocrinological peptides

Chairs: P.W. Schiller
Clinical Research Institute of Montreal
Montreal, Quebec, Canada

and

Xiao-Yu Hu
Lanzhou University
Lanzhou, China

Structure activity relationships (SAR) of corticotropin releasing factor

Jean Rivier, Catherine Rivier, Steve Sutton, Antonio Miranda, Steven C. Koerber, Jozsef Gulyas, Sabine L. Lahrichi, Anne Corrigan, A. Grey Craig and Wylie Vale

Clayton Foundation Laboratories for Peptide Biology, The Salk Institute, 10010 N. Torrey Pines Road, La Jolla, CA 92037, U.S.A.

Introduction

Corticotropin releasing factor (CRF) is a 41-peptide amide which stimulates the release of ACTH [1] among many other activities [2]. The actions of CRF are mediated through effective binding to a CRF receptor which has been recently characterized [3, 4]. The cloned cDNA of the CRF receptor encodes a 415-amino acid protein comprising seven putative membrane-spanning regions. Other members of this family of G protein-coupled receptors include receptors for growth hormone releasing factor, calcitonin and vasoactive intestinal peptide. It was also found that CRF's activities may be modulated by a 37 kDa CRF binding protein (CRF-BP) [5]. On the basis of C^{α}methyl substitutions of some amino acids in the sequence, which have led to more potent analogs and CD spectra showing significant α-helical content, CRF has been postulated to assume an α-helical conformation upon binding to its pituitary receptor [6]. A recent NMR study of hCRF in TFE/water (66/34, v/v) confirmed these results and identified a well defined α-helix between residues 6 to 36 with an extended N-terminus and a disordered C-terminus [7]. The first half of the α-helix was clearly amphipathic as recognized earlier from model experiments [8]. We have exploited this hypothesis in the design of a limited series of cyclic analogs and have taken into consideration the effects of side chain deletion (Alanine scan, [9]) as well as of changes in chirality (D-series scan, [10]) with the rationale that side chains necessary for binding could also be replaced by side chain bridges. In particular, we have used computer modeling to predict likely side chain bridging opportunities, and evaluated the effects of such replacements by correlating biological results with those derived from CD spectroscopy [8]. To further test this hypothesis, we have embarked on the synthesis of an i-(i + 3) bridge scan (Glu-X-X-Lys) of the whole CRF sequence. The CRF receptor and binding protein are also shown to have differential binding to a number of ligands [11].

Results and Discussion

In view of encouraging results with the introduction of lactam rings in both CRF [8] and GRF [12], we have systematically scanned the entire length of the CRF sequence with an i (i+3) Glu/Lys lactam bridge. Side chain to side chain cycles were introduced at the C-terminus of our antagonist standard (compound 1, Table 1), whereas cycles at the N-terminus were introduced in the full length molecule to yield CRF agonists. In an extensive study of the influence of bridge length (using a combination of Asp/Glu, Lys/Orn/Dbu), composition (Cystine bridge) and chirality of the bridge heads (LL, DD, DL and LD) Cyclo(20-23)[DPhe^{12},Lys^{23},Nle^{21,38}]hCRF$_{12-41}$ was shown to have optimized potency [8]. Because Glu and Arg are the native amino acids at positions 20 and 23, respectively, in hCRF, it is thought that the α-helix in that part of the molecule is stabilized by an ionic interaction between the side chains of these two residues. A similar ionic stabilization: Glu to Arg or Lys can be found in most other members of the CRF family, which includes sauvagine, sucker and carp urotensins I. Interestingly, three urotensins I (sole, maggy and flounder) have His at position 20 and Lys at position 23. Although all members of this family are essentially equipotent at releasing ACTH from dispersed pituitary cells in culture (EC$_{50}$ varying from 0.024 to 0.43), their relative affinity for the CRF-BP can vary by a factor of >1,000 while their affinity for the receptor falls within a factor of 2 (see below Table 2).

Table 1 shows that introduction of an i (i+3) Glu/Lys lactam bridge can have dramatic effects on potency. Compounds 10 and 13 are inactive at the doses tested. Five analogs (compounds 3, 4, 6, 7, 9) are less than 10% as potent as the parent analog 1. Three analogs (2, 11 and 12) are, respectively, 16%, 20% and 12% as potent as 1 respectively and compound 8 is about equipotent to the standard. As mentioned earlier, compound 5 is approximately three times as potent as the standard. Although in the past we have found little correlation between helicity as evidenced by CD and biological activity [8], we suspect that some of these more rigid peptides may show increased α-helicity as compared to the parent compound.

Because of our interest in understanding the mechanism of action of CRF at its receptor and of its interaction with the CRF-BP, we have also tried to identify which minimal structural components of CRF are required for binding to recombinant CRF-BP. We know that, upon binding to CRF-BP, CRF loses its ability to stimulate hypophyseal ACTH release (affinity of CRF for CRF-BP is ca. 20 times greater than that for the CRF receptor). We have found that CRF-BP binds hCRF and cUrotensin I with high affinity, sauvagine with moderate affinity, and oCRF with low affinity (Table 2). Unexpected was the observation that extension at the C-terminus or deletion at the N- and C-termini (Compounds 16-18) (which result in analogs with no affinity for the CRF receptor) yielded analogs that retained very high affinity for CRF-BP, suggesting completely different mechanisms of action. The CRF antagonist a-helical CRF(9-41) has high affinity for CRF-BP while [DPhe^{12},Nle^{21,38}]hCRF(12-41) has low affinity (data not shown) [11]. Because both of these peptides are commonly used as CRF receptor antagonists, we suspect that this observation should be taken into consideration in the interpretation of the *in vivo* data gathered with these analogs in situations where the CRF-BP is present.

Sequence of hCRF:

5 10 15 20 25 30 35 40
S E E P P I S L D L T F H L L R E V L E M A R A E Q L A Q Q A H S N R K L M E I I-NH$_2$

Table 1 *CRF antagonists: Glu to Lys (i, i+3) bridge scan*

	Compounds	Relative potency *In vitro*
1	$[DPhe^{12},Nle^{21,38}]hCRF_{(12-41)}$	1.0 Std
2	cyclo(17-20)$[DPhe^{12},Glu^{17},DLys^{20},Nle^{21,38}]hCRF_{(12-41)}$	0.16(0.05-0.5)
3	cyclo(18-21)$[DPhe^{12},Glu^{18},Lys^{21},Nle^{38}]hCRF_{(12-41)}$	0.1(0.03-0.3)
4	cyclo(19-22)$[DPhe^{12},Glu^{19},Lys^{22},Nle^{21,38}]hCRF_{(12-41)}$	0.04(0.01-0.08)
5	cyclo(20-23)$[DPhe^{12},Lys^{23},Nle^{21,38}]hCRF_{(12-41)}$	2.9(1.3-6.7)
6	cyclo(31-34)$[DPhe^{12},Nle^{21},Glu^{31},Lys^{34},Nle^{38}]hCRF_{(12-41)}$	0.01(0.001-0.07)
7	cyclo(32-35)$[DPhe^{12},Nle^{21},Glu^{32},Lys^{35},Nle^{38}]hCRF_{(12-41)}$	0.06(0.03-0.1)
8	cyclo(33-36)$[DPhe^{12},Nle^{21},Glu^{33},Lys^{36},Nle^{38}]hCRF_{(12-41)}$	0.8(0.4-1.7)
9	cyclo(34-37)$[DPhe^{12},Nle^{21},Glu^{34},Lys^{37},Nle^{38}]hCRF_{(12-41)}$	0.003(0.001-0.007)
10	cyclo(35-38)$[DPhe^{12},Nle^{21},Glu^{35},Lys^{38}]hCRF_{(12-41)}$	<0.001
11	cyclo(36-39)$[DPhe^{12},Nle^{21},Glu^{36},Nle^{38},Lys^{39}]hCRF_{(12-41)}$	0.2(0.08-0.4)
12	cyclo(37-40)$[DPhe^{12},Nle^{21},Glu^{37},Nle^{38},Lys^{40}]hCRF_{(12-41)}$	0.12(0.04-0.3)
13	cyclo(38-41)$[DPhe^{12},Nle^{21},Glu^{38},Lys^{41}]hCRF_{(12-41)}$	<0.001

Analogs were made using SPPS and purified using preparative HPLC [8]. Analogs were characterized using HPLC, AAA, CZE and MS [8]. Cyclizations were performed on the resin according to protocols described earlier [8]. Inhibition of CRF induced ACTH release by primary culture of rat pitutitary cells was the basis of the assay [8].

Table 2 *Binding affinities of CRF agonists*

	Compound	CRF-BP Ki (nM)	EC_{50} *in vitro* (nM)	Receptor Ki (nM)
14	oCRF	>1000	0.30	2.3
15	hCRF	0.17	0.43	3.2
16	hCRF-Cly-OH	0.24	7.6	>1000
17	hCRF(6-33)OH	3.5	3,000	>1000
18	hCRF(9-33)OH	11	71,000	>1000
19	carp Urotensin I	0.052	0.17	2.0
20	Sauvagine	17	0.024	1.2

K_i and EC_{50} values represent the mean and SEM of 3 or more separate assays. See Sutton et al. [11] for experimental details.

The fact that CRF-BP has low affinity for oCRF (K_i = 1110 ± 97 nM) as compared to hCRF (K_i = 0.166 ± 0.008 nM), and that both molecules are highly homologous between residues 3 to 36, which encompass the central domain (residues 9 to 33), suggests that the three different amino acids at positions 22 (Ala in hCRF for Thr in oCRF), 23 (Arg for Lys), and/or 25 (Glu for Asp) are critical for binding to CRF-BP in very subtle ways since the corresponding substitutions of these three amino acids are functionally very similar. It is conceivable that such an analog will have physiological and therapeutic implications. Indeed, hCRF(9-33)OH with high affinity for CRF-BP, low affinity for the CRF receptor and low potency as a CRF agonist *in vitro*, is capable of blocking the effect of CRF-BP [11].

Acknowledgements

This work was supported in part by NIH grant no. DK-26741, The Hearst Foundation, FAPESP, the Foundation for Research, California Division. Drs. W. Vale, C. Rivier and A. Craig are FR investigators. We thank L. Cervini, C. Miller, R. Kaiser, J. Porter, D. Pantoja, Y. Haas and S. Leonard for technical assistance and D. Johns for manuscript preparation.

References

1. Vale, W., Spiess, J., Rivier, C. and Rivier, J., Science, 213(1981)1394.
2. Taché, Y. andRivier, C., (Eds.), Corticotropin-Releasing Factor and Cytokines: Role in the Stress Response, Annals of the New York Academy of Sciences, vol. 697, The New York Academy of Sciences, New York, 1993.
3. Chen, R., Lewis, K.A., Perrin, M.H. and Vale, W.W., Proc. Natl. Acad. Sci., USA, 90(1993)8967.
4. Chang, C.P., Pearse, R.V., II, O'Connell, S. and Rosenfeld, M.G., Neuron, 11(1993)1187.
5. Potter, E., Behan, D.P., Fischer, W.H., Linton, E.A., Lowry, P.J. and Vale, W., Nature, 349(1990)423.
6. Hernandez, J.-F., Kornreich, W., Rivier, C., Miranda, A., Yamamoto, G., Andrews, J., Taché, Y., Vale, W. and Rivier, J., J. Med. Chem., 36(1993)2860.
7. Romier, C., Bernassau, J.-M., Cambillau, C. and Darbon, H., Protein Engineering, 6(1993)149.
8. Miranda, A., Koerber, S.C., Gulyas, J., Lahrichi, S., Craig, A.G., Corrigan, A., Hagler, A., Rivier, C., Vale, W. and Rivier, J., J. Med. Chem., 37(1994)1450.
9. Kornreich, W.D., Galyean, R., Hernandez, J.-F., Craig, A.G., Donaldson, C.J., Yamamoto, G., Rivier, C., Vale, W. and Rivier, J., J. Med. Chem., 35(1992)1870.
10. Rivier, J., Rivier, C., Galyean, R., Miranda, A., Miller, C., Craig, A.G., Yamamoto, G., Brown, M. and Vale, W., J. Med. Chem., 36(1993)2851.
11. Sutton, S.W., Behan, D.P., Lahrichi, S., Kaiser, R., Lowry, P., Potter, E., Rivier, J. and Vale, W.W., Endocrinology, in press.
12. Cervini, L.A., Corrigan, A., Donaldson, C.J., Koerber, S.C., Vale, W.W. and Rivier, J.E., In Hodges, R.S. and Smith, J.A. (Eds.) Peptides: Chemistry, Structure and Biology (Proceedings of the 13th American Peptide Symposium), ESCOM, Leiden, 1994, pp. 541–543.

Preparation and antinociceptive effect of enkephalin - poly(ethylene glycol) hybrids

Koichi Kawasaki[a], Mitsuko Maeda, Yuko Yamashiro, Tadanori Mayumi[b], Masakatsu Takahashi[c] and Hiroshi Kaneto[c]

[a]*Faculty of Pharmaceutical Sciences, Kobe Gakuin University, Ikawadani-cho, Nishi-ku, Kobe 651-21, Japan*

[b]*Faculty of Pharmaceutical Sciences, Osaka University, Yamadaoka, Suita-shi, Osaka 565, Japan*

[c]*Faculty of Pharmaceutical Sciences, Nagasaki University, Bunkyo-machi, Nagasaki-shi 852, Japan*

Introduction

Since poly(ethylene glycol) (PEG) is stable, has low toxicity, is bioinert and is only weakly immunogenic, it seems to be a promising candidate for a drug-carrier.

Recently we found that the hybrid formation of some oligopeptides with amino-PEG (aPEG) resulted in potentiation of the biological activity. We prepared PEG hybrids of laminin- and fibronectin-related peptides, (Tyr-Ile-Gly-Ser-Arg-Gly)-aPEG and (Arg-Gly-Asp)-aPEG, and examined their inhibitory effect on experimental metastasis in mice[1,2]. The inhibitory effect of (Tyr-Ile-Gly-Ser-Arg-Gly)-aPEG and (Arg-Gly-Asp)-aPEG was much more potent than that of the parent peptides, Tyr-Ile-Gly-Ser-Arg-Gly and Arg-Gly-Asp.

Following these studies, we prepared leucine enkephalin (Leu-Enk) analog-aPEG hybrids and examined their antinociceptive effect.

Results and Discussion

For preparation of the hybrids, PEG #4000(MW3000-3700) was converted to amino-PEG(aPEG) according to the procedure reported by Pillai and Mutter[3]. PEG#4000 was selected from various kinds of PEG because of our speculation that the bulky PEG portion is not large enough to hinder the binding of the peptide portion to its receptor. β-Endorphine, a potent analgesic containing enkephalin sequence, has a molecular weight of 3294, which is not much different from that of the LeuEnk-aPEG(#4000) hybrid. The (Leu-Enk)-aPEG hybrid was prepared by the solution method. The hybrid was purified by HPLC and the enkephalin content of the hybrid was 0.35 mmol/g. The hybrid was soluble in dimethylforamide, water, methanol, dichloromethane, and a 1:1 mixture of methanol and dichloromethane.

The antinociceptive effect of the hybrid administered intracerebroventricularly(i.c.v.) was examined by the tail-pinch method. The antinociceptive effect of the hybrid was compared with that of Leu-Enk. Prior to the assay, the analgesic effect of PEG and aPEG was examined and no effect was observed. LeuEnk did not produce any appreciable antinociception at doses up to 200 nmol/animal; however, the hybrid at doses of 30 and 100 nmol/animal exhibited a remarkable antinociceptive effect, and the effect of 100 nmol/animal of the hybrid is equipotent to that of 3 mg (about 10 nmol)/animal i.c.v. morphine. Thus the antinociceptive effect of the hybrid is more potent than that of LeuEnk in terms of their molar ratio.

The hybrid of [D-Ala^2]LeuEnk and aPEG(#4000) was also prepared by the solution method. The final product was purified by LH-20 column chromatography using a mixture of methanol/dichloromethane (MeOH/DCM) as an eluent, followed by RP-HPLC. The peptide content of the hybrid was 0.39 mmol/g.

Inhibitory effects of the hybrid and [DAla^2]LeuEnk on electrically induced contraction of the mouse vas deferens (MVD) are shown in Table 1.

Table 1 *Inhibitory Effect of Enkephalin Analogs and the aPEG Hybrid on the Electrically Induced Contraction of Isolated Mouse Vas Deferens (MVD)*

	MVD	
	IC_{50} (nM)	Retative Activities
Leu-Enk	12.00	1.00
[DAla^2]LeuEnk	6.08	1.97
[DAla^2]LeuEnk-aPEG*	0.05	240.00

* Calculated as [DAla^2]LeuEnk contained in the hybrid.

IC_{50} of the hybrid was calculated as the concentration of [DAla^2]LeuEnk contained in the hybrid. As shown in Table 1, the inhibitory effect of the [DAla^2]LeuEnk on MVD contraction was enhanced 120 times by hybrid formation with aPEG. Of LeuEnk, 12.0 nmol equals 7.1 mg; of [DAla^2]LeuEnk, 6.08 nmol is 3.7 mg; and of the hybrid, 0.05 nmol is 0.13 mg. It can thus be said that the effect of the hybrid was more potent than that of [DAla^2]LeuEnk even in terms of weight ratio.

The antinociceptive effect of the hybrid administered i.c.v. was examined by the tail-pinch method. [DAla^2]LeuEnk produced no prominent effect at doses up to 30 nmol/animal: however, at a dose of 10 nmol/animal exhibited a remarkable antinociceptive effect. The effect of 10 nmol of the hybrid was equipotent to that of 100 nmol of [DAla^2]LeuEnk. Thus it can be said that the antinociceptive effect of [DAla^2]LeuEnk was enhanced 10 times by hybrid formation with a PEG in terms of molar ratio.

In addition to the replacement of Gly^2 with DAla, *N*-alkylation of Phe was also reported as a strategy to enhance the opioid activity of enkephalin analogs.[4] Hybrids of [DAla^2]LeuEnk analogs containing *N*-methylated Phe [(Me)Phe] and D-Leu (DLeu) were therefore prepared. Hybrids of [DAla^2,(Me)Phe^4]LeuEnk analog: Tyr-DAla-Gly-(Me)Phe-aPEG, Tyr-DAla-Gly-(Me)Phe-Leu-aPEG and Tyr-DAla-Gly-(Me)Phe-DLeu-aPEG were prepared by the solution method.

Antinociceptive effects of the synthetic peptides and the hybrids are shown in Table 2.

Table 2 *Comparison of the Antinociceptive Effect of Authentic Peptides and their hybrids*

Drug	Dose/mouse	n	Antinociceptive effect (A U C)
Morphine	3 μg	28	618 ± 65
H-Tyr-DAla-Gly-(Me)Phe-OH	10 nmol	7	320 ± 63
	30 nmol	7	390 ± 94
H-Tyr-DAla-Gly-(Me)Phe-aPEG	10 nmol	6	349 ± 78
	30 nmol	6	394 ± 59
H-Tyr-DAla-Gly-(Me)Phe-Leu-OH	10 nmol	7	313 ± 92
	30 nmol	7	370 ± 88
H-Tyr-DAla-Gly-(Me)Phe-Leu-aPEG	10 nmol	7	885 ± 76**
	30 nmol	7	878 ± 81**
H-Tyr-DAla-Gly-(Me)Phe-DLeu-OH	3 nmol	7	83 ± 35
	10 nmol	7	113 ± 52
	30 nmol	7	126 ± 95
H-Tyr-DAla-Gly-(Me)Phe-DLeu-aPEG	10 nmol	6	803 ± 47**
	30 nmol	7	997 ± 32**

** : $P < 0.01$, compared with corresponding authentic peptide at the dosage.

H-Tyr-DAla-Gly-(Me)Phe-OH or H-Tyr-DAla-Gly-(Me)Phe-Leu-OH at doses of 10-30 nmol/animal administered i.c.v., produced the antinociceptive effect. However, the effect was less potent than that of morphine, 3 mg/animal. No appreciable antinociceptive effect of H-Tyr-DAla-Gly-(Me)Phe-DLeu-OH was found up to doses of 30 nmol/animal. Thirty nmol/animal of H-Tyr-DAla-Gly-(Me)Phe-aPEG, similar to its parent peptide, produced a weak antinociceptive effect and the effect was not comparable to morphine at dose of 3 mg/animal. On the other hand, H-Tyr-DAla-Gly-(Me)Phe-Leu-aPEG, 10 and 30 nmol/animal, and H-Tyr-DAla-Gly-(Me)Phe-DLeu-aPEG, 10 and 30 nmol/animal, compared with the authentic peptides, produced remarkable antinociception. The potency of these compounds was higher than that of 3 mg/animal of morphine, and the effect lasted at least 120 min.

In this study, the two hybrids (H-Tyr-DAla-Gly-(Me)Phe-Leu-aPEG and H-Tyr-DAla-Gly-(Me)Phe-DLeu-aPEG) produced remarkable antinociception compared with their parent peptides. The effect of the former at doses of 10 and 30 nmol/mouse and of the latter at doses of 10 and 30 nmol/mouse was more potent than that of 3 mg/mouse of morphine, and lasted at least 120 min. Unexpectantly, the LLeu epimer, H-Tyr-DAla-Gly-(Me)Phe-Leu-OH produced antinociception, whereas no appreciable antinociception by the DLeu epimer, H-Tyr-DAla-Gly-(Me)Phe-DLeu-OH was observed. However, when these peptides formed hybrids with aPEG, the hybrid with DLeu epimer produced a marked antinociceptive effect, and the effect of the hybrid with LLeu epimer was also intensified by the hybrid formation. One possible reason for the enhancement of antinociception by hybrid formation may be slower enzymatic degradation as compared with the authentic peptide. To examine the stability of PEG hybrids to enzymatic degradation, the LeuEnk-aPEG hybrid (H-Tyr-Gly-Gly-Phe-Leu-aPEG) was treated with chymotrypsin and the recovery of Tyr in a digest was compared with that of LeuEnk. Hydrolysis at the Tyr bond in the hybrid was slower than that in LeuEnk.

The low toxicity and low immunogenicity of PEG and the potentiation and prolongation of the antinociceptive effect of opioid peptides by hybrid formation with PEG imply that hybrids may be a useful adjunct in treating patients for chronic and severe pain.

References

1. Kawasaki, K., Namikawa, M., Murakami, T., Mizuta, T., Iwai, Y., Hama, T. and Mayumi, T., Biochem. Biophys. Res. Commun., 174(1991)1159.
2. Kawasaki, K., Namikawa, M., Yamashiro, Y., Hama, T. and Mayumi, T., Chem. Pharm. Bull., 39(1991)3373.
3. Pillai, V.N.R. and Mutter, J. Org. Chem., 45(1980)5364.
4. a) Schuman, R.T., Gesellchen, P.D., Smithwick, E.L. and Frederickson, R.C.A., "Peptides: Synthesis, Structure, Function ", ed. by Rich, D.H. and Gross, E., Pierce Chemical Co., Rockford, Illinois, 1981, pp.617-620. b) Kiso, Y., Miyazaki, T., Satomi, M., Inai, M., Akita, T., Moritoki, H., Takei, M. and Nakamura, H., "Peptide Chemistry 1981", ed. by Shioiri, T., Protein Research Institute, Osaka, 1982, pp.65-70.

BST refolding: The formation of dimers and aggregates

Jacob S. Tou, Bernard N. Violand, S. Brad Storrs and Michael R. Schlittler
Monsanto Company, 700 Chesterfield Village Pky N.,
St. Louis, MO 63198, U.S.A.

Introduction

Bovine somatotropin (bST) is a single chain polypeptide that contains 191 amino acids with two disulfide bonds (Fig. 1). It can be expressed in E. coil as an inactive specices in inclusion bodies [1]. One of the key steps in the production of active bST is the renaturation process. During this process, reduced bST is refolded and oxidized to the native state with the correct disulfide bond linkages. However, as a result of foldin/oxidation of reduced bst, several high molecular weight species are also generated as side products. This report describes the structures of two unique dimers gennerated during folding/oxidation of reduced bST. The formation of these two dimers and aggregates (high moecular weight species with undefined structure) as a function of three key refolding parameters is also discussed.

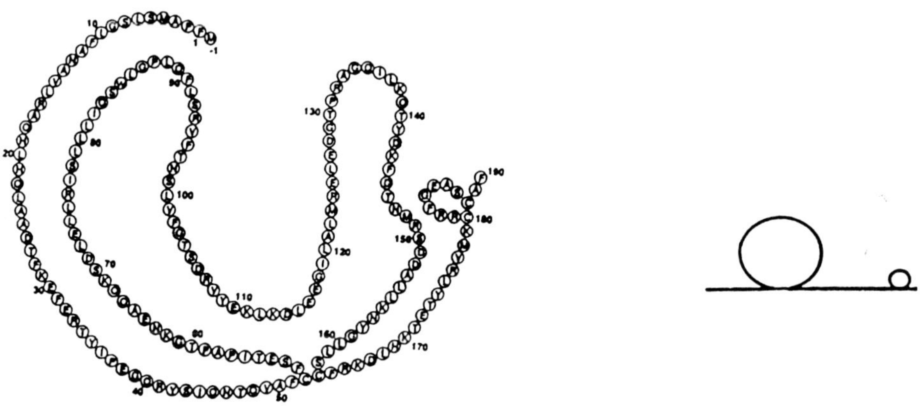

Fig. 1. Amino acid sequence of bST. The large and small circles represent the large and small disulfide loops, respectively.

Results and Discussion

Dimer structures

There arre two types of dimer structures that are derived from bST monnomer. The concatenated structural type is formed by two monomers that are held together by the interlocking of two disulfid loops (eg. Dimer B, Fig. 2). On the other hand, mispaired

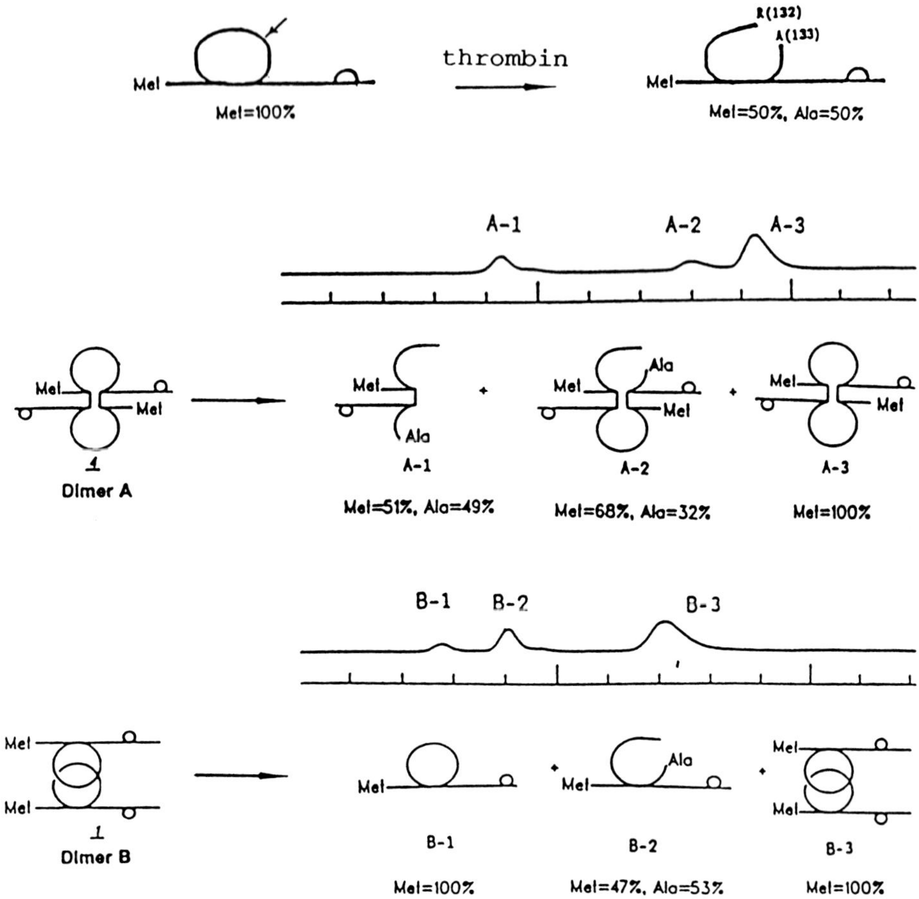

Fig. 2. Thrombin cleavage reaction. (Top) Thrombin can selectively cleave the amide bond of R(132)-A(133) of bST and generate a new N-terminal residue Ala. (Bottom) Partial thrombin cleavage products derived from the two dimer structures.

intermolecular disulfide linkages lead to covalent dimera (eg. Dimer A, Fig. 2). Previously we reported [2] that the structures of the two reefold dimers were unambiguously determined by tryptic peptide mapping, limited DTT reduction and thrombin cleavage studies. One structural was based on the study of the thrombin cleavage intermediates (Fig. 2). When treated with a limited amount of thrombin, Dimer B yielded two cleavage products along with an unreacted starting dimer. One of the cleavage products is bST monomer (B-1) while the other (B-2) is the thrombin-cleaved monomer. On the other hand, partial cleavage of Dimer A gave a thrombin-cleaved monomer (A-1), a new dimer (A-2) with only one of the two large loops cleaved, and the starting Dimer A. The assignments of these intermediates were based on their molecular weight as determined by SDS-PAGE and N-terminal amino acid sequencing.

Dimers/aggregates formation during refolding

There key refold parameters [3], urea concentration, pH and bST concentrations were examined for their impact on the formation of dimers and aggregates during renaturation. Fig. 3-a shows the effect ofurea concentration over a 2 to 7 M range on the formation of dimers and aggregates. The urea concentration has a profound influence on the formation of DDimer B and aggregates. The amounts of Dimer B increases at high urea concentration, although aggregate content was significantly higher at both high and low urea concentrations.

The effect of pfH on the generation of dimers/aggregates is shown in Fig. 3-b. As refold pH increases from 9 to 12, the amount of Dimer B increases while that of the aggregates decrease. Dimer A, howerver, is less affected by pH variations.

Protein concentration has a significant impact on the formation of both Dimer A and aggregates. As the protein concentration increases so do Dimer A and aggregates. However, the amount of DDimer B generated during refold seems to be less affected (Fig. 3-c).

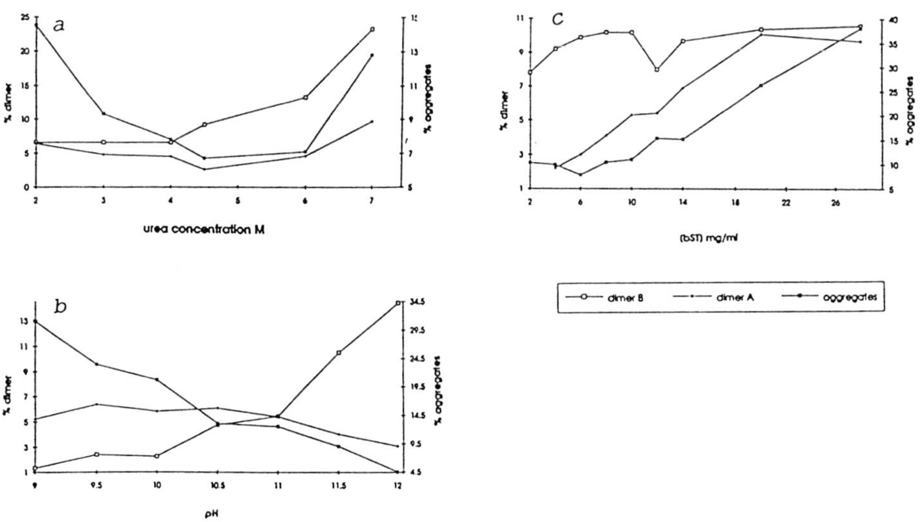

Fig. 3. Effect of (a) urea concentration, (b) pH, (c) protein concentration on the generation of dimers/aggregates during renaturation. For (a), the pH was constant at 11 and bST concentration was 9.5mg/ml. For (b), the urea concentration was 4.5 M and bST concentration was 8 mg/ml. For (c), the pH was constant at 11 and urea concentration was 4.5 M.

The formation of these high molecular weight species directly affects the recovery yield of the bST monnomer. Figs. 4a-c clearly show that dimers/aggregates were formed at the expense of bST monomer. The data presented here indicate that the optimal refold condition is at a urea concentration between 4-5 M, pH around 11.0, and at a protein concentration less than 10 mg/ml. Within these experimental parameters, one could obtain a maximal yield of bST monomer with a minimal amount of dimers/aggregates.

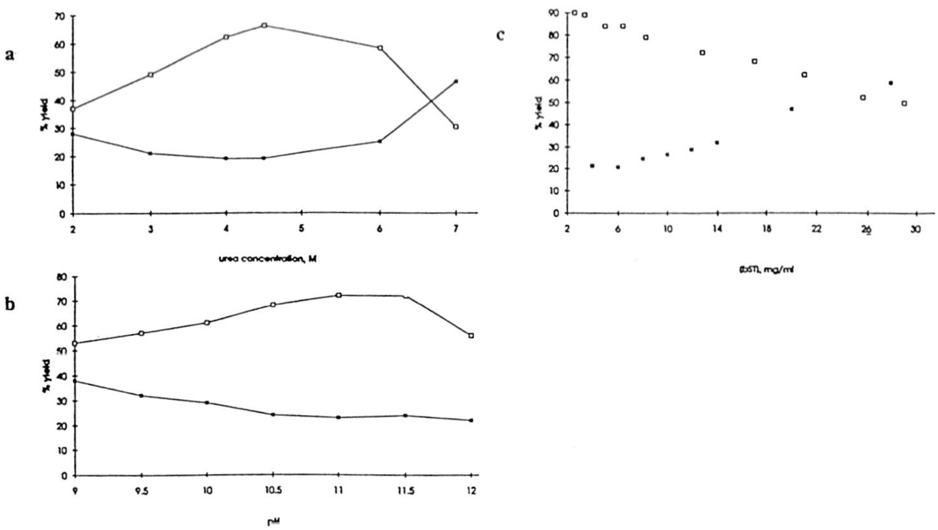

Fig. 4. Effects of (a) urea concentration, (b) pH, and (c) bST concentration on the formation of bST monomer and total dimers and aggregates. In all cases, dimers and aggregates were formed at the expenses of bST monomer.

The formation of these two dimers along with the absence of other mixed disulfide-linked bST dimers may imply the existence of specific protein-protein interactions during bST renaturation. The formation of the concentrated dimer (Dimer B) is of particular interest because of its topologically complex structure [4]. In fact, various renaturing conditions as described above may provide specific environments of controlling the protein-protein interactions [5]. The generation of monomer, dimers, and aggregates is the result of the different folding conformers and folding pathways induced by specific renaturation environments.

References

1. Bogosian, G., Violand, B.N., Dorward-King, E.J., Workman, W.E., Jung, P.E. and Kane, J.F., J. Biol. Chem., 264(1989)531.
2. Tou, J.S., Violand, B.N., Schlittler, M.R. and Jennings, M.G., J. Prot. Chem., 12(1993)237.
3. Storrs, S.B. and Przybycien, T.M., in Protein Refolding, American Chemical Society, Chapter 15, 1991, pp. 197-205.
4. Mao, B., Biopolymers, 30(1990)645.
5. (a) Brems, D.N., Biochemistry, 27(1988)4541: (b) Brems, D.N., Plaisted, S.M., Kauffman, E.W. and Hacel, H.A.,Biochemistry, 25(1986)6539.

Purification of a gonadotropin-releasing hormone-like peptide in human early placenta

Hong Yin[a], Chong-Li Zhang[a], Wei-Bin Shen[a], Ai-Min Zou[a], Charlean Miller[b] and Jean Rivier[b]

[a]*State key Laboratory of Reproductive Biology, Institute of Zoology, Academia Sinica, Beijing 100080, China*

[b]*The Clayton Foundation Laboratories for Peptide Biology, The Salk Institute, La Jolla, CA 92037, U.S.A.*

Introduction

Many works have proved that gonadotropin- releasing hormone (GnRH, or LHRH) can be synthesized by human placental trophoblast [1,2]. Action studies indicated that GnRH can regulate human chorionic gonadotropin (hCG) secretion and GnRH antiserum can affect mouse blastocyst implantation [3,4]. Therefore, human placental GnRH (hpGnRH) is regarded as an important potential regulator in the placenta. However, it is not clear whether hpGnRH is identical to mammalian hypothalamus GnRH (mGnRH). Tan and Rousseau's isolation work [5] and Seeburg and Adelman's definition of GnRH precursor mRNA in term placenta [6] indicated that hpGnRH is identical to mGnRH. On the contrary, studies on the placental GnRH receptor [7,8] and on the isolation from Gautron [9] suggested that hpGnRH is not identical to mGnRH. In order to clarify the extent of their similarity, we purified a GnRH-like peptide from human early placenta on the basis of our previous works [2,4].

Results and Discussion

Human early placental villi (6-9 weeks gestation) were extracted with methanol and isolated by Sep Pak C18 and twice by HPLC. Two main pools of immunoreactive GnRH (Fraction 4 and 7) were found in the first HPLC (Fig. 1). They presented very similar retention times (RT) in the second HPLC (Fig. 2, Fraction 36 and 22). Capillary electrophoresis (CE) analysis indicated that active Fraction 36 and Fraction 22 presented a peptide (migration time (MT) of 8.5 min, Fig. 3) which was not identical to that of either mGnRH (MT 17.9 min, from Sigma Co.) or [Hydroxyproline^{9}]GnRH (MT 19.3 min, from Dr. Rivier's Lab) (CE analysis grams not shown). The purity of this peptide from Fraction 36 was about 75 percent (Fig. 3).

Recently, Gautron reported a novel GnRH-like peptide, [Hydroxy-proline^{9}]-GnRH ([Hyp$_{9}$]GnRH) in mammalian hypothalamus [10]. In this study, we purified a hpGnRH whose characteristics are different from both mGnRH and [Hyp^{9}]-GnRH. This result indicated that human early placenta may be present its own GnRH-like peptide. The yield was about 0.2 µg/kg tissue, which was less than 25% of the concentration of GnRH measured directly by RIA [2]. Its sequence is being analyzed.

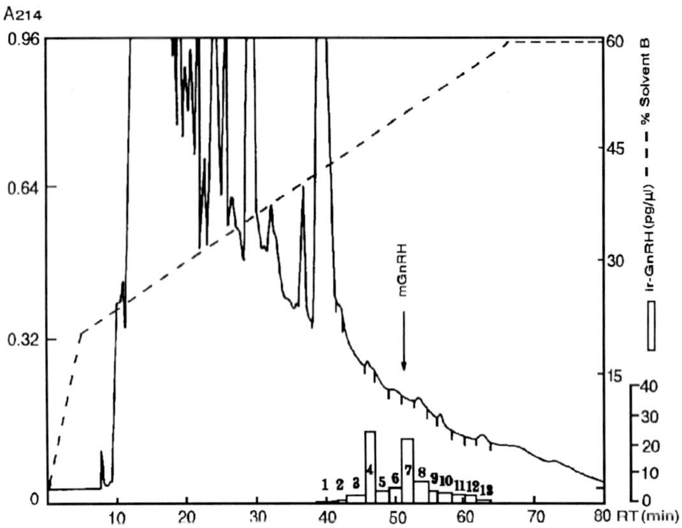

Fig. 1. The first HPLC of placental tissue extract.Conditions are: ODS column (10 x 250mm, 5μ), flow rate 1.0 ml/min; 0-60% solvent B gradient elution (solvent A: 20% MeOH/0.1% TFA, solvent B: 80% MeCN/0.1% TFA). 100g tissue extract loaded. Fractions collected and detected by GnRH RIA. The elution peak of standard mGnRH is indicated by the arrow.

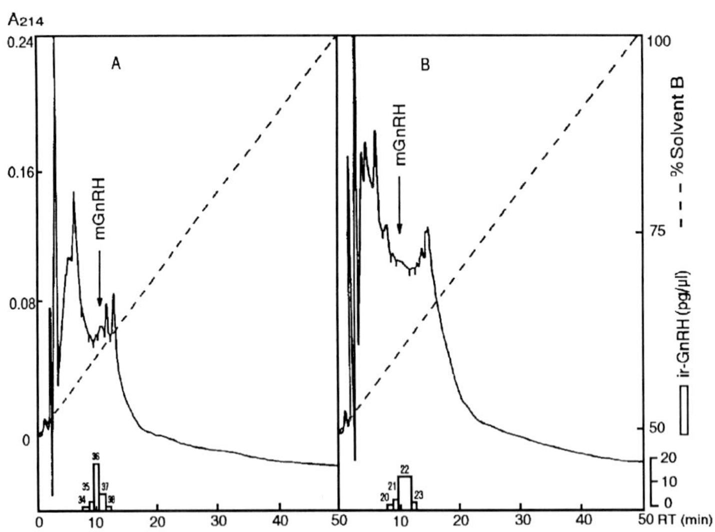

Fig. 2. The second HPLC of fraction 4 and 7 from the first HPLC-1. Conditions are: ODS column (4.6 x 250mm, 5μ), 50-100% solvent B gradient elution (solvent A: 0.1% TFA, solvent B: 50% MeCN/0.1% TFA). Half of fraction 4 (A) or fraction 7 (B) loaded. Fractions collected and detected by GnRH RIA. The elution peak of standard mGnRH is indicated by the arrow.

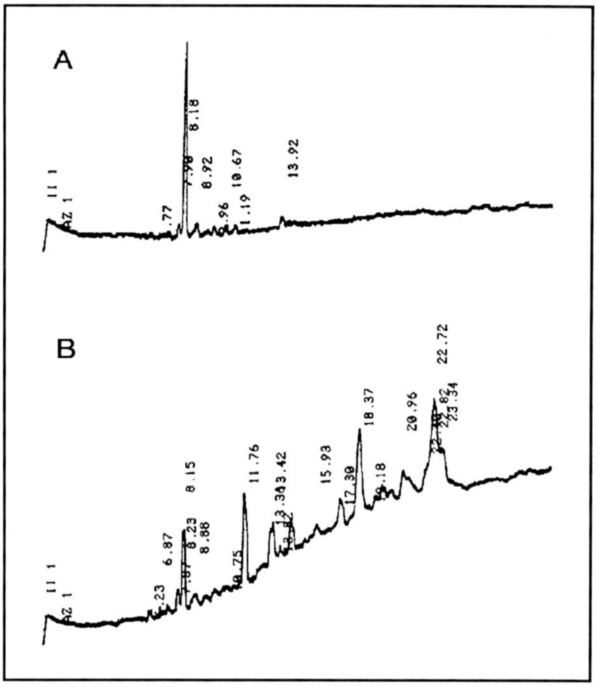

Fig. 3. Capillary electrophoresis of fractions 36 and 22 from the second HPLC. CE performed on a Beckman capillary electrophoresis P/ACE system 2050 with Spectra-Physics ChromJet SP4400 integrator, at 15 KV, 85 μA using 0.1 M sodium phosphate at pH 2.50. Silica-fused capillary cartridge was 50 cm x 75μm. Detection was performed at 214 nm, 0.013 AUFS and temperature was 30 °C. Sample dissolved in 10μl 0.1% TFA. (A): fraction 36; (B): fraction 22.

Acknowledgements

We thank Ms. Li-Hong Jiao who helped collecting tissues.

References

1. Khodr, G.S. and Khodr, T.M., Science, 207(1980)315.
2. Zhang, C.L., Cheng, L.R., Wang, H., Zhang, L.Z. and Huang, W.Q., Neuroendoc., 53(suppl.1,1991)77.
3. Khodr, G.S. and Khodr, T.M., Fertil. Steril., 30(1978)301.
4. Zhang, C.L., Cheng, L.R., Wang, H., Yin, H., Shen, W.B. and Huang,W.Q., Contraception, 46(1992)159.
5. Tan, L. and Rousseau, P., Biochem. Biophys. Res. Commun., 109(1982)1061.
6. Seeburg, P.H. and Adelman, J.P., Nature, 311(1984)666.
7. Currie, A.J., Fraser, H.M. and Sharpe, R.M., Biochem. Biophys. Res. Commun., 99(1981)332.
8. Qiu, X.D., Wang, H.Z., Gong, Y.T., Sheng, W.X., Shun, Z.D. and Zhou, W., Acta Biophys. Sinica, 22(1990)551.
9. Gautron, J.P., Pattou, E., Bauer, K., Rotten, D. and Kordon, C., Placenta, 10(1989)19.
10. Gautron J.P. Pattou E.,Bauer K. and Kordon C., Neurochem. Int., 18(1991)221.

Expression of inhibin/activin subunit mRNAs in corpus luteum during the human menstrual cycle

De-Jun Kong[a], Gui-Shen Lu[b], De-Xin Wang[b] and Chi-Ping Cheng[a]
[a]Department of Physiology, Harbin Medical University, Harbin 150086, China
[b]Institute of Materia Medica, Chinese Academy of Medical Sciences, Beijing 100050, China

Introduction

Inhibin is a glycoprotein hormone produced by the gonads and first purified in 1985. There are two inhibins based on a difference in their β subunits, inhibin A (α, βA) and inhibin B (α, βB). Interestingly, the β subunit of inhibin can form β subunit dimers which are called activin. There are activin A (βA, βB), B (βB, βB) and AB (βA, βB). Inhibin and activin are functional antagonists; inhibin suppresses, whereas activin stimulates FSH secretion [1].

It has been reported that the human corpus luteum (CL) is a major source of inhibin in the normal menstrual cycle. Gene expression of inhibin/activin α and βA subunits has been demonstrated in human granulosa-lutein cells [2], human CL [3] and primate CL [4, 5]. At any site where mRNAs for the α,βA and βB subunits are detected, inhibin A and B as well as activin A, B and AB may be produced. So far, however, expression patterns of inhibin/activin βB subunit mRNAs have not been well documented in human CL [2]. Therefore, little is known about the forms of inhibin or activin that are produced in human CL during the human menstrual cycle. In the present study, we report expression of inhibin/activin α, βA and βB in the human CL at various phases during the menstrual cycle.

Results and Discussion

mRNAs encoding the α subunit of inhibin/activin were detectable at a size of 1.6 kb in human CL. According to scanning densitometry, inhibin α subunit mRNA levels of human CL peaked in early luteal phase, declined gradually in mid and were markedly lower in late (Fig. 1). A 4.0kb transcript for βA subunit of inhibin/activin was seen in human CL. Expression of βA subunit mRNA appeared only in early luteal phase and decreased to undectable levels in mid and late phase during human menstrual cycle (Fig. 2). In contrast, human CL expression of βB subunit mRNAs, which were a size of 4.5 kb was found in early luteal phase, higher in mid than in early and peaked in late phase (Fig. 3). To our surprise, specific hybridization of the inhibin βB subunit probe was seen in total RNA from rat liver (Fig. 3D). The result will be discussed elsewhere.

We characterized the expression patterns of inhibin/activin α, βA and βB subunit mRNAs in human CL throughout the menstrual cycle. In the present study, the sizes of α and βA mRNAs of inhibin subunits in human CL were similar to that previously described [3] and βB subunit mRNAs were about 4.5 kb, similar to that of human placenta [6].

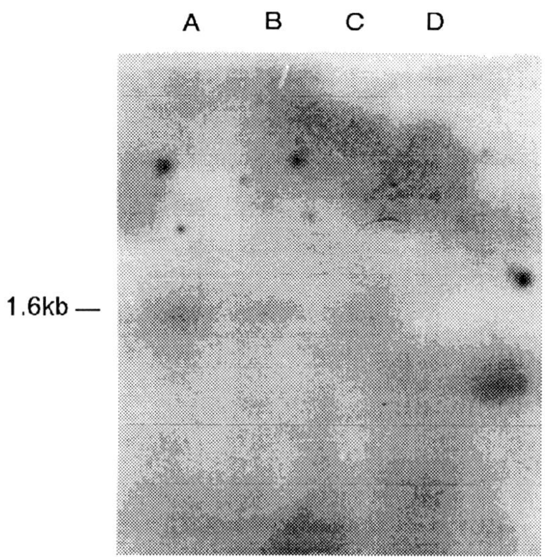

Fig. 1. Northern blot analysis of inhibin/activin α subunit mRNA levels in human CL during menstrual cycle. 10μg of total RNA from 3rd day (lane A), 8th day (lane B) and 15th day (lane C) human CL and 120μg of total RNA from rat liver (lane D) in 1.5% agarose-urea gel, transferred to nylon filters and hybridized to human inhibin α cDNA probe.

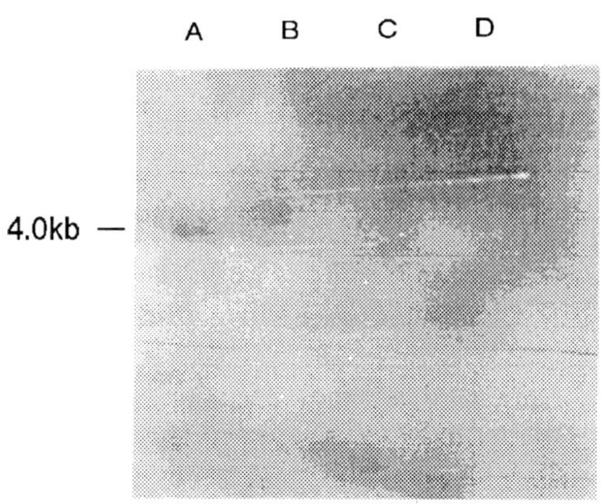

Fig. 2. Northern blot analysis of inhibin/activin βA subunit mRNA levels in human CL during menstrual cycle. 10μg of total RNA from 3rd day (lane A), 8th day (lane B) and 15th day (lane C human CL and 120μg of total RNA from rat liver (lane D) in 1.5% agarose-urea gel, transferred to nylon filters and hybridized to human inhibin βA cDNA probe.

We observed that expression of α subunit mRNAs of inhibin gradually declined from early to late luteal phases during the human menstrual cycle and that βA mRNAs were only detected in the early luteal phase. In contrast to α subunit, expression of βB subunit mRNAs can be found in the early luteal phase, are higher in mid than in early phase and peak in late phase. To our knowledge, it is the first time that such an expression pattern of βB (including βA) subunit mRNAs has been observed in human corpus luteum of early, mid and late luteal phase during the menstrual cycle. However, our results are strikingly similar to immunohistochemical results of sparse βA and relatively strong βB subunit immunoreactivity in human CL during menstrual cycle [7]. In contrast to our results, βB subunit transcripts in human CL have not been detected by in situ hybridization. This discrepancy may result from the low level of βB mRNA that can be detected by in situ hybridization. In fact, slightly specific hybridization with inhibin βB probe was found once by in situ hybridization in monkey luteal tissue [5]. Although little is known of the possible different properties of inhibin A and B, expression patterns of abundant α and βB subunit mRNAs in human CL suggest that inhibin B (α, βB dimer), rather than inhibin A (α, βA dimer), might be the major form of inhibin in human corpus luteum and that inhibin A, if any, is secreted only in early luteal phase. Thus, it is necessary to study further whether inhibin A and inhibin B have different effects on human ovary function. That inhibin α subunit mRNA is declining to low levels in late luteal phase while βB mRNA reaches a maximum may be interpreted to mean that human corpus luteum at late luteal phase may produce activin B (βB, βB dimer).

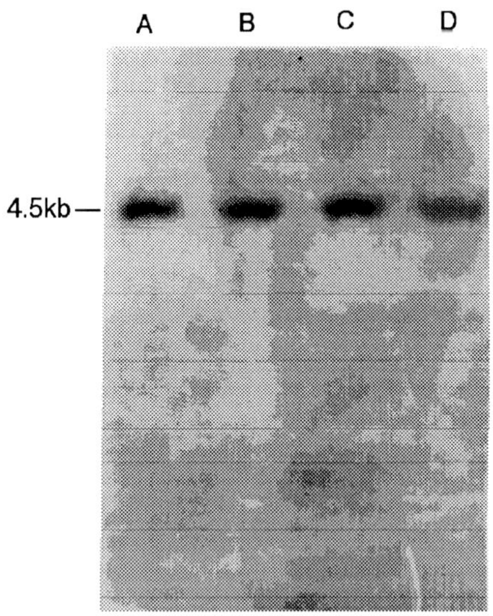

Fig. 3. Northern blot analysis of inhibin/activin βB subunit mRNA levels in human CL during menstrual cycle. 10μg of total RNA from 3rd day (lane A), 8th day (lane B) and 15th day (lane C human CL and 120μg of total RNA from rat liver (lane D) in 1.5% agarose-urea gel, transferred to nylon filters and hybridized to human inhibin βB cDNA probe.

The physiological effects of human ovarian activin are uncertain. It has been recently proposed that activin A may promote proliferation in cultured human granulosa-luteal cells [8] and that activin B may play a role in small antral follicles during earlier phases of primate follicular development [5]. We think that the function of activin B produced by human CL at late luteal phase may initiate the next cycle in both development of antral follicles and activity of the hypothalamus-hypophysis. The expression of β subunits may be regulated independently in human CL during menstrual cycle.

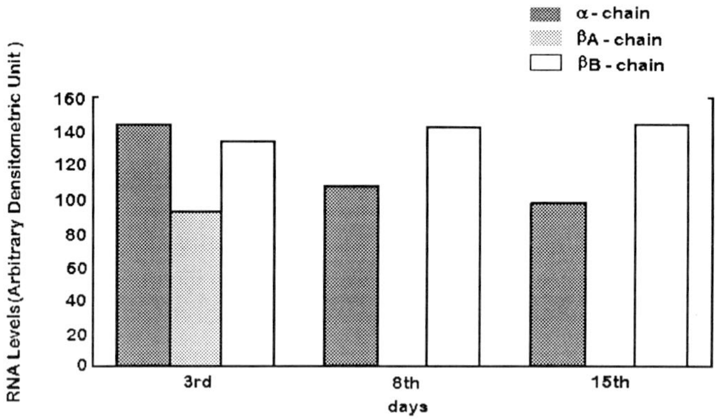

Fig. 4. Levels of inhibin α, βA and βB subunit mRNAs in human corpus luteum. Autoradiograms were scanned by densitometer and integration values (mean+SD, n=3) are shown as arbitral densitometic units.

Acknowledgements

We are grateful to Dr. Anthony J Mason for providing inhibin subunit α, βA and βB cDNA probes. We also thank Dr. Nicholas Ling and Dr. Shunichi Shimasaki (Department of Molecular Endocrinology, Whittier Institute for Diabetes and Endocrinology, La Jolla, CA, U.S.A.) for help in this work.

References

1. Burger, H.G., Reproductive Medicine Review, 1(1992)1.
2. Eramaa, M., Heikinbeimo, K., Tuuri, R., Hilden, K. and Ritvos, O., Mol. Cell. Endocrinol., 92(1993)R15.
3. Davis, S.R., Krozowski, Z., Mclachcn, R.I. and Burger, H.G., J. Endocrinol., 115(1987)R21.
4. Basseti, S.G., Winters, S.J., Keeping, H.S. and Zeleznik, A.J., Clin. Endocrinol Metab., 70(1990)590.
5. Schwall, R.H., Mason, A.J., Wilcox, J.N., Basset, S.G. and Zekeznik, A., J. Mol. Endocrinol., 4(1990)75.
6. Petraglia, F., Caura, L., Garuti, G.C., Abrate, M., Giardino, L., Genazzani, A. R., Vake, W. and Meunier, H., J. Clin. Endocrine. Metab., 71(1990)487.
7. Yamoto, M., Minami, S. and Nakano, R., J. Clin. Endocrinol. Metab., 73(1991)470.
8. Rabinovic, J., Spencer, S. J. and Jaffe, R. B., J. Clin. Endocrinol. Metab., 71(1990)1336.

Prediction of the three dimensional structure of activin

Jing-Chu Luo[a,c], Qiu-Yan Liu[b], Charles R. Coombers[b], Paul A. Bates[c] and Michael J.E. Sternberg[c]
[a]*College of Life Sciences, Peking University, China*
[b]*Dept. of Medical Oncology, Charing Cross and Westminster Medical School, London, U.K.*
[c]*Biomolecular Modelling Lab, Imperial Cancer Research Fund, London, U.K.*

Introduction

Activin and inhibin are known to be molecules involved in the regulation of follicle-stimulating hormone secretion by pituitary cells [1]. They are dimeric proteins composed of two β subunits (βa and βb) and an α subunit. Activin is a β subunit dimer. In vivo, activin and inhibin subunits are expressed in gonadal tissues, where they are known to regulate gonadal hormone production. They are also expressed in several other tissues, where their function is still unclear. It is now known that activin belongs to the supergene family of transforming growth factor β (TGF-β's) and that activin, like TGF-β, has a diverse set of functions. These include the control of proliferation, differentiation and development in variety of cell types, including breast cells [2], fibroblasts [3], nerve cells [4], germ-Sertoli cells [5], erythroleukemic cells [6], osteoclasts [7] and pituitary cells [8]. Activin has also been found to be a nerve cell survival factor [9]. Furthermore, activin plays a critical role in mesoderm induction during vertebrate development in, for example, Xenopus [10] and mouse [11] and in chick limb bud formation [12]. Various activin receptors, together with the TGF-β receptors, constitute a new class of ligand activated serine/threonine kinases. Sequence analysis suggests that the polypeptide chain of activin receptor consists of two domains which are linked by a transmembrane α-helix. The cytosolic domain encodes the serine/threonine kinase activity whereas the extracellular domain binds to the ligand. In order to further understand the interaction of activin and its receptor it would be useful to know the three-dimensional structure for each of the molecules. For example, such information would be useful in site directed mutagenesis experiments directed at understanding which parts of these molecules are involved in binding interactions. We report here a molecular model of activin-A which has been constructed by means of knowledge-based homology modeling. The main frame of the predicted structure is very similar to the parent structure of TGF-β2 while some difference can be found in the loop areas.

Results and Discussion

The crystal structure of TGF-β2 has been solved recently by two groups [13,14]. The striking feature of this molecule is the framework formed by eight cysteine residues. Interestingly, the topology of the disulfide cluster was also found in other growth factors such as nerve growth factor and platelet-derived growth factor-BB, which show low sequence similarity with TGF-b2. On the other hand, the monomer of activin-A shares about one third identity and the position of all cyteines is conserved (Fig. 1). The homodimer of TGF-β2 is linked by a disulfide bond between two cysteines at each subunit. The left-hand like monomer of this molecule consists of a long α-helix and two anti-parallel β-sheets (Fig. 2). The hydrophobic core of the dimer is formed by side chains of some of the residues on the α-helix and β-sheets, which is indicated with dark circles in Fig. 2.

```
TGF-β2     1        10         20          30          40          50        55
Activin-A  LDAAYCFRNVQDNCCLRPLYIDFKRDLGW-KWIHEPKGYNANFCAGACPYLW-----SSD
           ---GLECDGKVN-ICCKKQFFVSFK-DIGWNDWIIAPSGYHANYCEGECPSHIAGTSGSSL
              1        10         20          30          40         50    56

TGF-β2        60          70         80          90         100          112
Activin-A  TQHSRVLSLYNTIN--PEASASPCCVSQDLEPLTILYYI-GKTPKIEQLSNMIVKSCKCS
           SFHSTVINHYRMRGHSPFANLKSCCVPTKLRPMSMLYYDDGQNIIKKDIQNMIVEECGCS
             60         70         80          90         100        110   116
```

Fig. 1. Sequence alignment of activin-A monomer to TGF-β2. Sequence numbers are shown at the top and bottom. Identical residues are highlighted, conserved cysteines are underlined and deletions are denoted by dashes.

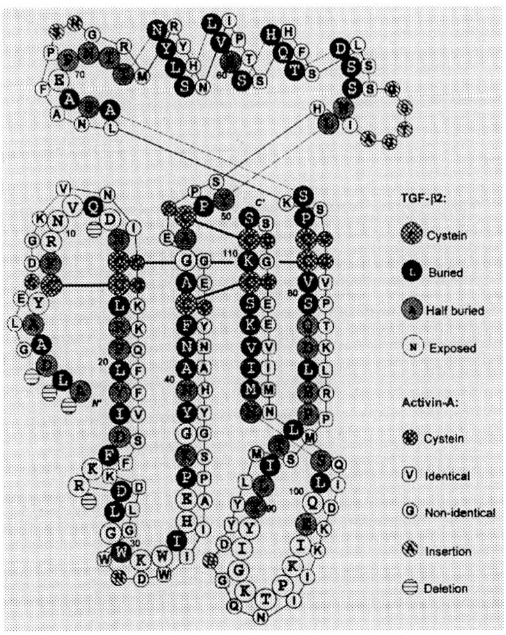

Fig. 2. Sequence alignment of activin-A to TGF-β2 based on the secondary structure framework which was generated from the crystal structure of TGF-β2. Residues of the parent structure are shown as big circles. Dark: buried; light-dark: half buried; plain: exposed. Residues of activin-A are aligned to TGF-β2 by either side. Small circles: non-identical; round-corner squares: identical residues. Insertions and deletions are indicated in different pattern.

The conserved cysteine cluster as well as the high percentage of secondary structure component both of α-helix and β-sheet provided the basis for predicting the three-dimensional structure of activin-A. Automatic sequence alignment was performed to obtain the initial step. Adjustment of some of the residues was then carried out based on the framework of the secondary structure of TGF-β2. Principles of protein conformation were considered in this step to yield a better result. Molecular graphics and molecular modeling programs were used to build up the model. Finally, energy minimization was employed to refine the model structure. The main framework of the predicted structure remains the same as TGF-β2, while conformation changes can be seen in the loop areas, due to sequence variety. This model provides some information which may be useful for further investigation of the functions of activin and its receptor in mutagenesis, signal transduction and other molecular biological studies. Further work on modeling the receptor and investigating the interaction between the ligand and receptor are underway.

Acknowledgments

The authors are grateful to Dr. L. Buluwela, Department of Biochemistry, Charing Cross and Westminster Medical School, for his critical discussion on the content of the manuscript. This work was partly supported by the Royal Society of Britain to J. Luo for his visit to the Imperial Cancer Research Fund. This work was also supported by the Medical Research Council (UK).

References

1. Vale, W., Rivier, J., Vaughan, J., McClintock, R., Corrigan,A., Woo, W., Karr, D., Spiess, J., Nature, 321(1986)776.
2. Liu, Q.Y., Gomm, J.J., Buluwela, L. and Coombes, R.C., Proceedings of America Association for Cancer Research 35, 85th Annual Meeting, San Francisco, Abs., (1994)1546.
3. Kojima, I., Ogata, E., Biochem. Biophys. Res. Commun., 159(1989)1107.
4. Hashimoto, M., Kondo, S., Sakurai, T., Etoh, Y., Shibai, H., Muramatsu, M., Biochem. Biophs. Res. Commun., 173(1990)193.
5. Mather, J.P., Attie, K.M., Woodruff, T.K., Rice, G.C., Phillips, D.M., Endocrinology, 127(1990)3206.
6. Miyamoto, Y., Kosaka, M., Eto, Y., Shibai, H., Saito, S., Biochem. Biophs. Rese. Commun., 168(1990)1149.
7. Sakai, R., Eto, Y., Ohtsuka, M., Hirafuji, M., Shinoda, H., Biochem. Biophs. Res. Commun., 195(1993)39.
8. Katayama, T., Shiota, K., Takahashi, M., Mol. Cell. Endocrinol., 69(1990)179.
9. Schubert, D., Kimura, H., Lacorbiere, M., Vaughan, J., Karr,D., Fischer, W.H., Nature, 344(1990)868.
10. Smith, J.C., Price, B.M., Van-Nimmen, K.D., Huylebroeck, Nature, 345(1990)729.
11. Lu, R.Z., Matsuyama, S., Nishihara, M.,Takahashi, M., Biol. Reprod., 49(1993)1163.
12. Chen, P., Yu, Y.M., Reddi. A.H.. Exp. Cell. Res., 206(1993)119.
13. Schlunergger, M.P., Grutter, M.G., Nature, 358(1992)430.
14. Sun, D.P., Piez, K.A., Ogawa, Y., Davies, D.R., Science, 257(1992)369.

Gonadotropin releasing hormone antagonists with acyl substitutions of 4-aminophenylalanine at positions 5 and 6

Guang-Cheng Jiang, Catherine Rivier, A. Grey Craig, Charleen Miller, John Porter, Anne Corrigan, Wylie Vale and Jean Rivier
Clayton Foundation Laboratories for Peptide Biology, The Salk Institute, 10010 N. Torrey Pines Road, La Jolla, CA 92037, U.S.A.

Introduction

GnRH antagonists are now recognized as potential drugs for the management or treatment of endometriosis, infertility, ovulation induction in women with chronic anovulation, precocious puberty, uterine myoma, ovarian hyperandrogenism and hirsutism, premenstrual syndrome (PMS), controlled induction of ovulation in *in vitro* fertilization programs, and may also be a promising lead in the treatment of breast and gynecological cancers. Most of these disorders were originally studied with long acting preparations of the superagonists which desensitize the gonadotrophs after approximately two weeks of treatment. An antagonist will likely displace the agonists in some situations because it avoids the initial up-regulation of the gonadotropin-gonadal axis, leads to rapid and predictable recovery, permits flexibility of the degree of gonadal suppression and can be used as a diagnostic test of gonadotropin dependent gonadal dysfunction. If GnRH antagonists are to be used successfully in humans, they need to be extremely potent, long acting and exhibit negligible side effects such as stimulating histamine release[1].

We have shown that although equally safe, Azaline A (Analog 1) was short acting while Azaline B (Analog 2) was considerably longer acting[1]. Novel Azaline B analogs were synthesized to further improve potency and solubility in aqueous buffers.

Results and Discussion

- Analogs were synthesized by SPPS methodology on a p-methylbenzhydrylamine resin.
- The protected peptide-resins were cleaved in anhydrous HF in the presence of a scavenger, precipitated, extracted and lyophilized.
- The crude peptides were purified by reversed-phase HPLC.
- The analytical techniques used for the characterization of the analogs included HPLC with two different solvent systems (acidic and neutral), amino acid analysis, optical rotation, capillary zone electrophoresis (CZE) and liquid secondary ion mass spectrometry (LSIMS). Results from these studies support the identity of the intended structures and demonstrated that most peptides were greater than 97% pure.
- Antiovulatory assay (AOA) results are expressed in terms of the dosage in micrograms/rat (rats ovulating/total number of treated rats).
- Measurement of circulating LH levels in castrated rats treated subcutaneously (200 mL) with the peptides was carried out over a period of 72 h as reported earlier [1].

Table 1 *Antiovulatory and castrated male rat assay results*

Analog	[Ac-DNal[1],DCpa[2],DPal[3],X[5,6],Ilys[8],DAla[10]]GnRH Description	AOA[a] μg/rat	#[c]	Duration of action[b]
1	[R,Lys[5](Atz),DLys[6](Atz)]-GnRH[b)] *Azaline A*	2.0	(1/10)	Short Acting
2	[R,Aph[5](Atz),DAph[6](Atz)]-GnRH *Azaline B*	1.0	(0/7)	Long Acting
3	[R,Aph[5],DAph[6]]-GnRH	1.0	(7/8)	
		2.0	(0/7)	
4	[R,Aph[5](For),DAph[6](For)]-GnRH	1.0	(3/9)	
		2.5	(0/8)	
5	[R,Aph[5](Ac),DAph[6](Ac)]-GnRH *Acyline*	1.0	(5/13)	Long Acting
		2.5	(0/7)	
6	[R,Aph[5](Ac-Gly),DAph[6](Ac-Gly)]-GnRH	0.5	(3/4)	Short Acting
		1.0	(5/8)	Intermediate
7	[R,Aph[5](Atz-Gly),DAph[6](Atz-Gly)]-GnRH	0.5	(3/4)	
		1.0	(5/8)	Long Acting
8	[R,Aph[5](Atz-bAla),DAph[6](Atz-bAla)]-GnRH	1.0	(9/18)	
		2.5	(0/6)	
9	[R,Aph[5](Atz-Gaba),DAph[6](Atz-Gaba)]-GnRH	0.5	(10/14)	Intermediate
		1.0	(1/8)	
10	[R,Aph[5](Atz-Ahx),DAph[6](Atz-Ahx)]-GnRH	0.5	(6/6)	
		1.0	(2/10)	
11	[R,Aph[5](Atz-Ala),DAph[6](Atz-Ala)]-GnRH	0.5	(4/4)	Short Acting
		1.0	(1/8)	
12	[R,Aph[5](Atz-DAla),DAph[6](Atz-Ala]-GnRH	1.0	(2/8)	

[a] Antiovulatory assay.
[b] Duration of action measured in the castrated male rat assay (50 mg/rat sc in 200 ml isotonic solution). Long duration of action = fully active after 72 hr; intermediate = fully active at 48 hr but not at 72.; short acting = fully active at 24 hr but not at 48 hr; Aph(atz) = 4-(N-5'-(3'-amino-1H-1',2',4'-triazolyl) phenylalanine.
[c] Number ovulating drug adminstration.

Because analog 2 was long acting and more potent than analog 1 in the AOA, we tried to pursue more systematically the study of analogs with simple acylated moieties on the 4-amino function of aminophenylalanines at positions 5 and 6, the rest of the molecule remaining constant with Ac-DNal-DCpa-DPal at the N-terminus and ILys-Pro-DAla-NH_2 at the C-terminus. We were first surprised that **3**, the parent underivatized analog, fully inhibited ovulation at 2 mg/rat and that the formylated (**4**), acetylated (**5**) and other closely related acylated analogs (**6-12**) were essentially equipotent in the AOA. It became evident that the bulk of the substitutions on the 4-amino function of phenylalanine was not influencing potency significantly in the AOA. Because of our early awareness that Azaline A (**1**) and Azaline B (**2**) had significantly different half lives *in vivo*, we tested most of the analogs presented in Table 1 in the castrated male rat. We found that drastic differences in duration of action resulted from seemingly insignificant changes in structure. Of particular interest was the observation that **5** was as efficacious as Azaline B and considerably easier to make. Interestingly, **5** seemed less potent than Azaline B both *in vitro* (not shown) and in the AOA. Its effectiveness in the castrated male rat, comparable to that of Azaline B, suggests that subtle and yet unidentified parameters, such as bioavailability (subtle differences in solubility), may influence the efficacy of an analog in the AOA. Other examples shown below suggest however that minor modifications of structure yield dramatic changes in biological efficacy (compare **6** to **7**, **8** and **10**), the possible result of subtle differences in bioavailability and affinity for both the receptor and other yet to be defined sites.

Acknowledgements

This work was supported by NIH under Contract NO1-HD-3-3171 with the Contraceptive Development Branch, Center for Population Research, NICHD and in part by the Hearst Foundation. We thank R. Kaiser, D. Pantoja, S. Johnson and Y. Haas for outstanding technical assistance.

References

1. Rivier, J., Porter, J., Hoeger, C., Theobald, P., Craig, A.G., Dykert, J., Corrigan, A., Perrin, M., Hook, W. A., Siraganian, R. P., Vale, W. and Rivier, C., J. Med. Chem., 35(1992)4270.

Study on molecular cloning of a G-protein coupled receptor candidate from rat hippocampus

Jun Guo, An-Wu Zhou, Ying Zhang, Xiang-Fu Wu and Yu-Cang Du
Shanghai Institute of Biochemistry, Academia Sinica, Shanghai 200031, China

Introduction

Signal transduction involving G-protein is a fundamental mechanism by which a wide variety of hormones or neurotransmitters regulate cellular functions [1,2]. Recent progresses in receptor biology has shown that a number of the receptors which interact with G protein, namely G-protein coupled receptor (GPCR), have a common structure characterized by seven hydrophobic transmembrane domains (TM). In these receptors, highly conserved sequence motifs often occur in the transmembrane regions TM2 and TM6 [3]. In this work we try to use these motifs to isolate cDNA encoding new receptors. Polymerase chain reaction (PCR) was used to amplify cDNA with degenerate oligonucleotide primers corresponding to sequences of TM2 and TM6, and a novel receptor-coding clone has been isolated from a rat hippocampus cDNA library.

Results and Discussion

A pair of oligonucleotides were synthesized by phosphoramidite method corresponding to TM2 and TM6 of FC5 receptor [3]:

5'-GGGGATCCACCAACATTCTGATCGTGAACCTCTCC -3'
5'-GGGCAGCCAGCAGACCGCGAAGGCGAC -3'

The template DNA from rat hippocampus cDNA library in the Lambda ZAPII vector (Stratagene, USA) was submitted to 45 cycles of PCR. A fragment as a band of about 500bp was isolated by agarose gel electrophoresis and then cloned into pUC19 plasmid.

```
GSQTAKAQRLVCDAIFWRGAVGDSRNGNGTLTNGDAGSNVDLSKELVNMIVASVTISL
TIQ
            TM2                                             TM3
TIKPGRSSTRWLTYVIADGIAQFIRDGAASQTLNQQVTRVICQLNARFSRAVECFTRG
ASG
                           TM4                                TM5
RLALPTATLVSRASTQYDARQNGLHLAPLDGPFAVCWLP
                                    TM6
```

Fig. 1. Deduced amino acid sequence from 500bp nucleotide fragment. Putative hydrophobic transmembrane domains are underlined.

A total of 11 clones were sequenced separately. Among them one clone (named as ZJ12, 496bp) was especially interesting. Its deduced amino acid sequence (as shown in Fig. 1) and the analysis of residue hydrophobicity indicated that this peptide fragment included five hydrophobic transmembrane domains from TM2 to TM6 (Fig. 2).

To obtain the complete primary structure of the putative receptors, the ZJ12 fragment was labeled as probe to screen for full-length clone, and one clone was selected. The work of analyzing this clone is presently being undertaken.

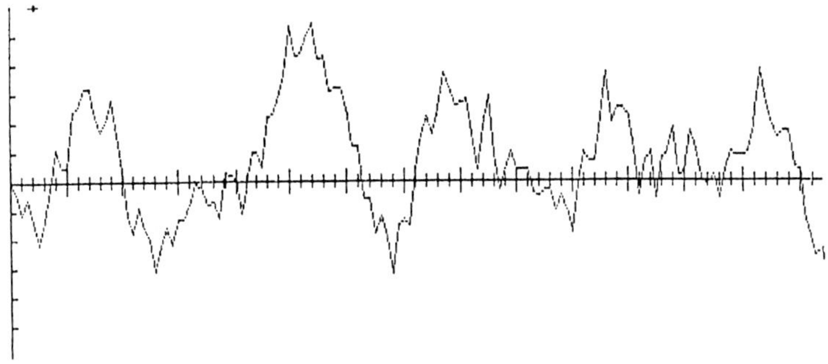

Fig. 2. The transmembrane domains identified by hydropathy analysis. Positive regions are relatively hydrophobic and negative regions are more hydrophilic.

At least several aspects of research remain to be done after cloning this putative G-protein coupled receptor: functional expression, searching for its ligand and quantitative distribution of the receptor or its mRNA, etc. This will help us to understand the signal transduction pathway mediated by G-protein coupled receptor.

References

1. Birnbaumer, L., Annu. Rev. Pharmacol. Toxicol., 30(1990)675.
2. Yamada, Y., Post, S.R., Wang, K., Tager, H.S., Bell, G.I. and Seino, S., Proc. Natl. Acad. Sci. USA., 89(1992)251.
3. Eva, C., Keinanen, K., Monyer, H., Seeburg, P. and Sprengel, R., FEBS letters, 271(1990)81.

The effects of pituitary adenylate cyclase activating polypeptide on guinea pig gallbladder *in vitro*

Mu-Xin Wei[ab], S. Naruse[c], T. Ozaki[b], K. Nokihara[d], E. Ando[d] and V. Wray[e]
[a]*National Institute for Physiological Sciences, Okazaki 444, Japan*
[b]*Nanjing Medical University, Nanjing 210029, China*
[c]*The Second Department of Medicine, Nagoya University School of Medicine, Nagoya 466, Japan*
[d]*Bio-application Laboratory, Shimadzu Co. Ltd., Kyoto 604, Japan*
[e]*Gesellshaft fur Biotechnologische Forschung, D-38124 Braunschweig, Germany*

Introduction

Pituitary adenylate cyclase activating polypeptide [PACAP] is a VIP-like neuropeptide isolated from ovine hypothalamus [1]. It exists in the ovine hypothalamus in two molecular forms, PACAP38 and PACAP27. PACAP strongly stimulates adenylate cyclase activity in the anterior pituitary and many other tissues. A dense network of PACAP-positive fibers was found in the hypothalamus and other areas of the brain [2]. Recently, PACAP-like immunoreactive nerve cell bodies and nerve fibers have been demonstrated in the gut wall of several mammals [3], and the specific PACAP binding sites were present in the digestive system [3,4]. These findings suggest that PACAP may have a regulatory role as a neurotransmitter or a neuromodulator not only in the central nervous system but also peripheral organs. It is known that VIP relaxes the gallbladder [5]. PACAP, on the other hand, evoked transient and tonic contractions of the gallbladder in conscious dogs [6]. In this study, we have tested the effect of PACAP38 and PACAP27 on isolated guinea pig gallbladder strips and compared it with that of VIP.

Results and Discussion

VIP relaxed gallbladder smooth muscle strips at concentrations ranging from 3×10^{-8}M to 3×10^{-7}M. The maximal inhibitory effect was observed in about 3 min. PACAP38 and PACAP27, on the other hand, contracted the muscle strips from 10^{-8}M to 3×10^{-7}M. The potency of PACAP38 was less than that of PACAP27 (Fig. 1). The peak response to PACAP38 was seen in about 7 min and to PACAP27 in about 3 min (Fig. 2).

The N-terminus of PACAP38 has high sequence homology with VIP[1]. At least two types of PACAP binding sites have been demonstrated in rat tissues. One (type 1), a dominate type in the hypothalamus, is highly specific for PACAP and the other (type 2), a dominant type in the lung and liver, has almost equal affinity for both PACAP and VIP. Many of the biological actions of PACAP, such as relaxation of vascular [7], bronchial and intestinal smooth muscle [2] and stimulation of exocrine pancreatic water and bicarbonate secretion [8], are very similar to those of VIP, suggesting that they are mediated by the latter type of receptors. It has been reported that PACAP38 evoked a transient and tonic contraction of the gallbladder in conscious dogs [6]. However, it had no effect on the resting tension of gallbladder smooth muscle strips *in vitro*. It appears that PACAP38 acts on the gallbladder via cholinergic mechanisms in dogs, as atropine inhibited the contraction induced by this peptide.

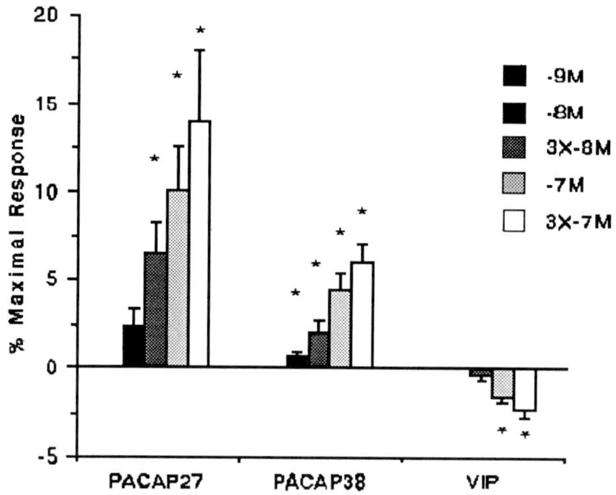

Fig. 1. The effect of PACAP38, PACAP27 and VIP on guinea pig gallbladder. Mean±SEM are given (n=12 PACAP38, n=3 PACAP27 and n=9 in VIP); Asterisks indicate significant (P<0.05) differences from the resting length.

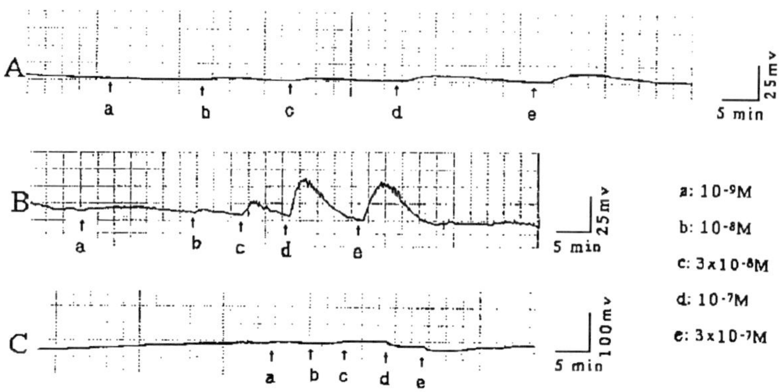

Fig. 2. The effects of PACAP38 (A), PACAP27 (B) and VIP (C) on isolated gallbladder smooth muscle strips of a guinea pig.

The present study demonstrated that PACAP38 and PACAP27 had concentration-related excitatory effects on guinea pigs gallbladder strips. In agreement with the previous study, VIP relaxed gallbladder strips in a concentration-related manner. The present results suggest that PACAP38 and PACAP27 can act on the gallbladder smooth muscle directly presumably via the activation of PACAP receptors (type 1). This mechanism may be dominant over the former mechanism (type 2) in the guinea pig gallbladder. It remains to be studied whether such receptors (type 1) are present in this organ. In the present study, PACAP38 was less active than PACAP27 in promoting the motility of the guinea pig gallbladder, which suggests that N-terminal 1-27 residues or imitations at position 27 are important in this response. In conclusion, the effect of PACAP38 and PACAP27 on motility of the guinea pig gallbladder was opposite to that of VIP.

Acknowledgements

This work was supported by Monbusho international scientific research program and a grant from the Ministry of Education, Science and Culture, Japan. Mu-xin Wei is a visiting scientist from Nanjing Medical University under the support of the government of Jiangsu Province, China.

References

1. Miyata, A., Ariimura, A., Dahl, R. R., Minamino, N., Uehara, A., Jiang, L., Cullar, M. and Coy, D.H., Biochem. Res. Commun., 164(1989)567.
2. Arimura, A., Peptides, 37(1992)287.
3. Sundler, F., Ekblad, E., Absood, A., Hakanson, R., Koves, K. and Arimura, A., Neuroscience, 46(1992)439.
4. Gottschall, P.E., Tatsuno, I., Miyata, A., Endocrinology, 127(1990)272.
5. Ryan, J. and Ryave, S., Am. J. Phyosiol, 234(1978)E44.
6. Mizumoto, A., Fujimura, M., Ohtawa, M., Ueki, S., Hayashi, N., Itoh, Z., Fujino, M. and Arimura, A.Reg., Peptides, 42(1992)39.
7. Naruse, S., Suzuki, T., Ozaki, T. and Nokihara, K., Peptides, 14(1993)505.
8. Naruse, S., Suzuki, T. and Ozaki, T., Pancreas, 7(1992)543.

Regulation of methionine-enkephalin on expression of interleukin-2 receptor α chain gene

Gang Li and Li-Chang Chen
Research Laboratory of Protein and Nucleic Acid, Hainan Medical College, Haikou 570005, China

Introduction

The immunological properties of opioid peptides have been extensively characterized. To date, however, little is known about their molecular mechanisms. It has been thought that there are certain functional relationships between opioid receptor and interleukin receptor [1]. Experiments therefore were designed to observe if methionine enkephalin (M-Enke), after binding to its receptor, could interfere with the expression of interleukin-2 receptor α chain (IL-2 R_α) gene. The results might help in understanding the regulative mechanism of M-Enk.

Results and Discussion

Reverse transcription (RT) of cDNA for IL-2 R_α and a double round of polymerase chain reaction (PCR) were performed in this experiment. Three primers were designed based on the sequence of IL-2 R_α cDNA [2] and synthesized in our laboratory. Primer 1 was used for cDNA synthesis, primer 2 and primer 3 were used for polymerase chain reaction(PCR) amplification (Table 1).

Table 1 *Sequences of primers used for reverse transcription and cDNA detection by PCR*

Primers	Sequence	Location[a]
Primer 1 (sense)	5'AGGAAGTCTCACTCTCAGGA3'	817-798
Primer 2 (anti-sense)	5'CTGGGACAACCAATGTCAATG3'	405-426
Primer 3 (sense)	5'CTTCACCTGGAAACTGACTGG3'	769-749

[a] On IL-2 R_α cDNA [2].

RNA used for RT-PCR was extracted from cultured lymphocytes. Peripheral blood lymphocytes were separated and cultured by a method previously described [3]. M-Enk was added to the culture to examine its effect on expression of IL-2 R_α. After two days incubation, mRNA in cultured lymphocytes was extracted using the guanidine-phenol-chloroform method with some modification [4]. cDNA for IL-2 R_α was synthesized by RT with primer 1 and a double round of PCR following RT with primer 2 and 3 was performed by modification of a previously described method [5].

The amplified products from RT-double PCR were visualized after agarose gel electrophoresis. Compared to the marker, the bands as shown in lanes 1 and 2 indicate an expected product of 365 base pairs (Fig. 1). Contents of the products were measured by ultraviolet spectrophotometer at 260 nm wavelength. The results show that the amount of the amplified product from lymphocytes cultured with M-Enk was only 87% of control, indicating that IL-2 R_α induced by Con A was probably inhibited by M-Enk.

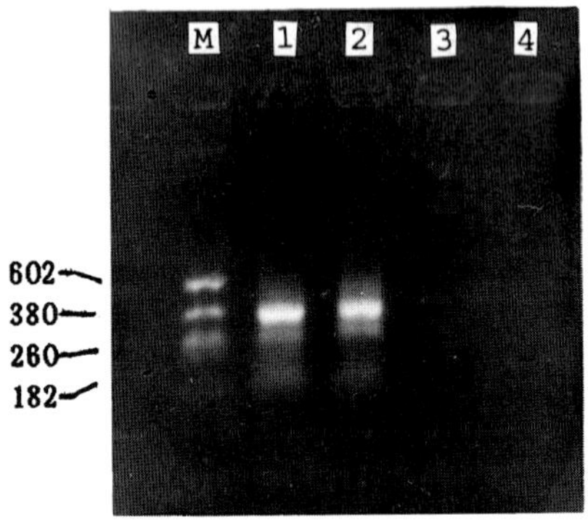

Fig. 1. Analysis of PCR products by agarose gel electrophoresis. Lane M, Marker; lane 1 and 2, amplified products; lane 3 and 4, control(PCR amplification reaction mixtures without primers).

Recent studies have shown that opioids can modulate cell-cell interaction of monocytes. On the other hand, it has been presumed that the receptor for opioids may function as the receptor for IL-2, and that there is at least a certain relationship between them. The gene encoding the human IL-2 R has been localized on chromosome 10 [6]. IL-2 and its receptor, which are important to cell growth and which are absent on resting T cells, may be expressed after cellular activation with mitogen (or an antigen) and are responsible for cell proliferation. Previous reports have indicated that M-Enk could act as inhibitor to suppress the proliferation of lymphocytes, but the mechanism of the inhibition is not clear. Therefore, it is presumed from the results that interfering with IL-2-mediated signal transduction may be involved in the suppressive effect of M-Enk on lymphocytes.

References

1. Loh, H.H. and Smith, A.P., Annu. Rev. Pharmacol. Toxical., 30(1990)123.
2. Leonard, W.J., Nature, 311(1984)626.
3. Li,G. and Fraker, P.J., Acta Pharmacol. Sin., 10(1989)216.
4. Chomczynski, P. and Sacchi, N., Analy Biochem., 162(1987)156.
5. Semple, M., Loveday, C., Weller, I. and Tedder, R., J. Med. Virol., 35(1991)38.
6. Leonard, W.J., Donlon, T.A., Lebo, R.V. and Greene, W.C., Science, 228(1985)1547.

Session V
Peptide immunology

Chairs: Jean Rivier
The Salk Institute
La Jolla, California, U.S.A.

De-Xin Wang
Institute of Materia Medica
Beijing, China

Yoshiako Kiso
Kyoto Pharmaceutical University
Kyoto, Japan

and

Xiao-Ji Xu
Peking University
Beijing, China

A peptide mimetic as antigen and immunogen

Arnold C. Satterthwait[a], Edelmira Cabezas[a], Julio C. Calvo[a], Shao-Qing Chen[a], Jia-Xiang Wu[a], Peng-Liang Wang[a], Yi-Ling Xie[a], E.A. Stura[a] and D.C. Kaslow[b]
[a]*Department of Molecular Biology, The Scripps Research Institute, La Jolla, CA 92037, U.S.A.*
[b]*NIAID, N. I. H., Bethesda, MD 20892, U.S.A.*

Introduction

Synthetic peptides have found important uses in immunology as antigens and immunogens [1]. However, their value is limited since peptides are often conformationally disordered in water and poor mimics of the ordered surfaces of native proteins. Consequently, affinities of peptides for antiprotein antibodies are often low. In addition, antipeptide antibodies often bind weakly if at all to native proteins limiting their use as vaccines. This leaves considerable room for improvement. Anfinsen's experiments [2] more than two decades ago suggested that affinities of peptides for antibodies could be improved as much as 10^4-fold by folding them. This effect has been confirmed with selected disulfide loops [3]. However, disulfide loops are few and far between in proteins and general approaches are needed to expand on these observations.

We have been exploring the use of covalent hydrogen bond mimics [4]. On average, every other amino acid in a globular protein engages in hydrogen bond formation between the main chain amide proton of one amino acid and the carbonyl oxygen atom of another (NH....O=CRNH). We have developed procedures for replacing structure defining hydrogen bonds on solid supports with a hydrazone link ($N\text{-}N{=}CHCH_2CH_2$) using multiple constrained peptide synthesis (MCPS) to form turns, loops and alpha helices [5]. Here we report on the enhanced antigenicity and immunogenicity of a malaria peptide folded with the hydrazone link.

Results and Discussion

The peptide of interest is from a neutralizing epitope on an EGF-like protein, Pfs25 [6], found during the sexual stages of *P. falciparum* malaria which infects hundreds of millions of persons annually. Both polyclonal and monoclonal antibodies to this protein completely block malaria transmission from human sera to the *Anopheles* mosquito (7). Consequently, vaccines based on this protein might be used to extend the life of anti-malarial drugs by blocking the transmission of escape mutants and breaking the cycle of malaria transmission. Several transmission blocking MAbs to this protein have been characterized [7]. One of these, Mab 4B7, binds weakly to a peptide, ILDTSNPVKT, that maps to a predicted beta-hairpin on the third of four EGF-like domains on Pfs25. This prediction is based on NMR structures for EGF-like proteins [8] and GORBTURN [9], an algorithm that predicts secondary structure and specific turn types. Beta-hairpins are characterized by a ladder of hydrogen bonds that can be replaced with covalent hydrogen bond mimics. Initially, a linear peptide and several loops (Table 1) were synthesized using MCPS [4]. Loop 1 was modeled on a GORBTURN prediction while Loop 2 was modeled on an EGF-like protein. In order to determine which loop best mimicked the corresponding loop on Pfs25, polyclonal rabbit antiserum to *P. falciparum* gametes was titered against linear and loop peptides in ELISAs (Fig. 1). The anti-gamete serum showed little or no reaction with control loops (not shown) or Linear 1. However, it reacted with each of the Pfs25 peptide loops, showing a preference for Loop 1. In contrast, antisera to improperly and/or incompletely folded yeast recombinant Pfs 25 [6] bound Linear 1 and Loop 1 peptide to a similar degree (Fig. 1), demonstrating that Linear 1 is exposed for reaction. MAb 4B7 also showed a strong preference for Loop 1 (titer = 64,000) compared with Linear 1 (<500). The preference of MAb 4B7 for Loop 1 was confirmed in competition ELISAs with a series of loops varying in size (8-12 amino acids from Pfs25) and cadence (overlaps of one amino acid within each size class).

Table 1 *Linear and Loop peptides used in this study*

Linear 1	Acetyl-G**ILDTSNPVKTG**GC-NH_2
Loop 1	**JILDTSNPVKTG**ZC-NH_2
Loop 2	**JILDTSNPVKTGV**GZC-NH_2
Loop (16-mer)	**JKC*ILDTSNPVKTGVC***SGZC-NH_2

The amino acid sequences in bold are from Pfs25. C* is Cys(Acm). Loops are formed by covalently joining J and Z in a hydrazone link: -N(CH_2CO-)N=CH-$CH_2CH_2CH_2$CO-.[3,4].

We believe these experiments have important implications. First, they show that antibodies to exposed loops, while clearly present in antisera, can go unrecognized by linear peptides. Since globular proteins often expose compact loops, an important class of antibodies may go undetected. Secondly, these antibodies can be detected with cyclic peptides. The cyclic peptide does not have to be a perfect mimic of the corresponding epitope on the protein since loops of different size and cadence react with antibodies better than the linear peptide. Alternative links might also be considered. Third, once a loop has been identified, it can be improved. Many of the tools available for peptide drug development might be applied to this process.

The improved affinity of antibodies for Loop 1 implies that it occupies a preferred conformation(s) in water. In fact, Loop 1 shows evidence of considerable conformational restraint at its "tip". Temperature coefficients for the backbone amide NH protons of Ser (3 ppb/degC) and Asn (1.5 ppb/degC) in 10% D_2O/H_2O are unusually low indicating sequestration of the amide protons from solvent and a defined structure. Further work is being directed towards an X-ray crystal structure determination of Loop 1 bound to MAb 4B7 (10). A comparison of the solution conformation of Loop 1 with its bound conformation may aid in understanding its improved affinity and provide a basis for rational vaccine design.

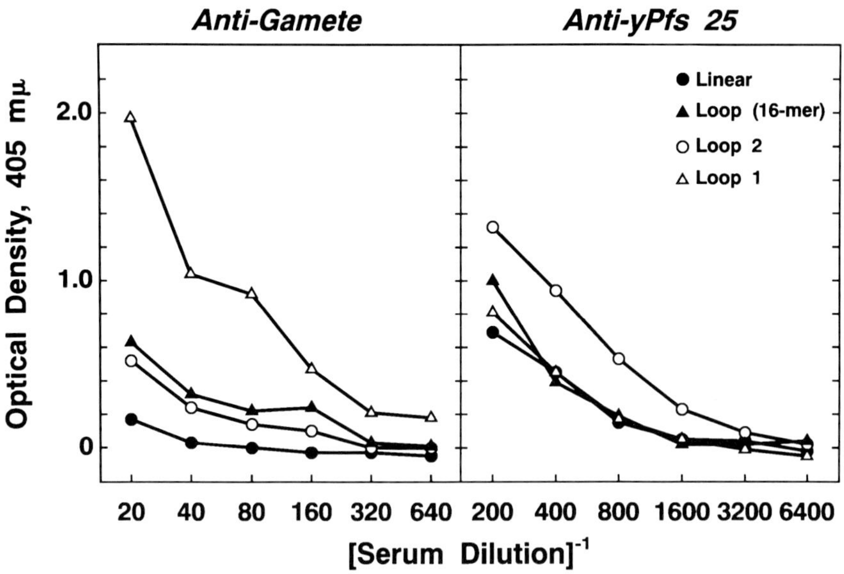

Fig. 1. ELISA titers of rabbit anti-gamete serum to Linear 1 and Loop peptides conjugated to maleimide-activated BSA (left panel). Peptides are described in Table 1. ELISA titers of mouse polyclonal anti-yeast recombinant Pfs25 serum to the same peptides. (right panel).

A systematic study of synthetic peptide vaccines requires covalently linked B-cell and T-cell epitopes for the stimulation of an immune response. To explore possibilities, we have taken advantage of two recent developments. Kaumaya and coworkers [11] have identified and explored colinear synthetic vaccines composed from B-cell and "promiscuous" T-cell epitopes that bypass genetic restriction in mice. We have used one of these T-cell epitopes for the syntheses of a 4-branched Linear 1-MAPS and a Loop 1-MAPS by applying Tam's modular approach to MAPS synthesis [12]. Linear 1 and Loop 1 were chemically ligated to a chloroacetylated "promiscuous" tetanus toxoid T-cell epitope, $ClCH_2CO$-[QYIKANSKFIGITEL], on a MAPS core. Both purified Linear 1-MAPS and Loop 1-MAPS showed single bands of expected mass (~15 kD) by high density polyacrylamide gel electrophoresis and exact mass by ion spray mass spectrometry.

Linear 1-MAPS and Loop 1-MAPS were used to immunize outbred mice and the antisera were analyzed (Table 2). Loop 1-MAPS antisera shows a slight preference for Loop 1 in ELISAs and can be distinguished from Linear 1-MAPS antisera. On the other hand, Loop 1 antisera differ from MAb 4B7 and anti-gamete antisera showing a strong reaction with Linear 1. This could reflect either the presence of anti-loop antibody that binds linear peptide and/or clipping of the loop by proteolysis to expose linear strands. However, it was Loop 1 antisera rather than the Linear 1 antisera that bound malaria gametes, suggesting that immunogenicity can be enhanced by folding the peptide.

Table 2 *ELISA titers of MAb 4B7 and polyclonal antisera from five mice immunized three times with Linear 1-MAPS and Loop 1-MAPS. Reaction of antibodies with P. falciparum gametes was observed in immunoflurorecence assays*

Antibodies	Antigens		
	Linear 1-BSA	Loop 1-BSA	Gametes
4B7 MAb, Ascites	<500	64,000	++++
1. Linear 1-MAPS	16,000	8,000	-
2. Linear 1-MAPS	16,000	8,000	-
3. Loop 1-MAPS	8,000	8,000	+
4. Loop 1-MAPS	4,000	8,000	-
5. Loop 1-MAPS	8,000	8,000	++

A strategy for rational vaccine development is apparent. It begins with the identification of a constrained peptide that shows improved affinity for neutralizing antibodies. This can be followed by optimizing this affinity using synthetic procedures augmented by NMR spectroscopy, molecular modeling and X-ray crystallography. Progress can be monitored by linking constrained peptides to T-cell epitopes in MAPS to give well characterized synthetic vaccines for immunization trials.

Acknowledgements

This work was supported by funding for malaria vaccine research by U.S.A.I.D. DPE-5979-A-00-1035-00. This is publication 8797-MB from the Scripps Research Institute.

References

1. Bersovsky, J., Science, 229(1985)932.
2. Sachs, D.H., Schechter, A.N., Eastlake, A. and Anfinsen, C.B., Proc. Nat. Acad. Sci. USA, 68(1971)1450.
3. Teicher, E., Maron, E. and Arnon, R., Immunochemistry, 10(1973)265.
4. Chiang, L.-C., Cabezas, E., Calvo, J.C. and Satterthwait, A.C., In Du, Y.-C., Tam, J.P. and Zhang, Y.-S. (Eds.) Peptides: Biology and Chemistry (Proceedings of the 1992 Chinese Peptide Symposium), ESCOM, Leiden, 1993, pp. 204–206.
5. Chiang, L.-C., Cabezas, E., Calvo, J.C. and Satterthwait, A.C., In Hodges, R.S. and Smith, J.A. (Eds.) Peptides: Chemistry, Structure and Biology (Proceedings of the 13th American Peptide Symposium), ESCOM, Leiden, 1994, pp. 278–280.
6. Kaslow, D.C., Quakyi, I.A., Sying, C., Raum, M.G., Keister, D.B., Coligan, J. E., McCutchan, T.F and Miller, L.H., Nature (London), 333(1988)74.
7. Barr, P.J., Green, K.M., Gibson, H.L., Barhurst, I.C., Quakyi, I.A. and Kaslow, D.C., J. Exp. Med. 174(1991)1203.
8. Kline, T.P., Brown, F.K., Brown, S.C., Jeffs, P.W., Kopple, K.D. and Mueller, L., Biochemistry (1990)7805.
9. Wilmot, C. and Thornton, J., J. Mol. Biol. 203(1988)221.

10. Stura, E.A., Kang, A.S., Stefanko, R.S., Calvo, J.C., Kaslow, D.C. and Satterthwait, A.C., Acta Crys. D. (1994) in press
11. DiGeorge, A.M., Wang, B., Kobs-Conrad, S.F. and Kaumaya, P.T.P., In Hodges, R.S. and Smith, J.A. (Eds.) Peptides: Chemistry, Structure and Biology (Proceedings of the 13th American Peptide Symposium), ESCOM, Leiden, 1994, pp. 732–733.
12. Defoort, J-P. Nardelli, B., Huang, W. and Tam, J.P., J. Peptide Protein Res. 40(1992)214.

An efficient method for preparing small synthetic peptide immunogens

Ying Chen, Jin Wang and Pu-Tao Shi
Shanghai Institute of Biochemistry, Academia Sinica, Shanghai 200031, China

Introduction

Antibodies raised against synthetic peptides have been widely used in various fields, such as studies of the relationship between structure and function of proteins and peptides, preparation of affinity chromatography columns for protein purification, synthetic vaccines and diagnostic kits, etc. However, it is difficult for micromolecular synthetic peptides to raise antibodies with high titers. To enhance their immunogenicity, small peptides therefore are usually conjugated to macromolecular carriers.

Generally, peptide-carrier conjugates are obtained by chemical conjugation by linking functional groups, such as amino, carboxyl, or thiol groups, on both sides. The most commonly used coupling agents are glutaraldehyde, carbodiimides, bis-diazotized benzidine (BDB), b-maleimido capric acyl-N-hydroxy succinimide ester (MCS), and m-maleimidobenzoyl-N-hydroxy succinimide ester (MBS). The conventional carriers are macromolecular proteins, such as bovine serum albumin (BSA), ovalbumin and tetanus toxoid (TT). Synthetic polypeptides such as poly (D,L-Ala,Lys) and polysaccharides such as dextran have been used as carriers. Besides the chemical conjugating method, physical adsorption methods in which polyvinylprolidone (PVP), alumina (Al_2O_3) and active charcoal act as carriers have been used. Bacteria have also been reported to act as adjuvants and carriers in place of Freund's adjuvant to prepare the antibodies to glycolipids [1] and natural glycoproteins [1]. In the present report, we utilized *Salmonellae typhi* to prepare synthetic peptide immunogens, immunized the rabbits and obtained high levels of high specific antisera.

Results and Discussion

Preparation of good immunogens is a prerequisite for obtaining high level antisera of high specificity. The usual method of raising anti-small peptide antibodies is to immunize animals with peptide-carrier conjugates. The immunogenicity of a peptide is also related to the properties and positions of the determinants on its surface, as well as its molecular conformation.

Although conventional chemical methods are widely used, they have several shortcomings. First, the conjugating reactions of peptides and carriers result not only in peptide-carrier conjugates, the purpose products, but also in peptide-peptide, carrier-carrier polymers in various ratios [3]. The conformations of polymers will be different from those of monomers. Thus, it is possible to create various new determinants. In addition, antisera obtained by chemical conjugation contain antibodies not only to peptides and carriers but also to coupling agents [4,5]. All of these will reduce the specificity of the antibodies.

Second, when the coupling agents conjugate peptides and carriers through linking their functional groups, such as amino groups, carboxyl groups, or thiol groups, they also chemically influence the peptide residues of the same groups (Table 1). This will lead to the modification of the molecular conformation and antigenicity of the peptides. Moreover, if the reacted residues are located precisely in the antigenic determinants, peptide antigenicity will be seriously reduced or even completely abolished [4].

Table 1 *Amino acid residues reported to be involved in the chemical coupling reactions most commonly used*

Coupling reagents	Modified amino acids
Glutaraldehyde	Lys, Cys, α-NH_2
Carbodiimides	Asp, Lys, Glu, α-NH_2, α-COOH
BDB	Tyr, Cys, His, Lys
MBS	Cys, Specific NH_2

Third, for immunogen preparation, it is necessary to add extra amino acids with functional groups to a few peptides without those groups, such as LHRH.

Physical adsorption for preparation of peptide-carrier complexes was considered, and PVP, Al_2O_3 and active charcoal have been used as carriers. However, the antisera titers thus obtained were very low.

We have prepared ten different peptide immunogens by physical adsorption on *Salmonella typhi* bacteria, immunized rabbits in routine way and obtained antibodies with high titers in short periods of time (Table 2).

Table 2 *Synthetic peptides and the results of immunizations*

Peptides	Amino acid sequences	Numers of AA residues	Ab titers (by ELISA)	Frequencies of boostings
Peptide 1	THRLAGLLSRSGGVV	15	25,000	4
Peptide 2	YDEIKKFVCGG	11	100,000	3
Peptide 3	FKCKYKQIFLGGV	13	25,000	3
Peptide 4	VDKLQKVCGG	10	35,000	3
Peptide 5	SSCFGGRIDRIGAQSGLGCNSFRY (S–S bridge between C3 and C19)	24	12,000	4
Peptide 6	RIGAQSGLGCNSFRY(NH_2)	15	12,000	4
Peptide 7	RRSSCFGGRID(NH_2)	11	10,000	4
Peptide 8	RPKPQQFFGLM(NH_2)	11	20,000	5
Peptide 9	FADKPVASGSNKLVQRG	17	15,000	4
Peptide 10	KLTFVCGSAMATGKFLLAG	19	20,000	3

We also investigated the specificity of the antibodies obtained. Fig. 1 shows the binding of rabbit anti-peptide 2 antibodies to itself, to vasopressin (VP), fragment of malaria surface antigens (MS), peptide 8 and peptide 3, and suggests good antibody specificity.

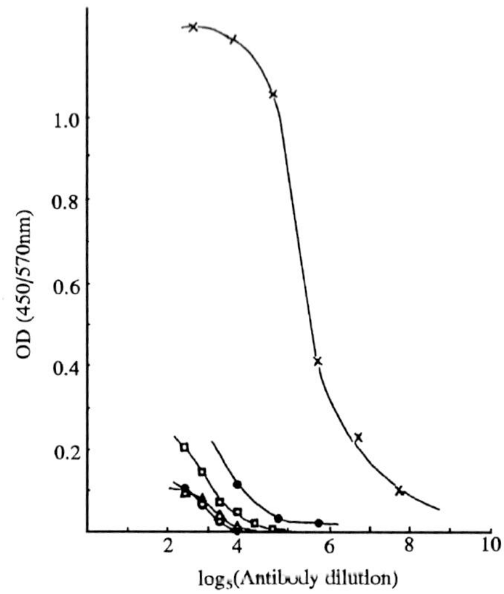

Fig. 1. Binding of Rabbit Anti-Peptide 2 Antibodies to ×*-Peptide 2,* △*-VP, o-MS,* □*-Peptide 8,* •*-Peptide 3.*

The results above show the feasibility of the method in which peptides are absorbed on *Salmonella typhi* bacteria to prepare small synthetic peptide immunogens and show that bacteria act as a carrier. Furthermore, using this mild physical adsorption method, the molecular structure of peptides will be natural. It leads to no polymer by-product, no new determinant, no chemical action on amino acid residues and no abolishment of antigenicity, and thus improves the specificity of the antibodies. *Salmonellae typhi* have some effects as *Mycobacteria* to generate and enhance immune responses. Briefly, this is a simple and efficient method for peptide immunogen preparation.

References

1. Higgins, T.J., Mol. Immunol., 2(1985)1265.
2. Bellstedt, D.U., Human, P.A., Rowland, G.F. and Van der Merwe, K.J., J. Immunol. Methods, 98(1987)249.
3. Lee, A.C.J, Powell, J.E., Tregea, G.W., Niall, H.D. and Stevens, V.C., Mol. Immunol, 17(1980)749.
4. Briand, J.P., Muller, S. and Van Regenmortel, M.H.V., J. Immunol. Methods, 78(1985)59.
5. Jones, G.L., Edmundson, H.M., L. Spenoer, L., J. Immunol. Methods, 123(1989)211.

Synthesis and conformation-activity relationships of MDP analogs

Ming-Zhu Zhang[a], Jie-Cheng Xu[a], Xiao-Ling Huang[a], Hou-Ming Wu[a], Xiao-Feng Wang[b], Xiao-Yu Li[b], Yi Zhang[c] and Shi-Yi Liu[c]
[a]*Shanghai Institute of Organic Chemistry, Chinese Academy of Sciences, Shanghai 200032, China*
[b]*Shanghai Institute of Physiology and* [c]*Shanghai Institute of Materia Medica, Chinese Academy of Sciences, Shanghai 200031, China*

Introduction

Ever since NAcMur-L-Ala-D-isoGln (muramyl dipeptide, MDP) was recognized as the minimal glycopeptide essential for the adjuvant activity of the mycobacterial cell wall, numerous MDP analogs have been synthesized in attempts to enhance their immunoadjuvant activity and to reduce pyrogenic side effects. However, only a few works on solution conformation have been done, although it would be useful for the design of new MDP analogs. In this paper, we report the synthesis of MDP analogs, and the study of the relationship between their conformation and activity.

Results and Discussion

A series of MDP analogs were synthesized by an improved procedure (Fig. 1). In addition, an efficient coupling reagent, Benzotriazol-1-yloxy-bis(pyrrolidino)Carbonium hexafluorophosphate (BBC), was used.

The solution conformations of some synthesized MDP analogs (which either make the molecule more flexible or rigid than the parent MDP molecule) have been studied in dried d6-DMSO by a variety of 1D, 2D NMR techniques at 600 MHz. These include ^{1}HNMR, TOCSY, COSY, NOESY and hydrogen-deuterium exchange. The results suggest the presence of two successive type II β-turns in MDP [γ-Abu]^{1}, MDP[Pro]^{1} and MDP-Oxalys, although the stability of these β-turns differ. However, only the second β-turn persists in MDP[Aze]^{1} (Fig. 2).

The immunoactivity of MDP analogs were evaluated by analyzing the proliferation of T cells and B cells, and the level of interleukin-1 (IL-1) release by macrophages *in vitro*. As shown in Table 1, these MDP analogs exhibit selective activity towards T cells, B cells and macrophages. By comparing activity with solution conformation, it was demonstrated that their immunoactivity probably was related to the second β-turn, while there is no apparent relationship between immunoactivity and the first β-turn.

Pyrogenicity was assayed by testing the temperature of middle back skin via intracerebroventricular infusion in rabbits. As shown in Table 2, MDP-Oxalys and MDP[Pro]^{1} are less pyrogenic than MDP and MDP[Aze]^{1}; while MDP[Pro]^{1}-L8 exhibited negligible pyrogenicity. Moreover, MDP[γ-Abu]^{1} was demonstrated to be devoid of pyrogenic side effects even when injected in large quantities. According to

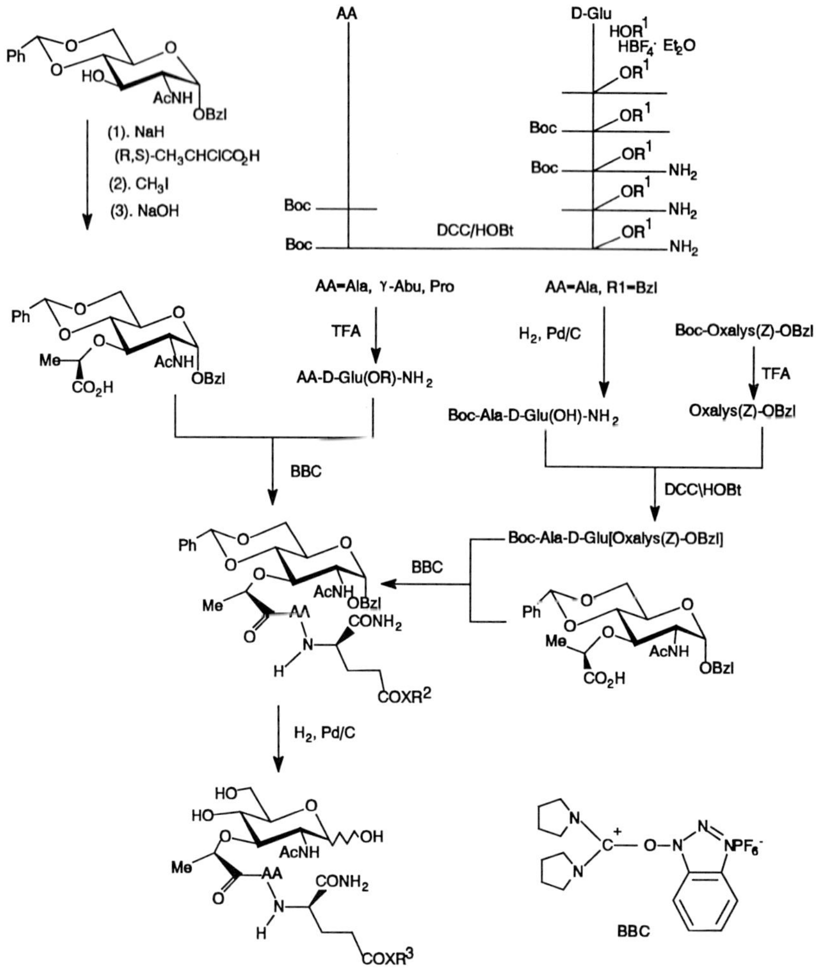

Name	AA	XR 3
MDP[g-Abu] 1	g-Abu	OH
MDP-Oxalys	L-Ala	L-Oxalys
MDP[Pro]1	L-Pro	OH
MDP[Aze] 1	L-Aze	OH
MDP[Pro]1-L8	L-Pro	OC_8H_{17}

g-Abu: $NH_2CH_2CH_2CH_2COOH$
Oxalys: $NH_2CH_2CH_2OCH_2CHNH_2COOH$
L-Aze: HN—COOH (azetidine ring)

Fig. 1. Synthesis of MDP analogs.

their conformation, we can infer that pyrogenic side effect can be abolished due to the high stability of the first β-turn. On the contrary, pyrogenicity may not be related to the second β-turn.

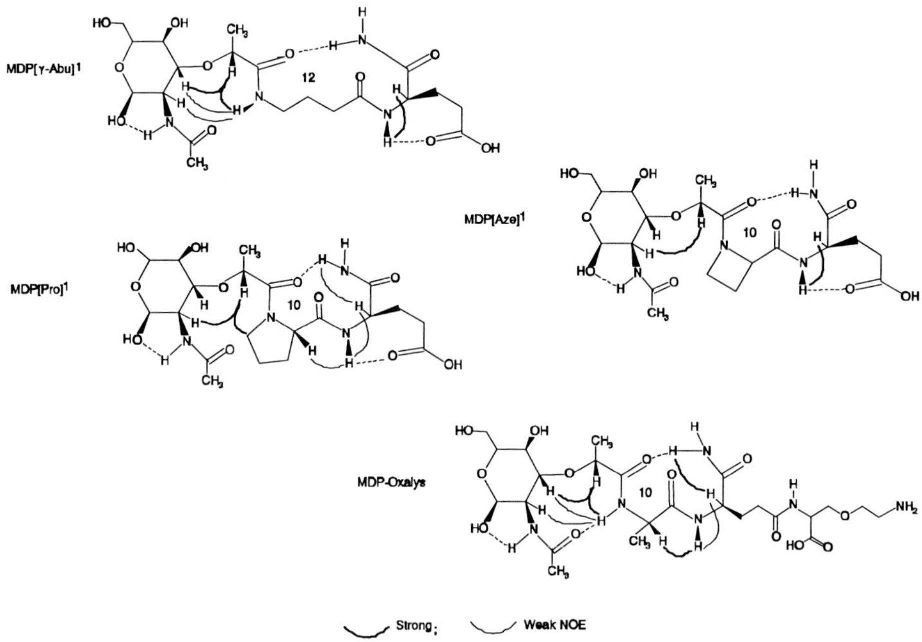

Fig. 2. Possible conformation of MDP analogs.

Table 1 *Effects of MDP analogs on T cell, B cell, and macrophage*

Conc.(μg/ml)	Con A(T)			LPS(B)			IL-1			Stability of First β-Turn	Stability of Second β-turn
	1	5	25	1	5	25	1	5	25		
MDP			↑↑	↑	↑↑↑	↑				**	**
MDP[γ-Abu][1]			↑	↑	↑↑	↑				***	**
MDP[Aze][1]			↑	↑	↑↑↑	↑↑					*
MDP-Oxalys				↑	↑↑↑	↑↑				**	**
MDP[Pro][1]		↑	↑↑			↓	↑			*	***
MDP[Pro][1]-L8	↑	↑	↑↑↑			↓				/	/

↑: $p<0.05$; ↑↑: p,0.01; ↑↑↑: $p<0.001$
* stable ** more stable *** most stable

Table 2 *Effects of synthetic MDP analogs (0.3ug/animal, i.c.v.) on middle back skin temperature in rabbits*

Name	Control	1~2 h	3~4 h	Pyro-genicity	Stability of first β-turn	Stability of second β-turn
MDP	36.14±0.50 (N=8)	36.56±0.42↑↑ (N=8)	37.34±0.53↑↑ (N=8)	**	**	**
MDP[Aze][1]	35.65±0.87 (N=6)	35.82±0.87 (N=6)	36.69±0.88↑↑ (N=6)	**		*
MDP[Pro]	36.07±0.82 (N=6)	35.99±0.68 (N=6)	36.65±0.73↑ (N=6)	*	*	***
MDP-Oxalys	35.15±0.70 (N=8)	35.15±0.72 (N=8)	35.57±0.84↑ (N=8)	*	**	**
MDP[Pro][1]-L8	35.20±0.96 (N=6)	35.12±0.83 (N=6)	35.62±0.85 (N=6)		/	/
MDP[γ-Abu][1]	35.42±0.74 (N=8)	35.33±1.05 (N=8)	35.54±1.07 (N=8)	no	***	**

↑: $p<0.05$ ↑↑: $p<0.01$ * stable ** more stable *** most stable

Summary

1. Some MDP analogs were synthesized by an improved procedure.

2. Study of solution conformation indicates the presence of two adjacent β-turns in MDP[γ-Abu][1], MDP-Oxalys and MDP[Pro][1], but only the second β-turn exists in MDP[Aze][1].

3. Immunoadjuvant activity seems to be related to the second β-turn, while the lack of pyrogenic side effects may be due to the comparatively higher stability of the first β-turn.

4. MDP[γ-Abu][1] and MDP[Pro][1]-L8 have potent immunoadjuvant activity, but they are not pyrogenic.

Acknowledgements

This study is supported by National Natural Science Foundation of China and State Key Laboratory of Bioorganic and Natural Products Chemistry.

References

1. Baschang, G., Tetrahedron, 45(1989)6331.
2. Chen, S.Q. and Xu, J.C., Tetrahedron Lett., 33(1992)647.
3. Sizun, P., Tetrahedron, 44(1988)991.

Studies on synthesis of HCV vaccine. I. Design, chemical synthesis and antigenicity of peptide fragments of HCV

Gang Liu[a], Zheng-Lun Liang[b], Meng-Shen Cai[a], Zhi-Hui Qin[a] and Hui Zhuang[b]
[a]School of Pharmaceutical Sciences and [b]School of Public Health, Beijing Medical University, Beijing 100083, China

Introduction

Hepatitis C Virus (HCV), the major causative agent of post transfusion non-A, non-B hepatitis (NANB), was first cloned and expressed by Choo et al [1]. From nucleotide sequence analysis, HCV has shown to be distantly related to the flavivirus and pestivirus families [2]. Failure to propagate HCV in tissue culture, the absence of a simple in vitro assay for infectious or cytopathic effects, and the lack of a susceptible small-animal model [3], were obstacles to the development of a hepatitis C vaccine. As a flavivirus pestivirus, HCV had a single major open reading frame (ORF) encoding a 3011-amino acid polyprotein that was secondarily cleaved into several structural and non-structural proteins. The structural region included three proteins: the nuclocapsid (C), envelop glycoproteins, E1 (gp33) and E2/NS1 (gp72) [4,5]. The non-structural region encoded protein NS2, NS3, NS4 and NS5. E2/NS1 corresponding to the gp53/gp55 envelope glycoprotein of the pestivirus and the NS1 of the flavivirus [5] that were known to elicit protective antibodies in hosts vaccinated with these proteins [6,7]. According to the amino acid sequence of HCV-BK [8], we designed sixteen peptide fragements with profiles of hydrophilicity, accessibility, and flexibility using the "Goldkey" computer program.

Results and Discussion

The following peptides were synthesized by the Merrifield Method with the acid-labile

P1:	AERSRSDQRPYCWHYPPPQCT
P3:	NSSGCPERMAQCRTIDKFDQ
P4:	DRPELS
P5:	NWTRGERCDLEDRDRPELS
P6:	PYCWHY
P7:	AESSRSDQRPYCWHY
P14:	CKPQRKTKRNTNRRPQYPEGRTWAQ
P15:	EDRDRPEL
P16:	VREGNSSR
P17:	GCVPCVREGNSSR
P18:	PQSFQVAHLHAPTGSGKSTKV
P19:	IVPDRELLYQEFDEMEECASHLPYIEQGMQLAEQFKQKALGL
P24:	APTGSGKSTKV
P25:	VSRSQRRGRTGR
P26:	DTHVTG
P27:	GSSAVDSGTATGPPDQASDDGDKGSDVESYSSMPE

tert-butyloxycarbonyl (BOC) group for temporary amino terminal protection and more acid-stable group for the protection of side chains with Merrifield Peptide Resin (2% cross-linked, 200-400 mesh, 1.04 mmol Cl/g)(purchased from SIGMA Company). DCC+HOBt or DCC+HOSu were used as coupling reagent. The protected target resin-bound peptides were cleaved with dried hydrogen fluoride (HF) and crude peptide were purified by column chromatography on Sephadex G-10, G-15 or G-25. Composition of peptides was confirmed by amino acid analyzer 835-50 of HITACH. Analytical HPLC of the peptides was performed by RP-HPLC (BONDAPAK C18 Column, 0.78x30cm). The synthetic peptides were used in ELISAs to detect antibodies to HCV in serum samples. First, upright and foursquare experiment was used for determing the coating quantity of peptide, diluted concentration of enzyme and samples. Wells of a 96 micowell Nunc plate received 100 μl carbonate-buffer (PH 9.6 0.05 M) containing 2.0 ng/μl peptides, and the plate was maintained for 1 hr at room temperature and 4°C overnight. The wells were washed five times with 0.9% NaCl Triton containing 0.05% Tween 20 and then blocked with 10% bovine serum in carbonate-buffer (PH 9.6 0.05 M) at 4°C for 6 hrs. After washing with 0.9% NaCl Triton, 10 μl of the serum samples (1:9) were added to each well and incubated for 0.5 hr at 37°C. Repeated washing and 1--10 μl (1:7000) mouse anti-hMwan horseradish peroxidase was added to each well and kept at 37°C for 0.5 hr. Final washing and color development of the bound peroxidase conjugate was achieved with substrate OPD solution. Absorbed OD values of each well were read at 492 nm.

As shown in Table 1, we found that synthetic peptide P14 corresponding to the C region ; P6 corresponding to the NS1 region and P19 corresponding to NS4 region showed very strong anti-HCV activity. In ELISAs, twenty eight sera were positive for both P14 and second-generation antigen ELISAs and seventeen sera were negative for both. Two sera were positive to second-generation antigen ELISAs and were negative to P14. One serum sample was positive to P14 and negative to second-generation antigen ELISA (see Table 2). The sensitivity of P14 therefore was 92.59%, and the specificity of P14 was 94.44%. The coincidences with second-generation antigen ELISAs was 93.33%. There were 2.2% (1/45) false positive tests for P14. When P14 was conjugated with BSA and mixed with Freund's Complete Adjuvant, rabbit were stimulated to generate anti-P14 antibody (1:2000).

According to the epitope profiles, we found 477-484 and 648-666 have the greatest possibility of containing linear epitopes. Seven peptides were designed and synthesized that corresponded to the amino acid resides in HCV-BK sequence: 475-495 (P1), 449-468 (P3), 658-663 (P4), 645-663 (P5), 484-489 (P6), 475-489 (P7), 655-662 (P15). The results of this ELISA were shown in Table 1. Peptide P6 was used to compare the ELISA results with second-generation assay. In a total of forty-four sera, twenty sera were reactive with P6 and thirty-four sera were reactive with second-generation assay (66.67% positive results). Fourteen sera that were reactive with second-generation assay were not reactive with P6. Ten sera nonreactive with second-generation assay were also not reactive with P6, indicating that P6 had no false positive results. Although we extended the amino acid sequences in search of a more effective peptide, their antigenicity decreased with the chain of peptide, such as P1 and P7, indicating that this region contained only a position of anigene-antibody binding. Another epitope 655-662 had low OD values as shown in Table 1. The same results were seen when we extended the amino acid sequence, such as P4, P5 and P15. In nonstructural region (NS3, NS4, NS5) of HCV, we defined P19 (NS3) as a linear immunogenic region. It had also 63.23% positive results when we tested 30 positive sera with ELISAs.

Table 1 *The results of ELISA of synthetic peptides*

	Sample									
Peptides	1	2	3	4	5	6	7	8	9	10
P1	0.13	0.12	0.08	0.25	0.11	0.11	0.12	0.13	0.07	0.10
P3	0.02	0.01	0.05	0.22	0.05	0.05	0.12	0.09	0.06	0.13
P4	0.01	0.08	0.02	0.02	0.07	0.11	0.13	0.09	0.06	0.06
P5	0.08	0.09	0.01	0.02	0.04	0.01	0.12	0.02	0.00	0.12
P6	0.54	0.17	0.25	0.68	0.75	0.10	0.82	0.06	0.15	0.43
P7	0.24	0.22	0.20	0.34	0.17	0.16	0.17	0.21	0.03	0.12
P14	1.22	1.19	1.23	1.06	0.80	0.67	0.89	1.02	1.00	1.13
P15	0.12	0.09	0.02	0.17	0.10	0.10	0.03	0.03	0.00	*N*
P16	0.05	0.09	0.05	0.07	0.03	0.04	0.04	0.08	0.05	0.02
P17	0.02	0.02	0.01	0.07	0.04	0.08	0.09	0.10	0.03	0.05
P18	0.12	0.03	0.08	0.11	0.20	0.10	0.11	0.12	*N*	*N*
P19	1.13	0.03	0.67	0.18	0.11	1.08	0.06	0.01	0.25	0.32
P24	0.10	0.00	0.06	0.04	0.06	0.12	0.07	0.04	0.17	0.00
P25	0.21	0.00	0.03	0.11	0.05	0.02	0.00	0.04	0.08	0.00
P26	0.00	0.04	0.03	0.02	0.08	0.09	0.03	0.12	0.03	0.05
P27	0.02	0.06	0.07	0.33	0.06	0.09	0.07	0.10	0.11	0.05

Table 2 *Comparative testing of 45 serum samples*

		Second-generation ELISA	
		+	-
P14	+	28	1
	-	2	17

+ and - represent positive and negative serum for anti-HCV Abs respectively

According to the results of all synthetic peptides, the profile of anti-HCV antibodies is shown in Fig 1.

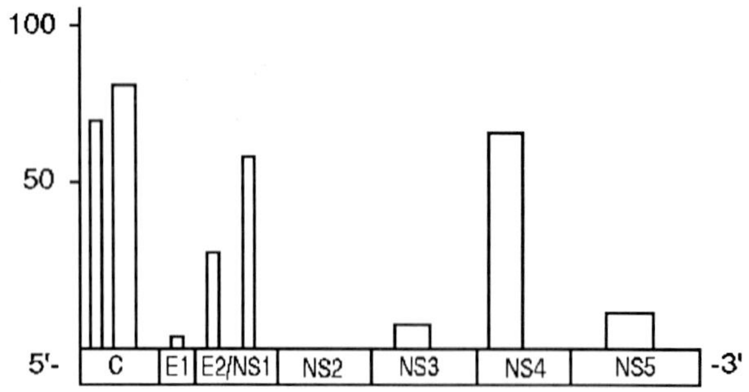

Fig. 1. The profile of anti-HCV antibodies in host.

Conclusions

1. The core region antigen (P14) has strongest activity in core region of HCV, with positive results of 93.33%.
2. There is a only a position Ag-Ab binding in HCV NS1 region (P6), 60.0% anti-HCV antibody was found with P6 ELISA.
3. On extending amino acid sequence of P6, the antigenicity decreased.
4. 63.23% of the positive sera can be detected by NS3 antigen (19) ELISAs.

Acknowledgements

We are grateful to Sun Tao of Institute in Basic Medical Sciences, AMMS, Beijing 100850, China and his colleagues for using their program "Goldkey" to do a parallel epitope analysis that gave a similar result as we got. The study was supported by a grant from Ministry of Public Health, China.

References

1. Choo, Q.L., Kuo, G., Weiner, A.J., Overby, L.R., Bradley, D.W. and Houghton, M., Science, 244 (1989)359.
2. Miller, R.H. and Purcell, R.H., Proc. Natl. Acad. Sci. USA, 87(1990)2057.
3. Farci, P. and Alter, H.J., Science, 258(1992)135.
4. Houghton, M.A., Weiner, A.J., Han, J., Kuo, G. and Choo, Q.L., Hepatology, 2(1991)381.
5. Hijikata, M., Kato, N., Ootsuyama, Y., Nakagawa, M. and Shimotohno, K., Proc. Natl. Acad. Sci. USA, 88(1991)2451.
6. Rumenapf, T., Stark, R., Meyers, G. and Thiel, H.J., J. Virol., 65(1991)589.
7. Schlesinger, J.J., Brandriss, M.W., Cropp, C.B. and Monath, T.P., J. Virol., 60(1986)1153.
8. Takamizawa A., J. Virol., 65(1991)1105.

Synthesis of antigenic peptides for another 26 kDa glutathione S-transferase of *Schistosoma mansoni*

Jia-Xi Xu and Meng-Shen Cai
School of Pharmaceutical Sciences, Beijing Medical University, Beijing 100083, China

Introduction

Among the candidates for an anti-schistosomal vaccine, the glutathione S-transferases (GSTs) are of major immunological interest. These GST isoenzymes appear to play a central role in the parasite detoxification system. Thus, the glutathione S-transferases have potential as components of a vaccine against schistosomiasis [1-4]. Immunochemical analyses have revealed that both *Schistosoma mansoni* and *Schistosoma japonicum* have at least two GST isoenzymes, of 26 and 28kDa(Sm26, Sm28, Sj26 and Sj28). Sm26 occurs as two different types (termed Sm26/1 and Sm26/2), which share 82% identity at the amino acid level [5]. In a previous paper, we reported the synthesis of antigenic peptide epitopes of Sm26/1[6]. In this paper, we studied the antigenic peptides of another GST isoenzyme, Sm26/2.

Results and Discussion

In order to find a strong immunogenic peptide for preparing an effective vaccine against Schistosoma, the epitopes of another 26kDa glutathione S-transferase of *Schistosoma mansoni* (Sm26/2) were predicted according to hydrophilicity by using Hopp and Woods's method [7], and according to flexibility and accessibility by the determination of the primary structure of Sm26/2 [8,9]. The peptides P26-39, P86-97, P139-153, P187-202 and P205-218 are epitopes of Sm26/2 antigen. Their sequences are shown below.

```
M1, P26-39,   ERYEERLYDRNDGD
M2, P86-97,   PKERAEISMLEG
M3, P139-153, HNTYLNGDHKTHPDF
M4, P187-202, PPIKNYLNSNRYIKWP
M5, P205-218, GWSATFGGGDAPPK
M6, P210-218, FGGGDAPPK
```

These antigenic peptides were synthesized by solid phase peptide synthesis method with the acid-labile tert-butyloxycarbonyl (Boc) group for temporary protection and acid-stable groups for side chain protection. The side chain-protected peptide-resins were treated by anhydrous hydrogen fluoride in the presence of anisole and thioanisole as scavengers and washed with cool ethyl ether. Peptides were extracted with 30% acetic acid and purified by gel filtration on Sephadex G15 and G25. They were checked for homogeneity by reverse-phase HPLC. The data of HPLC retention time and optical

Table 1 *The data of HPLC retention time and optical rotation*

Peptide	HPLC retention time	Optical rotation
p26-39	5.03[a]	-46(0.22)[e]
p86-97	5.83[b]	-44(0.90)[f]
p139-153	5.97[b]	-33(0.67)[f]
p187-202	5.96[b]	-37(0.09)[e]
p205-218	8.77[c]	-65(0.69)[f]
p210-218	10.15[d]	-69(0.64)[f]

HPLC: 10u YWG-C18 Column, 25cm×4.6mm, detection wavelength at 220nm(0.8AUFS), flow rate at 1mL/min, Eluent: A, 0.1%TFA-Water; B, MeOH, [a]A:B=80:20, [b]A:B=100:0, [c]A:B=60:40, [d]A:B=90:10. Optical rotation solvent: [e]Water, [f]20%HOAc/Water.

rotation are shown in Table 1. The compositions of the peptides were confirmed by amino acid analysis (shown in Table 2). They were conjugated to BSA (bovine serum albumin). The immunogenicity of all peptides and their BSA-conjugated products obtained with rabbit serum infected by *Schistosoma japonicum* have been determined by ELISA (enzyme-linked immunosorbent assay) method with goat anti-rabbit HRP (horseradish peroxidase) antibody. The results are shown in Table 3.

Table 2 *The amino acid analysis of synthetic epitope peptides*

Peptide	p26-39	p86-97	p139-153	p187-202	p205-218	p210-218
Asx	3.86(4)	-	4.26(4)	3.17(3)	1.02(1)	1.08(1)
Thr	-	-	1.78(2)	-	0.82(1)	-
Ser	-	0.86(1)	-	0.94(1)	0.89(1)	-
Glx	3.26(3)	3.12(3)	-	-	-	-
Gly	1.21(1)	1.18(1)	1.07(1)	-	4.39(4)	3.27(3)
Ala	-	1.08(1)	-	-	2.27(2)	1.18(1)
Met	-	0.86(1)	-	-	-	-
Ile	-	1.21(1)	-	1.76(2)	-	-
Leu	1.13(1)	1.16(1)	0.94(1)	1.10(1)	-	-
Tyr	1.79(2)	-	0.86(1)	1.81(2)	-	-
Phe	-	-	1.14(1)	-	1.10(1)	1.21(1)
Lys	-	1.21(1)	1.09(1)	1.81(2)	1.09(1)	1.16(1)
Arg	2.78(3)	0.98(1)	-	0.88(1)	-	-
His	-	-	2.88(3)	-	-	-
Pro	-	0.88(1)	0.79(1)	2.80(3)	1.78(2)	1.86(2)
Trp	-	-	-	a(1)	a(1)	-

a: Not determined.

As shown in Table 3, peptide-BSA conjugated products M1-BSA, M2-BSA, M3-BSA, M6-BSA and peptides M2, M6 showed higher immunogenicity. However, peptides M1, M3, M4, M5 showed general immunogenicity.

Table 3 *The immunogenicity of peptides and peptide-BSA with rabbit serum infected by Schistosoma japonicum*

Sample	M1	M2	M3	M4	M5	M6
P/N Value	1.93	2.30	1.88	1.88	2.09	2.25
Sample	M1-BSA	M2-BSA	M3-BSA	M4-BSA	M5-BSA	M6-BSA
P/N Value	3.31	3.23	3.59	1.22	1.27	3.31

Acknowledgement

We thank the Post Doctorate Fellowship for financial aid for J. X. Xu.

References

1. Henkle, K.L., Davern, K.M., Wright, M.D., Ramos, A.J. and Mitchell, G.F., Mol. Biochem. Parsitol.,40(1990)23.
2. Trottein, F., Kieny, M.P., Verwaerde, C., Torpier, G., Pierce, R.J., Balloul, J.M., Schmitt, D., Lecocq, J. P. and Capron, A., ibid., 41(1990)35.
3. Sher, A., James, S.L., Correa-Oliveira, R., Hieny, S. and Pearce, E., Parasitol., 98(1989)61.
4. Smith, D.B., Davern, K.M., Board, P.G., Tiu, W.U., Garcia, E.G. and Mitchell, G.F., Proc. Natl. Acad. USA, 83(1986)8703.
5. Wright, M.D., Harrison, R.A., Melder, A.M., Newport, G.R. and Mitchell, G.F., Mol. Biochem. Parasitol., 49(1991)177.
6. Xu, J.X. and Cai, M.S., The 2nd China-Canada symposium on Organic Chemistry, 4, Huangshan, Anhui, China, 1994.
7. Hopp, T.P. and Woods, K.R., Proc. Natl. Acad. Sci. USA, 78(1981)3824.
8. Janin, J., Nature, 277(1979)491.
9. Ponnuswamy, P.K. and Bhaskaran, R., Int. J. Peptide Protein Res., 24(1984)168.
10. Xu, J.X. and Cai, M.S., China Patent, CN94105973.1(1994).
11. Xu, J.X. and Cai, M.S., China Patent, CN94105974.X(1994).
12. Cai, M.S., Xu, J.X., Yang, J. and Qin, Z.H., China Patent, CN94105976.6(1994).

Studies on synthesis of HCV vaccine. II. Hepatitis C virus (HCV) immune selection

Gang Liu[a], Zheng-Lun Liang[b], Tao Sun[c], Meng-Shen Cai[a] and Hui Zhuang[b]
[a]School of Pharmaceutical Sciences and [b]School of Public Health, Beijing Medical University, Beijing 100083, China
[c]Institute of Basic Medical Sciences, AMMS, Beijing 100850, China

Introduction

HCV infection occurs frequently in individuals who have not had an obvious exposure to contaminated blood or blood products [1]. Most HCV infections are persistent despite the onset of a measurable immune response because HCV has the characteristics of rapid variability. The E2/NS1 amino terminus of HCV (amino acids sequence 384-414), which is hypervariable with respect to both nucleotide and amino acid sequence, has been termed the E2 HV domain [2,3] or HVR1 [4]. The E2 HV domain appears to be a rapidly evolving region of the HCV genome [3] and may contain linear neutralizing epitopes. For further immunogenicity studies on E2 HV of HCV, we selected three peptide fragments of the E2 HV region sequence of genotype-II and synthesized them with SPPS method.

Results and Discussion

The amino acid sequences are:

P2: VDGDTHVTGGAQAKTTNR (381-398) [5]
P9: STHVTGAVQGHSIRGTTSLFTSGPAQKIQ (384-412) [6]
P10: RTYTSGGTAGHTTSGITSLFSPGASQKIQ (384-412) [6]

These sequences were synthesized by the solid phase peptide synthesis method with acid-labile tert-butyloxycarbonyl (Boc) group for temporary amino-terminal protection and more-stable group for side protection. Peptide resins were cleaved by HF and crude peptides were purified with Sephadex-G25. Composition of the peptides was confirmed by HITACH amino acid analyzer 835-50. Analytical HPLC of the peptides was performed by RP-HPLC (BONDAPAK C18 Column, 0.78x30 cm). The target peptides were linked to a protein carrier (BSA) by glutaradehyde method with 5.0 mg peptide and 10.0 mg BSA in 0.01 M (pH 7.4) PBS and 1% glutaradehyde at room temperature for 20 hrs. The peptide/protein ratio of the conjugates was determined by amino acid analysis (P2-BSA=13; P9-BSA=18; P10-BSA=15) and the conjugates were then mixed with Freund's Complete Adjuvant. These mixtures were used to prepare anti-peptide antibody sera in rabbits for three months. Both anti-HCV antibodies and anti-peptide antibodies were detected with ELISAs shown in Table 1 and Table 2, respectively.

Table 1 *The anti-HCV antibody with synthetic peptides ELISAs*

	Sample									
Peptide	1	2	3	4	5	6	7	8	9	10
P2	0.25	0.42	0.40	0.50	0.50	0.11	0.03	0.08	0.72	0.48
P9	0.93	0.48	0.69	0.59	0.44	0.38	0.05	0.40	0.53	0.14
P10	1.18	0.09	0.27	0.63	0.49	0.20	0.60	0.49	0.18	0.52

Table 2 *Anti-peptide antibody with rabbit sera*

P2-BSA	P2-BSA	P2-BSA	P9-BSA	P9-BSA	P9-BSA	P10-BSA	P10-BSA	P10-BSA	
0.057	0.052	0.064	0.075	0.062	0.060	0.061	0.067	0.086	blank
0.074	0.076	0.064	0.071	0.195	0.206	0.134(1)	0.287(2)	0.295(3)	10-rabb.
0.071	0.058	0.063	0.069	0.162	0.178	0.121(1)	0.263(2)	0.337(3)	10-rabb.
0.115	0.138	0.052	0.072(1)	0.240(2)	0.431(3)	0.097	0.178	0.273	9-rabb.
0.058(1)	0.053(2)	0.070(3)	0.095	0.064	0.064	0.078	0.081	0.059	2-rabb.
0.065	0.066	0.083	0.102	0.081	0.075	0.109	0.055	0.074	norm.sreum

As seen in Table 1, we found all three peptides were moderately reactive with anti-HCV positive sera. They have weak antigenicity that coincides with HCV immune selection. Although E2 HV was the major target for host indentification and immunization, the hypervariability in the E2 HV domain may result in either humoral and/or T cell; mediated immune selection, as in the case of the V3 domain of HIV-1 gp120. Since HCV may generate escape mutants through antigenic variation [7], P2 did not elicit an immune response in rabbits when P2 was conjugated with BSA (see Table 2), indicating that not all E2 HV antigen could be identified by the host or else the host immune reaction was very weak. P9 and P10 can cross-react that may be caused by the homology of HCV [6]. The antibody dilution of both P9 and P10 was 1:1000.

Conclusions

1. There were anti-E2 HV antibodies in the immune host but they could not neutralize HCV.
2. These antibodies were not reactive with all E2 HV epitopes, which resulted in immune selection of escape mutants.
3. Some anti-E2 HV antibodies, possibly from same genotype, contained common structures in which the E2 HV epitopes could react with other anti-E2 HV antibodies.
4. Not all antigens of E2 HV were immunogenic or else its immunogenicity was too low to be functional.

Acknowledgements

The study was supported by a grant from Ministry of Public Health, China.

References

1. Alter, M.J., Margolis, H.S., Krawezynski, K., Judson, F.N., Mares, A., Alexander, W.J., Hu, P.Y., Miller, J.K., Gerber, M.A., Sampliner, R.E., Meeks, E.L. and Beach, M.J., New Engl. J. Med., 327(1992)1899.
2. Weiner, A.J., Brauer, M.J., Rosenblatt, J., Richman, K.H., Tung, J., Crawford, K., Bonino, F., Saracco, G., Choo, Q.L., Houghton, M. and Han, J.H., Virology, 180(1991)842.
3. Weiner, A.J., Geysen, H.M., Christopherson, C., Hall, J.E., Mason, T.J., Saracco, G., Bonino, F., Crawford, K., Marion, C.D., Crawford, K.A., Brunetto, M., Barr, P.J., Miyamura, T., Mchutchinson, J. and Houghton, M., Proc. Natl. Acad. Sci. USA., 89(1992)3468.
4. Hijikata, M. KAto, N., Ootsuyama, Y., Nakagawa, M., Ohkoshi, S. and Schimotohno, K., Biochem. Biophys. Res. Commun., 175(1991)220.
5. Takamizawa, A., Mori, C., Fuki, I., Manabe, S., Murakami, S., Fujita, J., Onishi, E.,Andoh, T., Yoshida, I. and Okayama, H., J. Virology, 65(3)(1991)1105.
6. Chen, W.R. and Tao, Q.M., National Medical Journal of China, 71(10)(1991)575.
7. Weiner, A.J., Christopherson, C., Hall, J.E., Bonino, F., Saracco, G., brunetto, M.,Crawford, K., Marion, C.D., Crawford, K.A., Venkatakrishna, S., Miyamura, T., Mchutchinson, J., Cuypers, T. and Houghton, M., J. Hepatology, 13(1991)S6.

Study on antigenic determinants of trichosanthin using peptide synthesis method

Hui-Qin Xie and Shan-Wei Jin
Shanghai Institute of Organic Chemistry, Chinese Academy of Sciences, Shanghai 200032, China

Introduction

Trichosanthin is a traditional Chinese medicine prepared from the root tuber of the perennial plant *trichosanthes kirilowii Maxim* (Cucurbitaceae) [1]. It is able to induce abortion in early and mid-trem pregnancy in women. In addition, it is an effective for treatment ectopic pregnancy, benign and malignant hydatidiform moles, as well as Choriocarcinoma. Recently trichosanthin has been reported as an anti-human immunodeficiency virus (anti-HIV) agent [2] and shows promise for treating AIDS in the USA. For a few patients, the main side effect of trichosanthin is an allergic reaction.

In order to reduce allergic reaction of trichosanthin, we are investigating the antigenic determinants of trichosanthin using the peptide synthesis method. In a previous paper, we reported that peptide AAGKIRENIPL, sequence 117-127 of trichosanthin, had definite antigenic activity [3].

Results and Discussion

Recently we have synthesized three peptides: GYRAGDTSYFFNEASATEAAKYVF-KDA (I), sequence 73-99; GYRAGDTSYFFNEA (II), sequence 73-86; SATEAAKYV-

Column: Lichrosome C18 (10.0mmIDx25cm)
Buffers: A, water(0.1% TFA); B, 60% MeCN (0.1% TFA)
Flow rate: 3ml/min
Gradient: 0-30-40min, 0-100-0%B
Detected: 230nm
Sensitivity: 0.5 AUFS/10mV

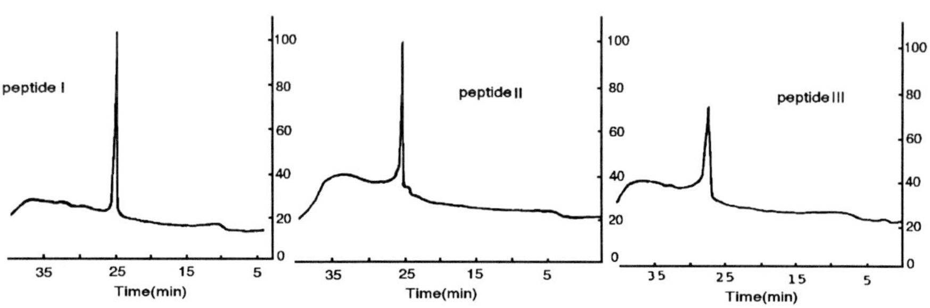

Fig. 1. RP-HPLC Separation of Peptide I,II, III.

FKDA (III), sequence 87-99 of trichosanthin with Merrifield solid phase method. These peptides were then purified with HPLC and checked by amino acid composition analysis as shown in Fig. 1 and Table 2. ELISA (enzymelinked immunosorbent assay) test of these peptides by reaction with rabbit anti-trichosanthin serum showed (Table 1) that only peptide (I) has definite antigenic activity. This result is quit interesting for exploring the relationship between its structure and function at molecular level.

Table 1 *Reactivity of anti-trichosanthin serum with trichosanthin fragments Antibody titers by ELISA OD450nm*

Exp[a]	Coating antigen	Rabbit anti-TCS Serum(1:2000)	Medium
	I-Crude	1.274(0.013)[a]	0.062(0.010)
	I-Pure	1.216(0.063)	0.055(0.002)
1	TCS-1	1.620(0.016)	0.100(0.037)
	TCS-2	1.309(0.152)	0.113(0.015)
	TCS-3	1.578(0.020)	0.111(0.001)
	Control	0.089(0.015)	0.052(0.002)
	I-Crude	1.117(0.034)	0.058(0.002)
	I-Pure	1.040(0.047)	0.076(0.018)
2	TCS-1	1.647(0.002)	0.084(0.011)
	TCS-2	1.612(0.001)	0.098(0.027)
	TCS-3	1.114(0.126)	0.068(0.013)
	Control	0.073(0.009)	0.059(0.010)
	II-Pure	0.138(0.006)[b]	0.073(0.003)
	III-Pure	0.126(0.009)	0.067(0.002)
1	TCS-1	0.901(0.003)	0.074(0.003)
	TCS-2	1.074(0.062)	0.080(0.002)
	TCS-3	0.923(0.043)	0.073(0.003)
	Control	0.086(0.009)	0.062(0.002)
	II-Pure	0.100(0.008)	0.088(0.022)
	III-Pure	0.119(0.003)	0.075(0.002)
2	TCS-1	0.837(0.012)	0.084(0.006)
	TCS-2	0.930(0.015)	0.098(0.017)
	TCS-3	0.825(0.036)	0.080(0.005)
	Control	0.080(0.001)	0.068(0.004)

[a] Exp 1: Coated antigen 1μg/well. Exp 2: Coated antigen 10μg/well.
[b] TCS-1: Trichosanthin from Dr. Ke Yi-Bao of Shanghai Inst. of Cell Biology.
[c] TCS-2: Trichosanthin from Wuhan Inst. of Biology Products.
[d] TCS-3: Trichosanthin from Jinsan Pharmaceutical Factory according to method of Shanghai Institute of Organic Chemistry.
[e] Numbers within parentheses indicated standard deviation.

Table 2 *Amino acid composition of peptides I, II, III*

	I	II	III
Asp	2.97(3)	2.05(2)	0.94(1)
Thr	1.77(2)	0.89(1)	0.93(1)
Ser	1.75(2)	0.94(1)	0.94(1)
Glu	1.86(2)	1.09(1)	1.00(1)
Gly	2.03(2)	2.27(2)	
Ala	6.09(6)	2.00(2)	4.51(4)
Val	1.00(1)		0.74(1)
Tyr	2.16(3)	1.43(2)	1.03(1)
Phe	2.68(3)	1.69(2)	0.91(1)
Lys	2.27(2)		1.84(2)
Arg	0.96(1)	1.00(1)	

References

1. Wang, Y., Qian, R.Q., Gu, Z.W., Jin, S.W., Zhang, L.Q., Xia, Z.X., Tian, G.Y. and Ni, C.Z., Pure and Appl. Chem., 58(1986)789.
2. McGrath, M. S., Hwang, K.M., Caldwell, S.E., Gaston, I., Luk, K.C. and Wu, P., Proc. Acad. Sci. USA, 86(1989)2844.
3. Wang, Y., Wu, J.X., Jin, S.W., and Xie, H.Q., In Du, Y.-C., Tam, J.P. and Zhang, Y.-S. (Eds.) Peptides: Biology and Chemistry (Proceedings of the 1992 Chinese Peptide Symposium), ESCOM, Leiden, 1993, pp. 104–106.

Study on analogs of the epitope peptides P141 of the *Schistosoma mansoni* P28 antigen

Jia-Xi Xu, Zhi-Hui Qin, Jun Yang and Meng-Shen Cai
School of Pharmaceutical Sciences, Beijing Medical University, Beijing 100083, China

Introduction

Among the potential vaccine antigens of *Schistosoma mansoni* that have been characterized, a 28kDa glutathione S-transferase antigen (abbreviated Sm28 or P28) appears to be very promising. This molecule was first isolated from Schistosoma mansoni adult worms [1] and recent studies have demonstrated that Sm28 is an excretory/secretory molecule with glutathione transferase activity. It is mainly synthesized in the protone phridial cells before being expressed at the surface of the parasite [2]. Since it has been identified and characterized, the protein has been cloned, sequenced and expressed in *Escherichia coli* [3].

The effectiveness of Sm28 as a vaccinating molecule was tested in various laboratory models and both the native form and the recombinant molecule can induce a high degree of protection after direct immunization of permissive hosts such as mice, hamsters and baboons, or semi-permissive hosts such as rats [4].

Results and Discussion

The *Schistosoma mansoni* P28 molecule is an antigen inducing protective immunity in various experimental models. Some synthetic epitopes, derived from the primary sequence of recombinant P28, were designed according to their hydrophilicity, mobility and accessibility profiles [4-8]. The epitope peptides P141 of the *Schistiosoma Mansoni* P28 antigen showed strong immunogenicity during Schistosoma infection in rabbits and rats with and without conjugation with carrier protein. It is probable that it can be incorporated into a synthetic peptide vaccine against Schistosoma. For the purposes of finding stronger immunogenic peptides, we have designed and synthesized four analogs of the epitope peptide P141, [Asn^1]-P141, [Gln^3]-P141, [Pro^5]-P141 and [Thr^6-Ser^7]-P141 using Chou-Fasman's conformational prediction[9] and value of amino acid hydrophilicity [10]. The peptide P141 and its analogs are shown below.

P141, H-Glu-Ser-Leu-Lys-Gly-Ser-Thr-Gly-Lys-Leu-Ala-Val-Gly-OH
[Asn^1]-P141, H-Asn-Ser-Leu-Lys-Gly-Ser-Thr-Gly-Lys-Leu-Ala-Val-Gly-OH
[Gln^3]-P141, H-Glu-Ser-Gln-Lys-Gly-Ser-Thr-Gly-Lys-Leu-Ala-Val-Gly-OH
[Pro^5]-P141, H-Glu-Ser-Leu-Lys-Pro-Ser-Thr-Gly-Lys-Leu-Ala-Val-Gly-OH
[Thr^6-Ser^7]-P141, H-Glu-Ser-Leu-Lys-Gly-Thr-Ser-Gly-Lys-Leu-Ala-Val-Gly-OH

ELISA results showed that analog [Gln^3]-P141 has higher antigenicity than P141, analog [Asn^1]-P141 has the same antigenicity as P141 and analog [Pro^5]-P141 has lower antigenicity than P141. However, analog [Thr^6-Ser^7]-P141 has no antigenicity.

References

1. Balloul, J.M., Grzych, J.M., Pierce, R.J. and Capron, A.J., J. Immunol., 138(1987)3448.
2. Balloul, J.M., Grzych, J.M., Pierce, R.J. and Capron, A.J., Mol. Biochem. Parasitol., 17(1985)105.
3. Balloul, J.M., Sondermeryer, P., Dreyer, D., Capron, M., Grzych, J.M., Pierce, R.J., Carvallo, D., Lecocq, J.P. and Capron, A., Nature, 326(1987)149.
4. Wolowczuk, I., Auriault, C., Bossus, M., Boulanger, D., Gras-Masse, H., Mazingue, C., Pierce, R.J., Grezel, D., Reid, G.D., Tartar, A. and Capron, A., J. Immunol., 146(1991)1987.
5. Wolowczuk, I., Auriault, C., Gras-Masse, H., Vendeville. C., Balloul, J.M., Tartar, A. and Capron, A., J. Immunol., 142(1989)1342.
6. Auriault, C., Gras-Masse, H., Wolowczuk, I., Pierce, R.J., Balloul, J.M., Neyrinck, J.L., Drobecq, H., Tartar, A. and Capron, A., J. Immunol., 141(1988)1687.
7. Mao, F., Doctoral Thesis, Beijing Medical University, P.R.China, 1991.
8. Cai, M.S., Mao, F., Xu, J.X., Yi, Y.Y. and Cheng, T.M., China Patent, CN94105038.6(1994).
9. Chou, P.Y. and Fasman, G.D., Biochem., 13(1974)222.
10. Kyte, J. and Doolittle, R.F., J. Mol. Biol., 175(1982)105.

Synthesis of antigenic peptides for 26 kDa glutathione S-transferase of *Schistosoma japonicum*

Jia-Xi Xu and Meng-Shen Cai
School of Pharmaceutical Sciences, Beijing Medical University, Beijing 100083, China

Introduction

Schistosomiasis is a chronic debilitating disease affecting human beings in 70 countries in several parts of the world. Schistosomiasis japonica is spread widely over Southeast Asia. It is estimated that 200-300 million people are infected by various species of schistosomes. In spite of the development of active and relatively safe drugs, prevention of rapid reinfection has remained a problem, requiring repeated drug applications over long periods. There is a clear need therefore for an effective vaccine inducing significant levels of protection against the invasive stage of the parasite [1].

Immunochemical analyses have revealed that the 26kDa glutathione S-transferase of *Schistosoma japonicum* (termed Sj26) can be immunogenic during Schistosome infection in mice [2-4]. It therefore has potential as component of a vaccine against *Schistosoma japonicum.*

Results and Discussion

The epitopes of the 26kDa glutathione S-transferase of *Schistosoma japonicum* Sj26 have been predicted according to their hydrophilicity using Hopp and Woods's method, and according to their flexibility and accessibility by the computer-aided determination of the sequences of Sj26, which are P23-46, P80-97, P136-153, P187-202 and P206-218. These sequences are as follows.

```
J1, P23-46,   YLEEKYEEHLYERDEGDKWRNKKF
J2, P80-97,   NMLGGCPKERAEISMLEG
J3, P86-97,   PKERAEISMLEG
J4, P136-153, RLCHKTYLNGDHVTHPDF
J5, P187-202, PQIDKYLKSSKYIAWP
J6, P206-218, WQATFGGGDHPPK
```

The epitopes synthesized using solid phase peptide synthesis method with Merrifield resin as carrier, α-t-Boc-amino acid assembly, and HOBt or HOSu/DCC as the coupling reagents. They were cleaved with dried liquid HF/anisole/thioanisole, extracted with 30% HOAc, purified on Sephadex G15 or G25 column and characterized by reverse-phase HPLC and amino acid analysis [5-7]. These epitopic peptides were conjugated to BSA by glutaraldehyde method.

Results generated from ELISA experiments showed that the synthetic peptides J2, J3 and their BSA-conjugated products have stronger antigenicity than J1, J4, J5 or J6 against rabbit serum infected by *Schistosoma japonicum*.

Acknowledgement

We thank the Post Doctorate Fellowship for financial aid for J.X. Xu.

References

1. Balloul, J.M., Sondermeyer, P., Dreyer, D., Capron, M., Grzych, J.M., Pierce, R.J., Carvallo, D., Lecocq, J.P. and Capron, A., Nature, 326(1987)149.
2. Henkle, K.J., Darven, K.M., Wright, M.D., Ramos, A.J. and Mitchell, G.F., Mol. Biochem. Parasitol., 40(1990)23.
3. Wright, M.D., Harrison, R.A., Melder, A.M., Newport, G.R. and Mitchell, G.F., ibid., 49(1991)177.
4. Smith, D.B., Darven, K.M., Board, P.G., Tiu, W.U., Garcia, E.G. and Mitchell, G.F., Proc. Natl. Acad. Sci. USA, 83(1986)8703.
5. Xu, J.X. and Cai, M.S., China Patent, CN94105973.1(1994).
6. Xu, J.X. and Cai, M.S., China Patent, CN94105874.X(1994).
7. Cai, M.S., Xu, J.X., Yang, J. and Qin, Z.H., China Patent, CN94105976.6(1994).

Immunochemical studies on pituitary adenylate cyclase activating polypeptide in guinea pig tissues

Eiji Ando[a], Kiyoshi Nokihara[a,b], Satoru Naruse[c], Michiko Suzuki[d] and Shigeru Kobayashi[c]

[a]*Biotechnology Instruments Department, Shimadzu Corp., Kyoto, Japan*
[b]*Tokyo University of Agriculture and Technology, Koganei, Tokyo, Japan*
[c]*University of Nagoya School of Medicine, Nagoya, Japan*
[d]*Yamanashi Gakuin College, Yamanashi, Japan*

Introduction

PACAPs were isolated as 38- and 27-amino acid residue peptide amides from ovine hypothalamus and designated PACAP38 and PACAP27. The latter is the N-terminus of the 38-form and has 68% sequence homology with VIP. PACAPs appear to stimulate adenylate cyclase in pituitary cells and this effect was at least 1000 times greater than that of VIP. Many of the biological actions of PACAPs are very similar to those of VIP [1]. As PACAPhas a high sequence homology with VIP, it is necessary to discriminate the PACAP-LI from VIP-LI. Additionally the amino acid sequence of human prepro-PACAP was deduced from the nucleotide sequence of cDNA [2]. PACAP-related peptide (PRP), prepro-PACAP(82-110), is considered to be a processed peptide having biological significance similar to PHM of prepro-VIP [3], with which it has 44% homology. We have efficiently synthesized PACAPs, their fragments and analogues using a novel simultaneous multiple peptide synthesizer with eight independent channels [4]. By the use of PACAP-related peptides, PACAP-specific antisera were generated and characterized. The specific RIA systems for PACAPs have been developed. By the RIA and histochemical techniques, PACAP-LI was demonstrated in guinea pig tissues, as we have already studied the physiological actions in guinea pigs [5,6] *in vitro*.

Experimental

Peptides were assembled by a stepwise procedure using a recently developed simultaneous multiple solid-phase peptide synthesizer, Shimadzu Model PSSM-8 [4] (Kyoto, Japan). For the present syntheses the Fmoc strategy was employed with the pre-activation of the N^{α}-Fmoc amino acids with PyBOP, HOBt and NMM in DMF. After simultaneous cleavage followed by precipitation from ether crude peptides were obtained, which were then purified by a single step of RPHPLC using 0.01 N HCl and acetonitrile as eluents. The resulting purified peptides were characterized by LSIMS using Shimadzu-Kratos CONCEPT II H (Manchester, UK) and by sequencing using a Shimadzu Model PPSQ-10.

Antiserum specific to PACAP38 (SS30) was obtained from rabbitsby immunization of N^{α}Cys-PACAP(24-38)-NH_2 as a hapten coupled to bovine thyroglobulin (BTG) with maleimidobenzoyl-N-hydroxy-succinimide ester. Conjugate of hPRP to BTG with water soluble carbodiimide was immunized to provide an antiserum KS30. To remove cross-reactions with carrier protein, KS30 was purified by affinity chromatography with a column containing peptide as a ligand on the TentaGel® by s novel raptd production method described previously by Nokihara and Ando [7]. Peptide wass iodinated by the chloramine-T method and the labeled peptide was purified with a Sep-Pak® cartridge. Antisera SS30 against PACAP(24-38)-NH_2 and RIN8922 against PACAP27 were used for the development of PACAP38- and PACAP27-RIAs, respectively, whereas the delayed method and the second antibody method for bound/free separation were successfully employed.

Guinea pig tissues were boiled in water and homogenized with acetic acid at a final concentration of 0.5 M. After centrifugation, the supernatant was applied on a C18 Sep-Pak® cartridge, eluted with 60% CH_3CN in 0.1% TGA and subjected to TFA. The supernatant of a stomach extraction was siparated on a reverse phase HPLC and each fraction was assayed.

Male Hartley strain guinea pigs were fixed by descending aorta perfusion with the Zamboni's fluid followed by the Bouin's fluid. Tissue sections (small intestine and stomach) were immunostained by the peroxidase anti-peroxidase method according to Kobayashi *et al* [8]. Specificity controls in the present immunohistochemical studies were performed by using SS30 (dilution, 1:2000) absorbed with PACAP38 (1.0 ug/ml).

Results and Discussion

PACAPs and related peptides, as well as their fragments listed below, were successfully synthesized by the efficient Fmoc strategy followed by cleavage in quantitative yields. HPLC profiles of several crude peptides, after cleavage, are shown in Fig. 1. The resulting crude peptides were easily purified by a single step of preparative RPHPLC and characterized by analytical HPLC, LSIMS, sequencing and amino acid analysis, and shown to be homogeneous. Peptide content was determined by amino acid analysis and HPLC.

The fragment PACAP(24-38)-NH_2 was iodinated, instead of PACAP38, as nonspecific binding of [^{125}I]PACAP38 was much higher than that of [^{125}I]PACAP(24-38)-NH_2, as a tracer. On the contrary, PACAP27 could be successfully iodinated and no significant nonspecific binding was observed. Two RIA systems were developed using antisera, SS30 and RIN8922, respectively. Antisera were characterized using different related peptides and their fragments, which included C-terminal deletion peptides such as PACAP(31-38)NH_2, PACAP(32-38)-NH_2, PACAP-(34-38)-NH_2, PACAP(35-38)-NH_2, PACAP(1-33)-OH, PACAP(1-34)-OH, PACAP(1-35)-OH, PACAP(1-36)-OH, PACAP-(1-37)-OH and PACAP(1-38)-OH. Table 1 indicates relative cross-reactivities of SS30 to the above peptides. No significant cross-reactivities were observed for the fragments: PACAP(1-15)-OH, PACAP(10-20)-OH, PACAP(18-27)-NH_2 and the structurally related peptides of the secretin-glucagon family such as human(h) secretin, PHM, hGIP, helodelmin, somatostatin, PHI, hGRP, hVIP, glucagon, hGRF, hPRP, guinea pig VIP, dogfish VIP up to 1 μg/ml. The recognition of SS30 requires the C-terminal pentapeptide amide. RIN8922, which is highly specific to PACAP27 and recognizes none of the PACAP fragments, PACAP38 or VIP. Antiserum KS30 was effectively purified as lgG specific to hPRP using the affinity column prepared by hPRP assembled on the TentaGel®-NH_2 without linker. The peptide coupled to the resin was characterized by comparison with the simultaneously prepared hPRP with the acid labile linker [7].

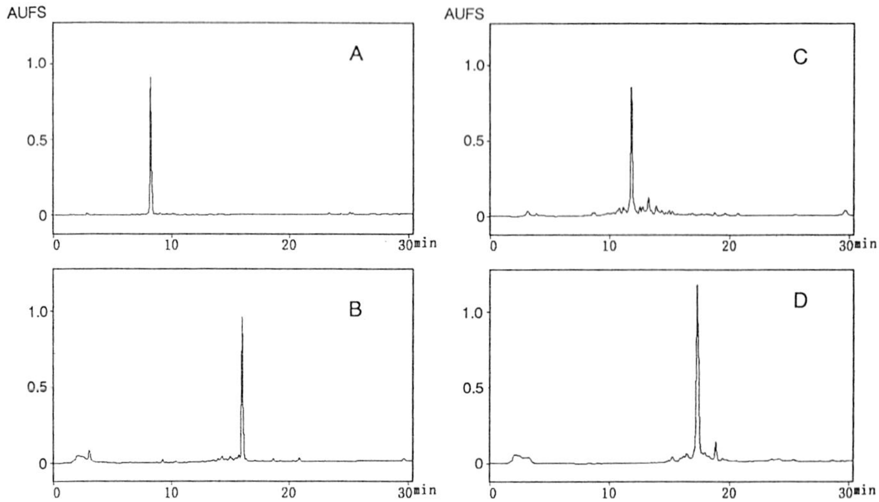

Fig. 1. RP-HPLC profiles of crude peptides after cleavage. Column: SynProPep® RPC18 (4.6×150 mm), eluent: 0.01 N HCl/acetonitrile (A; 95/5-65/35 in 30 min, B, C; 80/20-50/50 in 30 min, D; 90/10-60/40 in 30 min), flow rate 1.0 ml/min, absorbance 210 nm, A: PACAP(24-38), B: hPRP, C:PACAP38.

Table 1 *Relative cross-reactivity of antiserum SS30*

Peptide	% cross-reactivity
PACAP38	100
PACAP(31-38)-NH_2	9.3
PACAP(32-38)-NH_2	8.6
PACAP(34-38)-NH_2	6.6
PACAP(35-38)-NH_2	0.4
PACAP(1-33)-OH	<0.02
PACAP(1-34)-OH	<0.02
PACAP(1-35)-OH	<0.02
PACAP(1-36)-OH	<0.02
PACAP(1-37)-OH	<0.02
PACAP(1-38)-OH	0.25

After homogenizing various tissues of guinea pigs followed by centrifuging, the supernatant was treated with a Sep-Pak® cartridge and then subjected to RIA. Both

PACAP38-LI and PACAP27-LI were found as shown in Fig. 2. The two PACAP specific RIA systems allowed high concentrations of PACAP-LI to be shown in the various organs, especially in the brain. We have studied PACAP actions using guinea pig pulmonary artery [5] and gallbladder [6], *in vitro*. Similar amounts of immunoreactivity was found in the lungs as well as the stomach, but not in the gallbladder, although, when pooled gallbladder tissues were extracted, immunoreactivity was detected (details will be published elsewhere). As expected the diencephalon (mostly hypothalamus) contained higher concentrations of PACAP-LI. In all tissues PACAP38-LI were higher than that of PACAP27-LI. The distribution of PACAP38-LI and PACAP27-LI in the guinea pig are not similar to those in the rat [9].

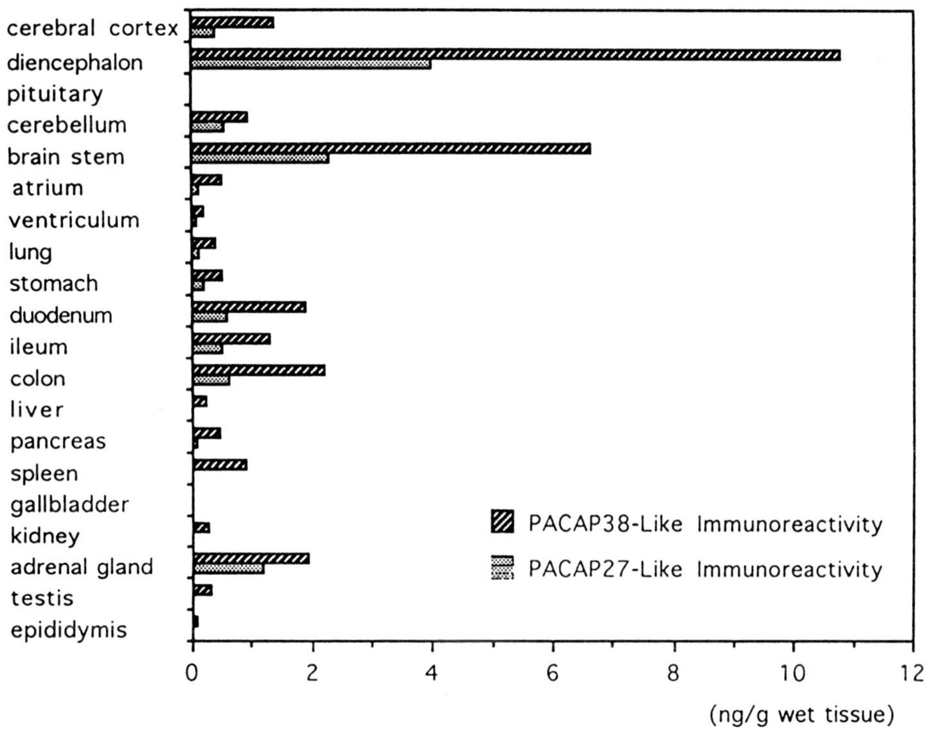

Fig. 2. PACAP38- and PACAP27-like immunoreactivities in guinea pig tissues. The results are mean values of n=5.

As we have studied the effects of PACAP on gastrointestinal blood flow and revealed that PACAP is a candidate for mediating neurally induced non-adrenergic and non-cholinergic gastric vasodilatation [10], the extracts obtained from the stomach were separated on a reverse phase HPLC column and each fraction was subjected to RIA. Major immunoreactive peaks for PACAPs were observed in the extracts of stomach and could be superimposed with the elution positions of those synthetic PACAP38 and -27, respectively (Fig. 3). Similar results were obtained from brain extracts.

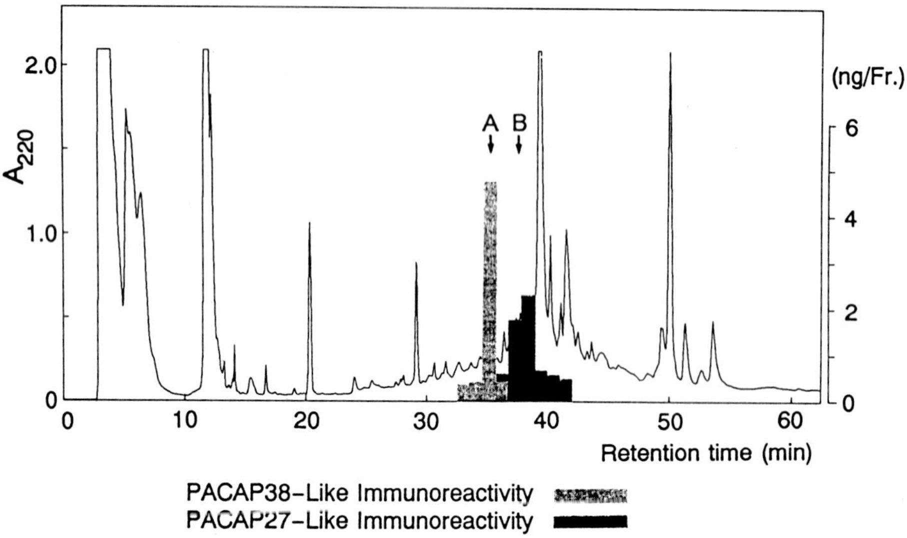

Fig. 3. RP-HPLC of extracts from the guinea pig stomch. A and B indicate the retention time of synthetic PACAP38 and PACAP27, respectively. Column: Vydac 218TP54 C18(4.6×250 mm), eluent: 0.1% TFA/acetonitrile containing 0.1% TFA; 100/0-40/60 in 60 min, flow rate 1.0 ml/min, absorbance 220nm.

Localization of PACAP-LI and PRP-LI in guinea pigs were studied by the immunohistochemical techniques. Immunostaining of the small intestine and stomach was carried out by the use of antibodies, SS30 against PACAP38 (dilution ×1500), RIN8922 against PACAP27 (dilution ×1000) and IgG (40 ng/ml) derived from KS30 against hPRP, respectively. In the specificity control described in the experimental, no immunostaining of tissue elements was obtained when the antisera were replaced by the pre-absorbed antiserum. However, the positive immunostaining reappeared when the antiserum (1:2000) was pre-incubated with PACAP38 (0.13 g/ml). Fig. 4. shows immunohistochemical demonstration of PACAP-LI in guinea pig small intestine. Either PACAP38 or PACAP27 immunoreactive nerve cell bodies were localized mainly in the central portion of the myenteric ganglia, whereas in the submucosal ganglia no ganglion cell bodies were immunopositive. Strongly immunopositive nerve terminal varicosities were found in both the myenteric and submucosal ganglia. Co-existence of immunoreactive PACAP38 and PACAP27 were apparent: there was a tendency that the PACAP38 immunoreactive nerve cell bodies were also positive for PACAP27, and *vice versa*. The hPRP-LI were dimonstrated in almost all nerve cell bodies of the enteric nervous system examined. Projections of neurons also contained hPRP-LI, however this reaction in the neuronal projections was remarkably weaker than that in the nerve cell bodies. Non-nerve cells, probably mast cells, showed a strtongly positive reaction to hPRP. They were scattered in the connective tissue of the submucosal layer. PRP may also be released to the nerve terminal. Co-existence of PACAPs and PRP, which are predicted to be generated from the same precursor, was not confirmed. Similar results were obtained from guinea pig stomach (Fig. 5).

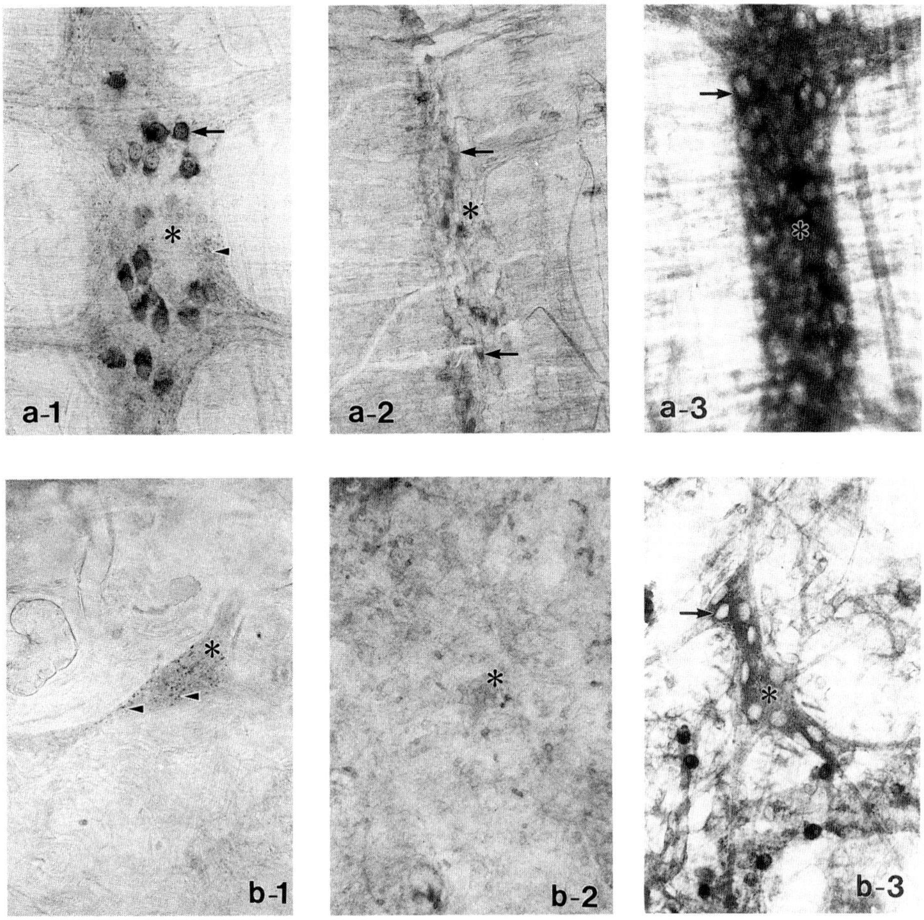

Fig. 4. Immunohistochemical demonstration in guinea pig small intestine. a: myenteric plexus; b: submucous plexus; a-1 and b-1: SS30, a-2 and b-2: RIN8922; a-3 and b-3: lgG derived from KS30. Immunopositive cell bodies (arrows) and nerve terminal varicosities (arrowheads) are clearly observed. Asterisks indicate ganglia. X340

Conclusion

The use of a novel simultaneous peptide synthesizer rapidly provides high quality peptides, particularly of analogues and fragments. A fragment peptide can provide the specific antibody of the whole biological active peptide with low non-specificity and can allow easier labelling. Using synthetic peptides, PACAP-specific antisera were generated, characterized and two PACAP specific RIA systems have been developed. PACAPs and PRP exist not only in brain but also in enteric nervous systems. The results suggest the possibility that both PACAP38 and PACAP27 coexist in tissues and have a physiological role as neuromodulator or neurotransmitter. The molecular forms of PACAP, including prepro-PACAP in the different tissues and physiological actions of PRP remain to be revealed. The present antisera and RIA are useful for further biological and physiological studies.

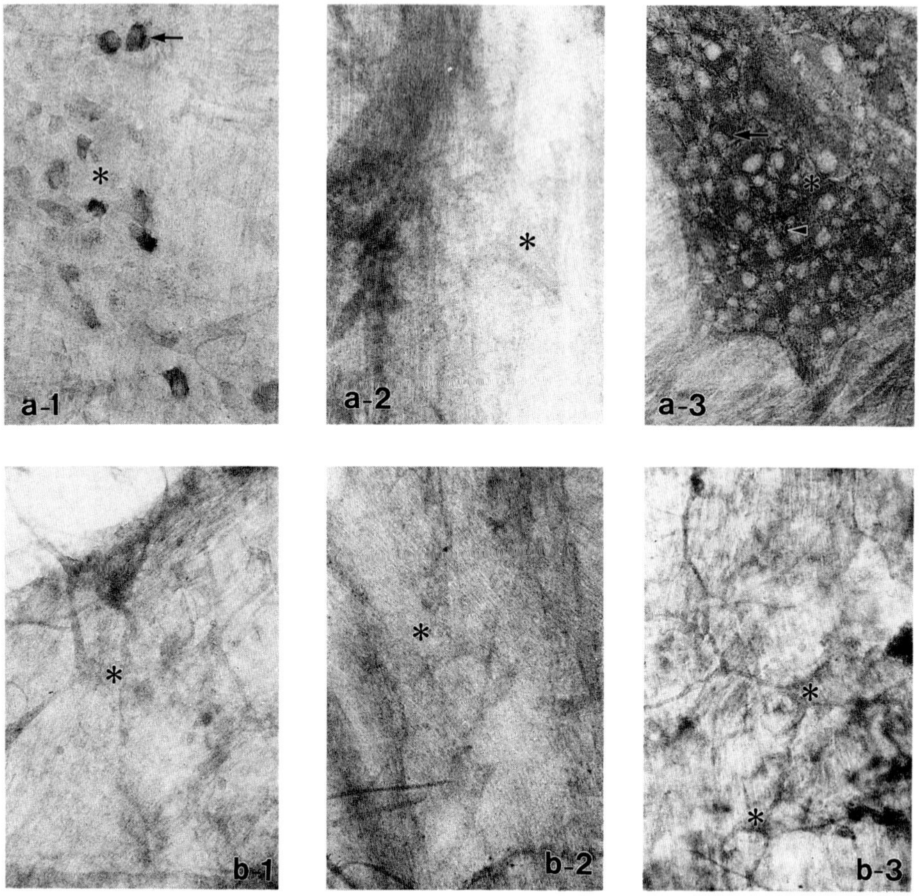

Fig. 5. Immunohistochemical demonstration in guinea pig stomach. a: myenteric plexus; b: submucous plexus; a-1 and b-1: SS30, a-2 and b-2: RIN8922; a-3 and b-3: IgG derived from KS30. Immunopositive cell bodies (arrows) and nerve terminal varicosities (arrowheads) are clearly observed. Asterisks indicate ganglia. X340

Acknowledgements

The authors wish to thank M. Shimizu (Peninsula Laboratories Inc., CA) for a gift of RIN8922, Dr. V. Wray (Gesellschaft für Biotechnologische Forschung, Germany) for his helpful discussions. A part of this work was supported by the Monbusho International Scientific Research Program from the Ministry of Education, Science, and Culture, Japan.

References

1. Arimura, A., Reg. Peptides, 37(1992)287.
2. Ohkubo, S., Kimura, C., Ogi, K., Okazaki, K., Hosoya, M., Onda, H., Miyata, A., Arimura, A. and Fujino, M., DNA and Cell Biol., 11(1992)21.
3. Itoh, N., Obata, K., Yanaihara, N. and Okamoto, H., Nature, 304(1983)547.
4. Nokihara, K., Yamamoto, R., Hazama, M., Wakizawa, O., Nakamura, O., in Innovation and Perspectives in Solid Phase Synthesis, Proceedings 2nd Symposium on Solid Phase Synthesis (R. Epton-Ed.), Intercept Ltd., Andover, 1992, p.445.
5. Naruse, S., Sazi, T., Ozaki, T., Nokihkara, K. and Wray, V., Biomed. Res., in press.
6. Wei, M., Naruse, S., Nakamura, T., Nokihara, K. and Ozski, T., Biomed. Res., in press.
7. Nokihara,K. and Ando,E., in Peptide Chemistry 1993, Proceeding 31st Japan Symposium on Peptide Chemistry (Y. Okada-Ed.), Protein Research Foundation, Osaka, 1994, p.25.
8. Kobayashi, S., Suzuki, M. and Yanaihara, N., Archiv. Histol. Japon., 48(1985)27.
9. Arimura,A., Somogyvari-Vigh, A., Miyata, A., Mizuno, K., Coy, D.H. and Kitada, C., Endocrinology, 129(1991)2787.
10. Naruse, S., Nakamura, T., Nokihara, K., Ando, E. and Wray, V., in VIP, PACAP and Related Regulatory Peptides, Proceedings 1st VIP-PACAP Symposium (G. E. Rosselin-Ed.), in press.

Diagnosis of hepatitis C virus infection by enzyme-linked immunosorbent assay using synthetic peptide antigens

Bing Xu[a]**, Jun-Shen Zhan**[b] **and Qing Dong**[a]
[a]*Star USA Inc., New York, NY 11733, U.S.A.*
[b]*Sichuan Star Biotechnological Engineering Industry Co. Ltd., Chengdu 610041, China*

Introduction

Hepatitis C virus (HCV), a member of the flavivirus family has been shown to be the major etiologic agent of vital non-A,non-B hepatitis. HCV causes persistent infections in most cases and leads to the development of chronic hepatitis and liver cirrhosis in about 50% and 10% of cases, respectively. A significant proportion of patients with liver cirrhosis will also develop primary hepatic cell carcinoma. HCV is prevalent at rates of 0.36% to 1.0% among blood donors in industrialized countries and at rates of 2.5% to 6.4% in the populations of developing countries. HCV constitutes the major cause of chronic liver diseases throughout the world.

The immunogenic epitopes in hepatitis C virus (HCV) core protein and envelop protein have been defined by Enzyme-Linked Immunosorbent Assay (ELISA) using several synthetic peptides with 18-35 amino acid residues. The peptides were assayed with 100 serum samples from anti-HCV-positive blood donors and another 200 serum samples from anti-HCV-negative blood donors, as determined by commercial assays. About 98% of the anti-HCV-positive samples reacted with a combination of peptides covering residues of HCV core protein and correlated almost exactly with genotypic analysis of HCV sequences amplified from the samples by polymerase chain reaction (PCR). We found that human antibodies to HCV core proteins are mainly directed to linear determinants and can easily be reproduced using synthetic peptides. These defined immunodominant regions and this potential application as clinical diagnosis tools for HCV infection will be discussed.

Presently, the most common assay is the NS-4 derived recombinant protein C100-3 is as antigen on the solid phase in an indirect ELISA. The sensitivities and specificity of the assay is not satisfactory for early determination.

We were interested in further characterizing immune recognition of HCV, and to identify the immunodominant regions with HCV protein by using short synthetic peptides.

Results and Discussion

Peptides employed in this test were synthesized by the solid-phase-peptide-synthesis method with Fmoc-amino acid strategy. The crude products were purified by high performance liquid chromatography (HPLC), and then used for ELISA.

To identify the immunogenic region on HCV structural proteins, we synthesized six peptides from the N-terminus of the putative HCV structural protein (Fig. 1).

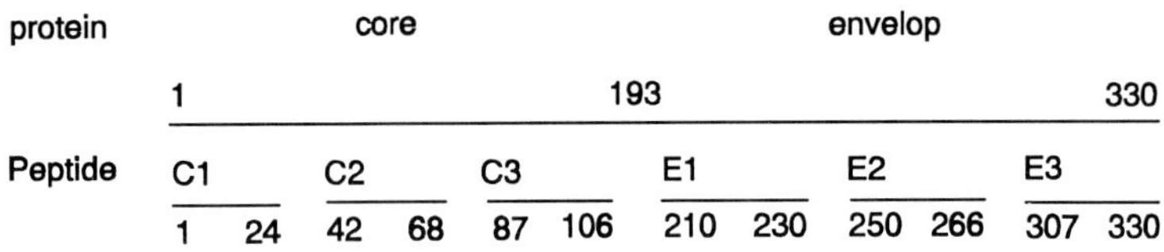

Fig. 1. Location of HCV core and envelop protein, and synthetic peptides.

A total of 100 HCV antibody-positive serum samples were selected from blood donors at a blood bank by commercial assay according to the manufacturer's manual (Ortho Diagnostic Systems, Raritan NV). Another 200 anti-HCV negative samples were obtained from blood donors.

ELISA were performed on 96-well microtiter plates which were prepared by the following method: the peptide solution ($NaHCO_3$ buffer, pH9.5) was put into each well of the plate, and then blocked with 2% bovine serum albumin for 2h at room temperature. Sera were added to the coated microtiter plates incubated at 37°C for 1h, then incubated with goat antibodies to human immunoglobulin G conjugated with horseradish peroxidase at 37°C for another 30 min. These plates were washed and developed.

In a competitive ELISA, serum specimens pre-incubated with 10 times excess of peptides or PBS at 37°C for 30 min were added to microtiter wells coated with peptides. A serum specimen showing more than a 30% decrease of absorbance in the competitive ELISA was judged to be positive.

Table 1 *Activities of six epitope peptides for 100 anti-HCV ELISA positive blood sample*

No. of sample	PCR	No. of Positive Samples						
		C_1	C_2	C_3	E_1	E_2	E_3	C_{100}
100	96	65	70	98	0	12	28	100

We were able to test peptide activities against the antibodies in the 100 anti-HCV positive serum samples by ELISA with 96-well microtiter plates coated with the synthetic peptides as antigens. The results are shown on Table 1. Peptide epitopes C_1, C_2, and C_3 derived from the core protein of HCV, and E_2, and E_3 from the envelop protein were reacted with the HCV positive antibodies in sera in different degrees. In the core protein group, C_1, C_2 and C_3, which corresponded to the N-terminus half of the core protein, reacted with antibodies in 65%, 70% and 98% of 100 serum samples, respectively. In particular, peptide C_3 reacted with antibodies in all but two serum samples. In the envelope protein portion, peptide epitopes E_2 and E_3 reacted with 12 and 28 samples of 100 HCV positive serum samples, respectively. Peptide epitopes E_2 and E_3 correspond to the N-terminus as well as the adjacent of hydrophobic C-terminal region of the envelop protein.

These results indicate that the N-terminal half of the core protein is the most immunogenic. We detected the HCV genome in 60 of the 100 HCV positive samples by RT-PCR, in 5' non coding region. The remarkable correlation was observed between these two assays.

On the basis of sequence similarities between HCV and flavivirus, the first 380 amino acid sequence of N-terminal port of open reading frame of HCV genome is believed to be the structural protein of the viral particle.

In present study, we defined the 5 peptide regions ($C_{1\text{-}3}$ and $E_{2,3}$) corresponding to immunogenic region on the N-terminal half (C_3) of the core protein that corresponds to the most immunogenic virus region. This region contained immunogenic epitopes that reacted with antibodies in nearly all serum samples judged to be positive by commercial kit test.

Session VI
Peptide libraries

Chairs: Kit S. Lam
University of Arizona
Tucson, Arizona, U.S.A.

and

Zhen-Kai Ding
Beijing Institute of Pharmacology and Toxicology
Beijing, China

Library of libraries: A novel approach in synthetic combinatorial libraries

Nikolai F. Sepetov[a], Viktor Krchňák[a], Magda Staňková[a], Kit S. Lam[b], Shelly Wade[a] and Michal Lebl[a]
[a]*Selectide Corporation, 1580 E. Hanley Blvd., Tucson, AZ 85737, U.S.A.*
[b]*Arizona Cancer Center and Department of Medicine, University of Arizona, College of Medicine, Tucson, AZ 85724, U.S.A.*

Introduction

The goal of screening peptide libraries for pharmacological purposes is not necessarily to find the most active peptide, but to identify structural features of the peptides critical for biological activity i.e. to find a motif which can be used subsequently for developing a drug. Therefore a *library of motifs* can give as much or more useful information as a *library of peptides.*

Results and Discussion

There are two types of library-based screening for new pharmacologically important leads: parallel and iterative (for the review of library techniques see e.g. [1,2]). The parallel approach is based on the principle of generating the complete, or as complete as possible, multiplicity of unique structures in a single library, which is then screened to identify the solid phase supports on which active compounds were synthesized. The compositions of the positive test compounds are then determined from information on the solid phase support. Since the number of unique compounds which must be synthesized to obtain a complete library increases exponentially, synthesis of complete libraries longer than pentapeptide is becoming impractical. This is the main disadvantage of the parallel approach. On the other hand it is accepted that in a long peptide not all amino acid residues are equally important for activity and only certain residues at certain positions (motif) are critical. Consequently it is not necessary to synthesize all possible structures, but instead to synthesize only compounds containing all possible motifs but not all structures sharing the same motif.

For a particular target it is impossible to predict *a priori* how many residues are important for activity and how many residues must be defined in order to observe activity in a particular screening assay. In the iterative approach to the combinatorial libraries a number of libraries are synthesized, each initially containing one or two sequencial positions which are defined, with the remaining positions completely randomized. The most active library in a particular biological test is selected, beginning an iterative process of synthesis and screening of additional libraries until all portions of the active molecule are defined. In other words, the first step of the iterative process consists of screening libraries with very short motifs containing only one or two sequencial residues. The "Library of Libraries" (library of motifs) approach is a powerful symbiosis of both approaches: parallel and iterative. To generate a single library of motifs we need to randomize only a few residues in a long peptide and also to randomize the positions of these randomized residues. In all nonrandomized positions we need to have some "average" amino acid, side chains of which may not participate in binding with the macromolecular target, but which may properly orient the critical residues. In our experiments we used a mixture of 19 natural amino acids to fill the "average" amino acid positions. We believe that it is reasonable to synthesize libraries with motifs containing three or four residues. This allows us to reduce dramatically the size of complete libraries and yet still expect sufficiently high activity for detection in screening.

We have synthesized several "Libraries of Libraries" containing three residue size motifs. The libraries were screened using anti-β-endorphin antibodies, streptavidin and thrombin. For the synthesis of the first library we used 19 proteinogenic (without Cys) amino acids in randomized positions and a mixture of the same amino acids in nonrandomized positions. In another library 78 amino acids were incorporated at randomized positions and a mixture of 19 proteinogenic amino acids of either L or D configuration were used to fill nonrandomized positions.

Motifs found in the library constructed from 78 natural and unnatural amino acids in the screening for streptavidin binders are given in Fig. 1. Similarity to the natural motif His-Pro-Gln is clearly visible. The imidazole ring of His is indispensable for binding as well as the secondary amine in the next position. Glutamine can be replaced by another residue with hydrogen bonding potential. Aromatic residue probably finds an additional beneficial interaction in the binding site.

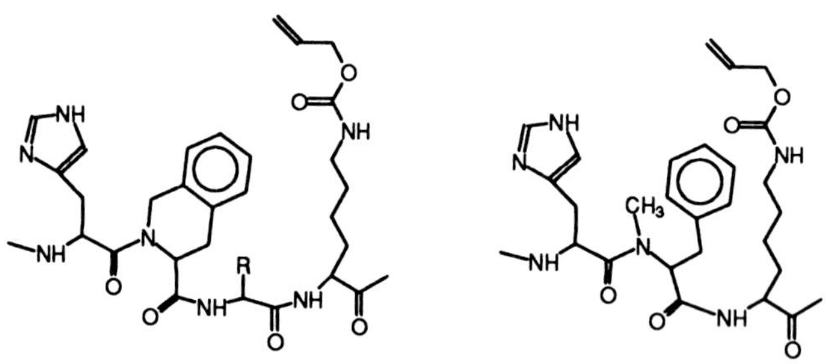

His-Tic-D-X-Lys(Alloc)   His-(Me)Phe-Lys(Alloc)

Fig. 1. Structure of streptavidin ligand motifs.

References

1. Gallop, M.A., J. Med. Chem., 37(1994)1233.
2. Gordon, E.M., J. Med. Chem., 37(1994)1385.

Peptide library screening based on the one-bead one-peptide concept

Kit S. Lam[a] **and Michal Lebl**[b]
[a]*Arizona Cancer Center, Department of Medicine, University of Arizona, Tucson, AZ 85724, U.S.A.*
[b]*Selectide Corporation, 1580 E. Hanley Blvd., Tucson, AZ 85737, U.S.A.*

Introduction

Combinatorial library methods are valuable tools for drug discovery and basic research. There are three general methods to generate and screen huge peptide libraries (over 10^6 peptides): (i) the biologic peptide libraries using filamentous phage [1], plasmids [2], and polysomes [3], (ii) combinatorial peptide libraries with iterative process [4, 5] and positional scanning [6], or (iii) combinatorial peptide library based on the one-bead one-peptide concept or the Selectide process [7].

In the Selectide proces, we use a "split synthesis method" [7,8] to generate random peptide libraries such that each resin-bead expresses only one peptide entity [7]. These huge libraries are then mixed with an enzyme-receptor complex. The beads that interact with the enzyme-rcceptor complex turn color in the presence of substrate. The color beads are then physically isolated for microsequencing [7,9]. More recently, with a modified functional screening method, peptide substrate motifs for cAMP-dependent protein kinases and c-src protein tyrosine kinases ($p60^{c\text{-}src}$) were also determined. In this article, the general method as well as examples of some of the applications of the Selectide process will be discussed.

Peptide Library Synthesis

Peptide libraries were synthesized with a split synthesis approch using standard Fmoc chemistry [7,9]. Polystyrene beads grafted with polyoxyethylene (TentaGel S, Rapp Polymere, Germany) were used as solid-phase for the peptide synthesis. Details of the library synthesis method were described previously [7,9]. In our standard peptide library synthesis, all 19 eukaryotic amino acids except for cysteine were used in each randomization step. Cysteine was excluded to eliminate undesirable crosslinking by disulfide bond. Linear peptide libraries as long as 15-mer had been synthesized and screened. After the desired length of the library was achieved, the N^{α}-Fmoc and side chain protecting groups were removed, and the beads were thoroughly washed with an aqueous buffer prior to screening [7,9].

Peptide Library Screening for Binding

In this method, the positive beads were identified by their ability to interact with a tagged-acceptor molecule. The acceptor molecule (e.g. receptor, enzyme, virus, or even small molecules) can be tagged directly by an enzyme [7,9], a fluorescent probe [10], a radionuclide [11], or a chromophore [12]. Alternatively, a secondary reagent with the above reporter molecule can also be used. For example, biotinylated-acceptor molecule can be the primary reagent, and streptavidin-alkaline phosphatase conjugate the secondary reagent. In the enzyme-linked detection scheme, the beads that interacted with the acceptor-enzyme conjugate turned color upon addition of substrate. The color beads were then physically isolated under a dissecting microscope and subjected to microsequencing as described [7,9].

We have applied this library screening method to various biological systems such as monoclonal antibodies (specific for continuous [7], and discontinuous [13] epitopes), streptavidin [7,14], avidin [14], MHC-class I molecules (B2 and A2) [15], and thrombin.

Peptide Library Screen for Post-translational Modification

In this method, the positive beads were identified by their ability to be enzymatically modified (e.g. phosphorylation) by an enzyme (e.g. protein kinase). To identify peptide substrate motif for cAMP-dependent protein kinase, we incubated a random heptapeptide library with cAMP-dependent protein kinase and [γ-^{32}P]ATP [16]. The bead-library was then washed thoroughly and immobilized on a glass plate with 1.5% agarose (w/v). The [^{32}P]-labelled beads were then localized by autoradiography. Using this method we identified peptide substrates for both cAMP-dependent protein kinase wd p60$^{c\text{-}src}$ protein tyrosine kinase (Table 1). Potentially, this can be used as a general method for identification of linear substrate motifs for other post-translational modifications of proteins provided that modifying enzymes and radiolabelled donors are available. If radiolablelled donors are not available, one may still be able to detect the covalently modified sites by antibodies that recognize these modified sites.

Table 1 *Identification of peptide substrate for protein kinase*

(I) cAMP-dependent protein kinase	RRYSV
	IIRRKSE
	SQRRFST
	YRRTSLV
(II) p60$^{c\text{-}src}$ protein tyrosine kinase	YIYGSFK

Identification of Peptides that Interact with Small Molecules

To test whether the combinatorial library method can be used to identify binding motif for small molecules, we used indigo carmine as a small molecule target [17]. Indigo carmine was chosen because it is a small water soluble organic dye molecule (M.W. = 466.56, Fig. 1). Since it is intrinsically colored, no additional reporter molecule is needed. The peptide libraies were incubated with 10 μM indigo carmine in the presence of 0.26 M NaCl (to reduce non-specific ionic interaction) and 0.1% Triton X-100 (to reduce non-specific hydrophobic interaction). After an hour, the color of a few beads became intensively blue. Some blue beads were isolated for microsequencing. The amino acid sequence of these beads is shown in Table 2. All ligands isolated had a genaral motif of "+OOO+" where '+' represents either Lys or Arg and 'O' represents relatively hydrophobic or aromatic amino acids. The two flanking basic amino acids are highlighted in boldface in Table 2.

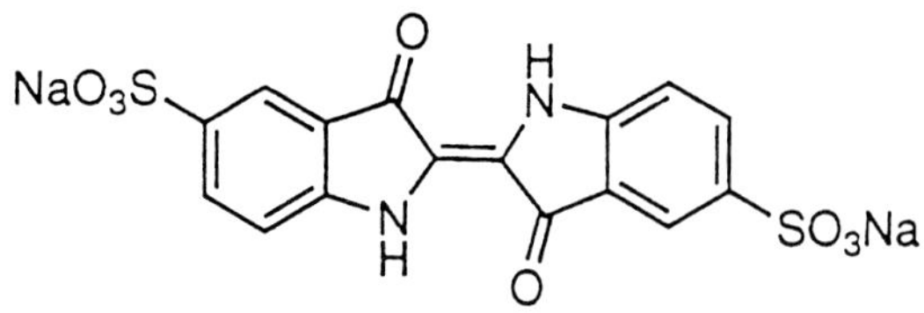

Fig. 1. Chemical structure of indigo carmine.

Table 2 *Identification peptides that interact with indigo carmine*

Libraries	Peptide
xxxxxxxx	akw**k**wvy**r**, **i**k**i**vy**r**fr y**k**vvy**r**is, vk**k**mvi**k**f
XXXXXXX	LT**K**LVL**K**, Y**K**VVY**K**L, VT**K**IIF**K**

Perspectives

This article summarizes principles and methods of combinatorial peptide library screening based on the one-bead one-peptide concept (Selectide process). Only a few examples are given. Over the last few years, there has been enormous progress in various aspects of the Selectide process. Using double-releasable linkers, peptides can be released from the beads for solution-phase assay [18]. Libraries with cyclic structures, unnatural amino acids and non-peptide moieties were synthesized and tested. Structural determination of small organic non-peptide ligand \using mass spectroscopic or peptide-encoding methods was accomplished [19]. Non-peptide libraries with rigid or flexible scaffolds (e.g. steroid, Kemp-triacid) were also developed. New coupling chemistries for combinatorial are currently under intensive investigation from many laboratories [20].

As an alternative to the synthetic approach described in this paper, the Selectide bead-library can be synthesized biochemically using enzymes. With the appropriate glycosyl-tranferases and precursors, oligosaccharide libraries can be synthesized. A similar approach can also be applied to polyketides.

Although our major focus has been on biological problems, we believe the Selectide process can also be applied to other disciplines such as materials science. For example, one could develop new material with desired photoelectric, physical, electronic, or electromagnetic propertis provided that an appropriate detection scheme can be developed. Combinatorial library method is an emerging field and will likely be useful for investigators from many disciplines. Table 3 summarizes some potential applications of the Selectide Process.

Table 3 *Some potential applications of the Selectide process*

1.	Basic research (identification of binding ligands, protein-peptide and protein-protein interaction, molecular recognition)
2.	Drug-discovery (lead discovery and optimization of leads, receptor binding, biological activity)
3.	Materials science (optical, photochemical, photoelectrical, electronic and electromagnetic properties, molecular switch)
4.	Molecular assembly (binding pairs with high affinity)
5.	Artificial enzyme (e.g. cyclized peptides with/without co-factors as side chains)

Acknowledgements

This work was supproted by Selectide Corporation and by NIH grants CA-17094, CA-57723, and CA-23074. K.S.L. is a scholar of the Leukemia Society of America.

References

1. Scott, J.K. and Smith, G.P., Science, 249(1990)386.
2. Cull, M.G., Miller, J.F. and Schatz, P.J., Proc. Natl. Acad. Sci. USA, 89(1992)1865.
3. Kawasaki, K., (1991) International Patent Application WO 91/05058.
4. Geysen, M.H., Rodda, H.M. and Mason, T.J., Mol. Immun. 23(1986)709.
5. Houghten, R.A., Pinilla, C., Blondelle, S.E., Appel, J.R., Dooley, C.T. and Cuervo, J.H., Nature, 354(1991)84.
6. Pinilla, C., Appel, J.R., Blanc, P. and Houghten, R.A., BioTechniques 13(1992)901.
7. Lam, K.S., Salmonm, S.E., Hersh, E.M., Hruby, V.J., Kazmierski, W.M. and Knapp, R.J., Nature, 354(1991)82.
8. Furka, A., Sebestyen, F., Asgedom, M. and Dibo, G., Int. J. Peptide Protein Res., 37(1991)487.
9. Lam, K.S. and Lebl, M., Methods, A companion to Methods in Enzymology 6(1994).
10. Needels, M.C., Jones, D.G., Tate, E.H., Heinkel, G.L., Kochersperger, L.M., Dower, W.J., Barrett, R.W. and Gallop, M.A., Proc.Natl. Acad. Sci. USA, 90(1993)10700.
11. Kassarjian, A.., Schellenberger, V. and Turck, C.W., Pept. Res., 6(1993)129.
12. Boyce, R., Li, G., Nestler, H.P., Suenaga, T. and Still, W.C., J. Am. Chem. Soc., 116(17)(1994)7955.
13. Lam, K.S., Lebl, M., Krchňák, V., Lake, D.F., Smith, J., Wade, S., Ferguson, R., Ackerman-Berrier, M. and Wertman, K., In Hodges, R.S. and Smith, J.A. (Eds.) Peptides: Chemistry, Structure and Biology (Proceedings of the 13th American Peptide Symposium), ESCOM, Leiden, 1994, pp. 1003–1004.
14. Lam, K.S. and Lebl, M., Immunomethods, 1(1992)11.
15. Smith, M.H., Lam, K.S., Hersh, E.M., Lebl, M. and Grimes, W.J., Mol. Immunol., (1994) in press.

16. Wu, J., Ma, Q.N. and Lam, K.S., Biochemistry, 33(1994)14825.
17. Lam, K.S., Zhao, Z.G., Wade, S., Krchnak, V. and Lebl, M., Drug Development Research, 33(1994)157.
18. Salmon, S.E., Lam, K.S., Lebl, M., Kandola, A., Khattri, P.S., Wade, S., Pátek, M., Kocis, P., Krchnák, V., Thorpe, D. and Felder, S., Proc. Natl. Acad. Sci. USA, 90(1993)11708.
19. Nikolaiev, V., Stierandova, A., Krchnak, V., Seligman, B., Lam, K.S., Salmon, S.E. and Lebl, M., Pept. Res., 6(1993)161.
20. Bunin, B.A. and Ellman, J.A., J. Am. Chem. Soc., 114(1992)10997.

Drug lead discovery and systematic analoging using the multipin technique

Jian-Xin Wang, Stuart Rodda, Gordon Tribbick, Andrew M. Bray, Robert M. Valerio and N. Joe Maeji
Chiron Mimotopes Pty. Ltd, PO Box 1415, Rosebank MDC, Clayton, Vic 3169, Australia

Introduction

An emerging trend in pharmaceutical research is the implementation of the concept and practice of synthetic combinatorial libraries, which started a decade ago as a result of breakthroughs in parallel synthesis of peptides or peptide mixtures using solid phase peptide chemistry [1-3]. Among several techniques, the multipin system [1] has unique features that make it a useful tool for synthesis and screening chemical libraries. In this article, we will discuss some applications of the multipin technique in drug lead discovery and, particularly, in the systematic analoging of peptide drug leads.

Results and Discussion

Drug leads derived from protein sequences

Many endogenous polypeptides and proteins are involved in regulation of important biological functions by interaction with their specific cellular receptors. Protein-protein interactions are now considered as major targets for new drugs. If the binding site involves one or several small linear portions of the protein ligand, it then becomes possible to apply the multipin technique to systematically scan the sequence for active fragments. Although the isolated binding fragments may have significantly lower affinity than the parent protein, they can provide initial structural information for further drug development through subsequent analoging processes.

The interaction between the coat glycoprotein gp120 of human immunodeficiency virus (HIV) and human T helper cell membrane protein CD4 is an example of such macro-molecular interaction. To identify the predominant binding region on the CD4 molecule, 428 overlapping octapeptides scanning the entire sequence of CD4 was synthesized on pins. Screening of the peptides singled out a unique region (residues 81-91) at the C-terminus of domain 1 as the major binding fragment for gp120 (Fig. 1) which was recognized as part of a binding groove of CD4. Extensive structural replacements were carried out on the core peptide SDTYCE and several strong binding analogs were developed [4]. Among them, one analog showed an *in vitro* inhibitory effect on HIV reproduction at a concentration of 50 μg/ml (data not shown).

Drug leads screened from peptide libraries

For receptors whose endogenous ligands are either unknown or non-peptidic, a combinatorial peptide library provides a source for discovering potential peptidic drug leads. In a model study, a peptide library was applied to a monoclonal antibody which recognizes a linear peptide DVPDYA. A complete hexapeptide library was synthesized containing all possible combinations of one or two defined residues. This library had more than 6000 pools of peptide mixtures. Results obtained revealed a comprehensive binding profile of the antibody with the highest binding to the peptide mixture XXXXYA (where X = mixture of 19 L-amino acids). A statistic algorithm was used to analyse the patterns of the binding peptide mixtures and to deduce the epitope/mimotopes using the structural information. Eighteen deduced hexapeptides, centered on "dominant" residues as determined from the binding to the library mixture are shown in Table 1. Peptides with similar sequences to the native epitope had the strongest binding, whereas those with little similarity failed to show any significant binding.

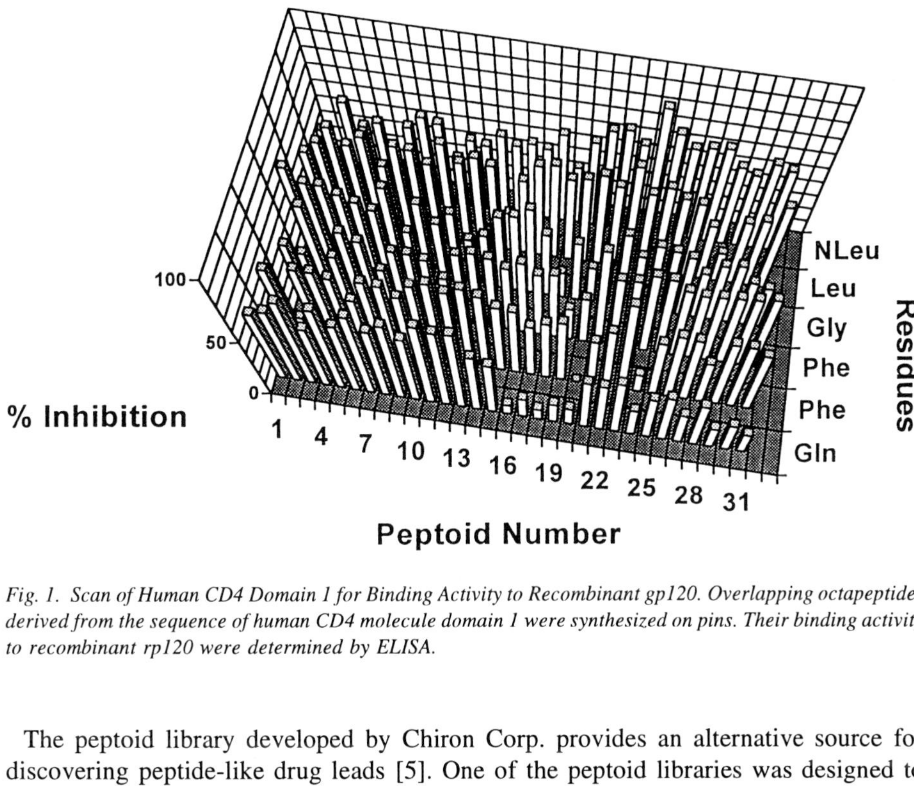

Fig. 1. Scan of Human CD4 Domain 1 for Binding Activity to Recombinant gp120. Overlapping octapeptides derived from the sequence of human CD4 molecule domain 1 were synthesized on pins. Their binding activity to recombinant rp120 were determined by ELISA.

The peptoid library developed by Chiron Corp. provides an alternative source for discovering peptide-like drug leads [5]. One of the peptoid libraries was designed to contain structural elements of ligands to 7-transmembrane G-protein coupled receptor (7-TM/GPCR) family. This library contains approximately 5000 compounds subgrouped into several mixture pools. Compounds with high binding affinity towards these receptors were quickly discovered after deconvolution of the mixtures. For example, one compound, CHIR-2279, was found to have a low nanomolar affinity for α_1-adrenergic receptors. Another compound, CHIR-4531, was found to have a low nanomolar affinity to the μ-opiate receptors [5].

Table 1 *Epitope/mimotopes predicted for an antibody which recognizes an epitope DVPDYA*

Epitope/Mimotopes Predicted	Binding (mOD)
DVPDYS	2971
DVPDYA	2899
DYRDYA	2728
DARDYA	2675
CIPDYA	2289
WVPDYA	2040
WVMDYA	1135
YAWMYA	200
WARMRR	165
RDRWGA	156
RHWRYA	151
RCWNRR	145
CFPWYA	141
YARIWR	140
YWQWYA	115
AYVDQA	113
CCYNWA	99
CFPWYA	96
CDYDGY	81
YAWEWY	68

Systematic analoging

The marriage of the concept of the conventional structure-activity relationship (SAR) study with the combinatorial library approach has created a new strategy that could be described as "systematic analoging" (SA) for drug lead optimization. In a typical SA of peptide leads, any structural elements available: D- and L-amino acids, non-natural amino acids, N-substituted glycines etc. will be inserted into each position of the lead compound in a systematic fashion. This will generate a large set of analogs with all possible combinations of such structural modifications. Structural information required for the optimal activity may then be extracted from the database by computer-aided drug modelling and statistical analyses.

We have conducted a systematic study on the stereo-requirements of substance P (SP) using the multipin system [6]. In this case, a complete set of 512 analogs with systematic replacement(s) of 9 L-amino acid residues by their D-isomers was synthesized. Their approximate receptor binding affinities (IC_{50}) were determined by an iterative process using a competitive radioligand binding assay. The results obtained from the screening have provided a comprehensive database for quantitative structural analysis and molecular modelling for SP. One analysis was conducted on 189 peptides with IC_{50} values less than 5 μM using a modified Free-Wilson model. Parameters obtained give a quantitative description of the predominant contribution from the C-terminal residues (Gln[6] to Met[11]) and small contributions from the N-terminal residues (Arg[1] to Gln[5]) to the overall binding activity of SP.

Recently. we synthesized and tested a set of N-acetylated hexapeptides analogous to the C-terminal region of SP, ie. **Ac-Q-F-F-G-L-J-NH_2**(J = Nleu). For this study, each position within the hexapeptide was systematically replaced with a series of 61 unusual amino acids, including a set of novel N-substituted glycines. A total of 366 novel analogs was prepared. Their binding potency was assessed at a concentration of 1 μM by a competition binding assay with a recombinant murine neurokinin (NK1) receptor expressed on the Sf9 cell surface membrane. Part of the results are shown in Fig. 2. Several promising peptide/peptoid chimeras in that group have estimated affinities about 150-300 nM, which was a 6 to 10 fold improvement over the binding affinity of the parent peptide (1.8 μM, data not shown in Fig. 2).

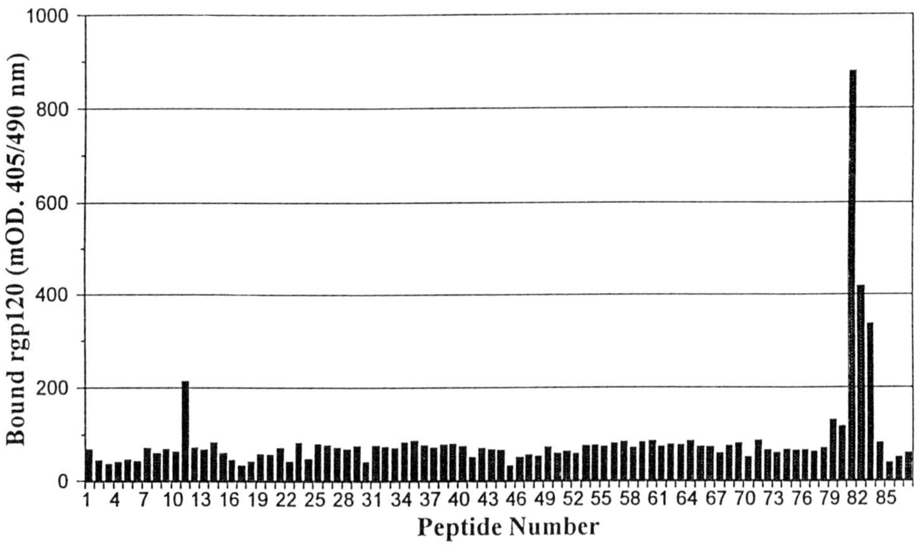

Fig. 2. Random Analoging Using Peptide/Peptoid Chimeras. The 6-mer SP C-terminal fragment (J = Nleu) was systematically replaced at each position by 31 non-natural amino acids or peptoids. Their binding potency to NK1 receptors were measured by a competitive binding assay at a concentration of 1 μM. Results are expressed as percentage of controls.

Summary

Recent developments in synthetic chemical libraries and high throughput screening have led the way towards building a drug discovery highway. As one of the original multiple synthesis designs, the multipin technology has undergone continuous development in the last few years to meet requirements for new drug development. Although the chemistry has expanded from peptide to non-peptide synthesis, whereas the system remains simple and convenient for synthesizing various libraries. The format has made it particularly suitable for many high throughput assays, including both solid phase and solution phase assays.

References

1. Geysen, H.M., Meleon, R.H. and Barteling, S.J., Proc. Natl. Acad. Sci. USA, 81(1985)3998.
2. Houghten, R.A., Proc. Natl. Acad. Sci., 82(1985)513.
3. Geysen, H.M., Rodda, S.J. and Mason, T.J., Mol. Immunol., 23(1986)709.
4. Wang, J.X., DiPasquale, A.J. and Geysen, H.M., Abstract of 16th Annual Lorne Conference on Protein Structure and Function. 1991, pC41.
5. Zuckermann, R.N., Martin, E.J., Spellmeyer, D.C., Stauber, G.B., Shoemaker, K.R., Kerr, J.M., Figliozzi, G.M., Goff, D.A., Siani, M.A., Simon, R.J., Banville, S.C., Brown, E.G., Wang, L., Richter, L.S. and Moos, W.H., J. Med. Chem., 37(1994)2678.
6. Wang, J.X., DiPasquale, A.J., Bray, A.M., Maeji, N.J., Spellmeyer, D.C. and Geysen, H.M., Int. J. Peptide Protein Res., 42(1993)392.

Author index

Subject index